The Natural History of Sexuality in Early America

The Natural History of Sexuality in Early America

GRETA LAFLEUR

Johns Hopkins University Press
Baltimore

© 2018 Greta LaFleur
All rights reserved. Published 2018
Printed in the United States of America on acid-free paper

Johns Hopkins Paperback edition, 2020
2 4 6 8 9 7 5 3 1

Johns Hopkins University Press
2715 North Charles Street
Baltimore, Maryland 21218-4363
www.press.jhu.edu

The Library of Congress has cataloged the hardcover edition of this book as follows:

Names: LaFleur, Greta, 1981– author.
Title: The natural history of sexuality in early America / Greta LaFleur.
Description: Baltimore : Johns Hopkins University Press, [2018] | Includes bibliographical references and index.
Identifiers: LCCN 2018000285 | ISBN 9781421426433 (hardcover : alk. paper) | ISBN 9781421426440 (electronic) | ISBN 1421426439 (hardcover : alk. paper) | ISBN 1421426447 (electronic)
Subjects: LCSH: Sex—United States—History. | Sex customs—United States—History. | Sexual ethics—United States—History.
Classification: LCC HQ18.U5 L34 2018 | DDC 306.70973—dc23
LC record available at https://lccn.loc.gov/2018000285

A catalog record for this book is available from the British Library.

ISBN-13: 978-1-4214-3884-9
ISBN-10: 1-4214-3884-4

Special discounts are available for bulk purchases of this book. For more information, please contact Special Sales at specialsales@press.jhu.edu.

Johns Hopkins University Press uses environmentally friendly book materials, including recycled text paper that is composed of at least 30 percent post-consumer waste, whenever possible.

CONTENTS

Acknowledgments vii

Introduction: Toward an Environmental Theory of Early Sexuality   1

1   The Natural History of Sexuality   32

2   The Complexion of Sodomy   63

3   "Egyptian Lusts" at the Gallows   103

4   Botanical Sexuality and the Colonial Landscape   137

5   "Negro Hill" and the Sexuality of Space   164

Epilogue: Thinking Sex—without the Subject   189

Notes   207
Bibliography   261
Index   279

ACKNOWLEDGMENTS

It would be difficult to write a book about the legacies of environmentalist thinking without a consideration of the environment in which I wrote it and, indeed, of the people and places who have given form and content to my world over the past six years.

At Johns Hopkins University Press, I thank Elizabeth Demers, Greg Britton, and Matt McAdam, who guided me through the production of the manuscript, for their leadership and insight, and Lauren Straley for her hard work on ferrying it through the print process. I also thank the anonymous readers who offered me such excellent and detailed feedback on the manuscript. This book is better for their efforts and reflection.

At the University of Pennsylvania, Nancy Bentley, Max Cavitch, David Kazanjian, and Heather Love advised the dissertation that provided the foundation for this book, and I thank them for their time, generosity, and careful feedback. Also at Penn, Suvir Kaul and Chi-ming Yang offered enormously useful suggestions and guidance. And to my fellow graduate students—fellow travelers and indefatigable companions through the misery and the mirth—thanks to all of you for your support and company. My time at Penn would not have been the same without Dave Alff, GerShun Avilez, Rachel Banner, Marina Bilbija, Julia Bloch, Claire Bourne, Ashley Cohen, Paige Contreras-Gould, Megan Cook, Sarah Dowling, Steph Elsky, Jonathan Fedors, James Fiumara, Ari Friedlander, David Gardner, Matt Goldmark, Chris Hunter, Emily Hyde, Jen Jahner, Benjy Kahan, Adrian Khactu, Jeehyun Lim, Phil Maciak, Melanie Micir, Christen Mucher, Justine Murison, Will Nessly, Emily Ogden, Rosemary O'Neil, Courtney Rydel, Poulomi Saha, Jill Shashaty, Emma Stapely, Michelle Strizever, Mecca Sullivan, Chris Taylor, Thomas Ward, Emily Weissbourd, Sunny Yang, and Jason Zuzga. Cal Biruk deserves special mention here. We spent both college and graduate school together, not to mention many legendary beach trips and holiday sojourns to The Feve in Oberlin, Ohio, and it has been wonderful to grow from teenagers into adult scholars alongside one another.

Moving from Philadelphia to take a position in the English Department at the University of Hawai'i was an exciting, if daunting, move, and I am so grateful for the warm welcome and vibrant intellectual community that met me there. Deep thanks to Cristina Bacchilega, Ned Bertz, Steve Canham, Urvashi Chakravarty, Kim Compoc, Monisha Das Gupta, Joy Enomoto, Cindy Franklin, Candace Fujikane, Tina Gerhardt, Ku'ualoha Ho'omanawanui, Craig Howes, Chuck Lawrence, Christine Lipat, Laura Lyons, Brandy Nalani MacDougall, Mari Matsuda, Maiana Minahal, Cheryl Naruse, Njoroge Njoroge, Craig Santos Perez, Tagi Qolouvaki, Rich Rath, Suzanna Reiss, Matt Romaniello, Ayu Saraswati, Susan Schultz, S. Shankar, Valerie Wayne, and John Zuern for their warmth and intellectual fellowship. While at UH, I was also lucky enough to participate in Chuck Lawrence's Race, Gender, and Culture Junior Faculty Seminar, which was one of the best things I did during my first year there. Special thanks are due to Chuck and Mari for opening up their home to all of us. Val Wayne and Matt Romaniello's Early Modern Studies Faculty Working Group was another highlight of my time in Honolulu, and I learned much from the thinking we did and the feedback I received in our meetings.

I moved to Yale University in 2013 and was welcomed by the hardworking intellectual enthusiasm of Vanessa Agard-Jones, Daphne Brooks, George Chauncey, Jean Cherniavsky, Michael Denning, Kate Dudley, Anne Eller, Crystal Feimster, Marta Figlerowicz, Joe Fischel, Ron Gregg, Inderpal Grewal, Cassi Hartblay, Briallen Hopper, Matt Jacobson, Katie Lofton, Dan Magaziner, Joanne Meyerowitz, Tavia Nyong'o, Eda Pepi, Steve Pitti, Joanna Radin, Anthony Reed, Jill Richards, Evren Savci, Alicia Schmidt-Camacho, Susan Shand, Caleb Smith, Rasheed Tazudeen, Dillon Vrana, Michael Warner, and Sunny Xiang. Special thanks are due to my American Studies junior colleagues Albert Laguna, Laura Barraclough, and Dixa Ramirez, who have been the most incisive, strategic, and hilarious company through faculty meetings and BBQ shark lunches alike. The development of this book was girded by fellowships from the Fund for Lesbian and Gay Studies, the A. Whitney Griswold Faculty Research Fund, and the Frederick W. Hilles Publication Fund at Yale University, and the manuscript benefited greatly from that generous support. Part of chapter 4 appeared in an earlier version in my article "Precipitous Sensations: Herman Mann's *The Female Review*, Botanical Sexuality, and the Challenge of Queer Historiography," *Early American Literature* 48, no. 1 (2013): 93–123.

This project itself would not have been possible without fellowship support from archives all over the United States that made it possible for me to spend months at a time steeping in the primary sources—most of them ephemera and thus often only accessible through archival research—that became the central objects of inquiry for the book. I thank the John Carter Brown Library at Brown University, the Massachusetts Historical Society, the William Andrews Clark Library at the University of California at Los Angeles, and the American Antiquarian Society (especially Ashley Cataldo and Paul Erickson) for the gift of time to tarry in those remarkable collections. Additionally, I want to express my deep gratitude to the staff and other scholarly community at the Huntington Library, who made my almost two years of research there exciting, productive, and refreshingly easy.

The very best things in the world eschew institutions. Since starting the process of turning a dissertation topic into a book, I have been lucky to be included in a number of different reading and writing groups that have importantly shaped my thinking and writing. Thanks to the broad and changing membership of the Southern California Americanists Group, and especially Michelle Chihara, Bert Emerson, Chris Hunter, Thomas Koenigs, Sarah Mesle, and Sharon Oster. The Early Americanist Writing Group that Caroline Wigginton wrangled in 2013 has read every single draft of every single chapter of this book, and I thank Angie Calcaterra, Abram von Engen, Travis Foster, Michele Navakas, Wendy Roberts, Kacy Tillman, and Caroline Wigginton for their collegiality, time, careful feedback, and good cheer. Similarly, GerShun Avilez, Maxe Crandall, Kyla Schuller, and Sarah Schulman, of Schulman's Queer First Book writing workshop, helped me conceptualize my project as a *book* for the first time, and their feedback on the entire manuscript in its early stages was immeasurably valuable. Carrie Hyde and I traded chapters several times, and I thank her for her thinking and her crisp feedback. Brian Connolly has been a cherished colleague (and, at times, commiserator) in the field of early American studies and has generously been offering me his historically precise and theoretically virtuosic feedback on my work since my first chapter workshop in 2009. I cannot thank him enough for the time and effort he has poured into my writing. Megan Cook graciously offered her expertise in book history at several hundred important moments in the revision of the manuscript, and I thank her for continuing to text and email me back. In February of 2016, Katy Chiles, Katie Lofton, Valerie Traub, and Michael Warner generously agreed to workshop my manuscript. Their feedback and guidance helped me to revise the book into its final form and get it under contract. Thanks also to the Americanist Faculty Workshop at Haverford College, the Americanist Research Colloquium at UCLA, the Washington Area Early Americanists Group, the Local Americanists Group at the University of Maryland, the Queer Studies Workshop at Wesleyan University, and the Omohundro Institute of Early American History and Culture, who invited me to workshop chapters of this project and to benefit from the rigorous thinking of their participants.

David Lobenstine's developmental editing helped render this book the best version of itself; I cannot thank him enough. Yale PhD students Sarah Robbins, Isabel Ortiz, and Kelsey Henry poured hours into the editing and proofing process, and I thank them for their meticulous, painstaking work. I also thank the graduate students who participated in my spring 2015 "Introduction to Biopolitics" course, my spring 2017 "History of Race" course, and the Yale Women's, Gender, and Sexuality Studies Program workshop that read my introduction in spring 2017. Their insights helped sediment my thinking, and their feedback energized the revision process.

Like any book, this one evolved slowly, over the course of years, and during that time several people offered mentorship at critical moments that—perhaps unbeknownst to them—fundamentally shaped my thinking, my work, and my career. Jordan Stein reached out to me while I was still a graduate student, after my first-ever conference presentation, expressing his enthusiasm for my project and offering to read my work; Bethany Schneider stepped in and offered truly necessary (and, as it turns out, deeply *right*) advice at no less than three absolutely definitive junctions in my trajectory as a scholar, and I cannot thank her enough for that prescient, well-timed, and good-natured Leo guidance; and the inimitable Katie Lofton has offered advice, support, and a game plan at more moments than I can count, since my arrival to Yale. Their efforts have been so very meaningful. They have provided a model of what good mentorship can achieve, and I strive to emulate their example with my students.

As anyone who has ever written a book would surely concur, it is an arduous process that places an enormous amount of strain not only on the writer but on anyone unlucky enough to stroll into their life during the many years that it takes to finish one. But if my friends were unlucky to encounter me in the ugliest moments of my writing process, I was unbelievably fortunate to encounter them. My years in Philadelphia were made sweeter by the presence of Anne-Marie Christensen, Becka Gorelick, Bernadette Walker, Dexter Rose, Eileen Carlin, Ezra Berkeley Nepon, Hazey Fairless, Heath Williams, Hye-won Gehring, Michelle Posadas, Niki Paul, Ollie Dooling, Rachel Marcus, Rath Radelman, Sarah Jamison, Star Morris, Tara Rubinstein, and Tristan Guarini. It was also in Philadelphia in 2005 that I met Poulomi Saha, without whose enduring, geographically flexible friendship I am certain that I would not have finished this book. Her presence, over the past six years, at the other end of the phone calls to Carlisle, Providence, or El Cerrito; on the other end of the video chat; or at the other end of the table, has delighted me in good times and kept me afloat during bad ones.

It was during my time in Honolulu that I started spending time in California queer communities, and as the years passed, I increasingly grafted my life to the people and places I frequented there. The people who became my West Coast friends and family offered me unprecedented warmth and a soft landing when I needed it most and sustained me with immeasurable light-heartedness, adventurousness, and fun. I have been so changed by the geography of that state and by the landscape of their kindness and care. This support was indispensable to both the initial revisioning of this project and its eventual completion, and I offer the sweetest and most immense gratitude to my West Coast sisters, brothers, and siblings, including Adele Failes-Carpenter, Allison Harris, Amanda Ream, Ari Banias, Avery Ucker, Becca Cohen, Connie Hockaday, Ebah Sunlei, Elena Hillard, Elan Margolis, Finn Paul, Hannah Horovitz, Heather Lukes, Isa Knafo, Ivan Ramos, Juana Maria Rodriguez, Kathleen Frederickson, Karen Tongson, Kyla Wazana Tompkins, Laurel Braitman, Leon Hilton, Lex Vaughn, Lisa Lindberg, Liz Hodgson, Marcy Coburn, Maxe Crandall, Maya Bookbinder, Megan Auster-Rosen, Meg Panichelli, Molly McGarry, Nick Mitchell, Nicki Green, Onya Hogan-Finley, Patty Ahn, Rayna Matthews, Sal Nicolazzo, Sarah Kessler, Sophia Poirier, Travis Harvey, Will Fisher, and Uri McMillan. As many (if not all) of you have heard me exclaim, two or more drinks in, I love being a gay adult, and you all are what makes gay adulthood such an enduringly excellent experience. And to Lis Goldschmidt, for the wonderful, uniquely transformative love you have poured so liberally into our connection: thank you. This book may not have been started had you not been on the other end of the phone calls and long flights to San Francisco.

New York has fed me differently, but no less fully. To Adrienne Sneed, AJ Murphy, Alice Eisenberg, Allison Palmer, Anna Wolk, Buzz Slutsky, Cassie Wagler, Emilie Connolly, Janet Neary, Jeanne Vaccaro, Jordan Stein, Kyla Schuller, Laura Mauldin, Lena Afridi, Mara Mills, Sam Feder, Sarah Blackwood, Ted Kerr, and Claudia Secor-Watkins: I'm so happy we're all here. Emma Heaney deserves special mention here for her abrupt arrival to my life and for her insistence that we write together; I have so many unwritten odes to her sharp thinking and tenacious friendship.

Anyone whose research is contingent on archival access knows that this kind of itinerant research has both benefits and drawbacks; living for a few months at a time in multiple locations is exciting, but it can be a lonely and draining endeavor. I have spent more time traveling than not, since restarting this project in earnest, and I have been lucky to have the best possible company as I have moved from archive to archive and city to city. To the friends, lovers, colleagues, and otherwise energetic folks who have offered suggestions, thoughts, activities, invitations, and directions to the nearest gay bar during the many years of my untoward mobility: that emotional and intellectual energy can be found all over the pages that follow. Writing a book can be an isolating process, but you all saw to it that it was never a lonely one. Thank you for sharing your joy with me and for reflecting mine back to me. To my many queer tour guides, socialites, artists and art experts, and party monsters, including Alexis Mitchell, Allyson Mitchell, Ben Reed, Brandie Taylor, Caitlin Manicom, Carmella Fleming, Chiming Yang, Christina Leon, Cecilia Berkovic, Cooper Dion, Christina Quarles, Chelsea Starr, Deidre Logue, Elizabeth Steeby, Erin Ikeler, Hannah Mittenberg,

Hester Blum, Isabel Osgood-Roach, Jess Lee, Lee Watts, Marie Suzor-Morin, Nathalie Kouri-Towe, Nicole Georges, Pete Coviello, Ruthie Doyle, Sarena Sairan, Sasha Wortzel, Sharlene Bamboat, Sophia Rose, Susanna Boehme, Steph Markowitz, TJ Burk, Wallace Blocker, and Zoe Walsh: thank you, for all of it. I owe special gratitude to Gisèle Suzor-Morin, for her company, and sweet consistency; her arrival to my life, in the year leading up to the completion of this project, has truly been a gift.

Deep thanks are also due to my family of origin. My mother, Anne LaFleur, has offered unmitigated support and kindness through the inevitable highs and lows of these years of writing: I am so grateful to have her as a parent. To the rest of my family: my father, Kurt Lynn; my siblings, Zulay, Flor, Jon, Tonie, Margot, and Elvis; my aunts Sarah, Liz, and Anne, and my uncles, Vin and Bob; my cousins Rachel, Olivia, and Kevin; and my grandmother, Catherine LaFleur, who died during the final revisions of this book: your love means, and has always meant, so much. And last, but by no means least, my companion in life, Marge—exactly as old as this project itself—has made every move, every transition with me, and her quiet company and insatiable squirrel bloodlust has made even the hardest moments of this work easier.

Finally, Bryn Kelly and Amanda Harris each left us as I was finishing this project. Writers both, their sudden departures have pressed on me, as I have struggled to put pen to paper in a world that was ill-equipped to receive them. The loss of each of these radiant sisters is felt in the blank spaces of the pages that follow and in the many pages I have been unable to write since they left. I dedicate this book to them—to the artists they were and to each of their complicated, devastating refusals of the world as it is.

The Natural History of Sexuality in Early America

INTRODUCTION

Toward an Environmental Theory of Early Sexuality

In 1597, William Davis set sail aboard the *Francis of Saltash*, an English ship charged with delivering a cargo of "fish, and herrings, and such-like commodities" to Civitavecchia, a port town on the western Italian coast.[1] Arriving safely in Italy to make its delivery, the ship proceeded to Algiers, where Davis, a veteran diarist like many sailors of his era, carefully recorded his impressions of the "Turks" and Islamic culture that he observed there. Davis's narrative, after its initial printing in London in 1614, was reprinted frequently throughout the eighteenth century as part of a massive collection of travel writings culled from the library of the Earl of Oxford and compiled as *A Collection of Voyages and Travels*. Davis's narrative demonstrates his familiarity with contemporary travel writing and the ethnographic conventions that characterize the genre. Davis's accounts of his encounters with people of other cultures, like the other travel writings in *A Collection of Voyages and Travels*, frequently offer his impressions of the dress, diet, behavior, and cultures of the places that Davis visits; he notes, for example, that the women in Civitavecchia are "very lewd and wicked, for even in that ancient city of *Rome* there are many thousands of lewd living women that pay monthly unto the pope for the sinful use of their wicked bodies."[2] His remarks on the "Turks" of Algiers, however, are markedly more succinct. Of them, Davis writes, "These *Turks* are goodly people of person, and of a very fair complexion, but very villains in mind, for they are altogether sodomites."[3]

Davis's bald assessment of the "Turks" he encounters as "sodomites" is something of a narrative surprise, arriving as it does with little context and even less explanation. But more perplexing still is that Davis offers no account of witnessing or even hearing about sodomy while in Algiers. How then do we, as twenty-first-century readers, understand Davis's characterization of "Turks" as "sodomites"? How can someone be a sodomite without engaging in sodomy?

As Davis's narrative progresses, however, it becomes clear that his identification of "Turks" as "sodomites" turns not on any sort of behaviors he witnesses in Algiers, but rather on his observation of the specificities of Algerian food, customs, and worship. It is not sexual inclination but mode of living that earns "Turks" the title of sodomites, and Davis's detailed descriptions of the extravagant cushions they sit on, and the relative nakedness in which they sleep, provide additional evidence of the fundamental differences between the sodomitical "Turks" and Davis's own status as an Englishman and a Christian. Organized into an Orientalizing expression of both judgment and wonder, Davis's assessment of "Turks" as sodomites points to the early modern rhetorical strategy of describing ethnic or cultural variety through the idiom of sexual aberrance, of presenting sexual behavior—sodomy, in this case—as a metaphorical gloss for the profundity of difference. Few historians of sexuality have addressed this early modern rhetorical tendency, but those who have tend to read "sodomy" as a metaphor or index for a broad range of differences or disapprovals and have not considered the specific and pervasive association between sodomy and religious or ethnic difference.[4]

Just over a century later, the Swedish scientist Carl von Linné (Carolus Linnaeus) published his *Systema Naturae*, or *Systems of Nature*, in which he introduced a broad taxonomic system for the categorization of the organic world. First published in Latin, the *Systema Naturae* grew in size as it grew in popularity, swelling from a publication of just 12 large pages in its first Leiden edition to a two-volume edition of more than 1,300 pages by its tenth edition in 1758, twenty-three years later.[5] While today, Linnaeus is primarily thought of as a botanist and the founder of the system of binominal nomenclature that would be adopted as the international Latinate standard for the classification of living organisms, Linnaeus's study of animal orders made as significant an impact on the scientific culture of his own time as his study of plants. Indeed, the *Systema Naturae* has earned the sustained attention of both historians of science and historians of race, because Linnaeus includes humans in his taxonomy of the animal world and because he was one of the first naturalists to divide humans into regional taxa—*Europaeus, Asiaticus, Americanus,* and *Afer*—that organized these "types" of humans according to phenotypic, cultural, and even characterological qualities.[6] Historians of race have since located in Linnaeus's taxonomy of *Homo sapiens* an early scientific attempt to schematize "human variety" that anticipated nineteenth-century theories of racial difference, a question that would emerge as one of the central, if not the most important, concerns of natural historical inquiry in the eighteenth-century Atlantic world.[7]

The 1735 *Systema Naturae* does not offer a particularly detailed elaboration of what Linnaeus termed the "varieties" of *Homo sapiens,* simply listing the four categories *Europaeus albus, Americanus rubescens, Asiaticus fuscus,* and *Africanus niger,* all of which etymologically correspond to continental regions. But as Linnaeus revised this work over the course of twenty-three years, the breadth and detail of the *Systema Naturae* increased, and so too did the descriptions of human kinds:

AMERICANUS   a. reddish, choleric, erect.
Hair black, straight, thick; *Nostrils:* wide; *Face:* harsh, *Beard* scanty.
Obstinate, merry, free.
Paints himself with fine red lines.
Regulated by customs.

EUROPAEUS   b. white, sanguine, muscular.
Hair flowing, long. *Eyes* blue.
Gentle, acute, inventive.
Covered with close vestments.
Governed by laws.

ASIATICUS   c. sallow, melancholy, stiff.
Hair black. *Eyes* dark.
Severe, haughty, avaricious.
Covered with loose garments.
Ruled by opinions.

AFER   d. black, phlegmatic, relaxed.
Hair black, frizzled. *Skin* silky. *Nose* flat, *Lips* tumid.
Women without shame. *Mammae* lactate profusely.
Crafty, indolent, negligent.
Anoints himself with grease.
Governed by caprice.[8]

Alongside phenotypic descriptions of skin color, hair texture and color, and humoral constitution, Linnaeus includes a series of terse, authoritative assessments of the characterological qualities of each "variety" of *Homo sapiens:* the *Americanus,* or Indian, is "obstinate," "merry," and "free"; the *Asiaticus* is "severe," "haughty," and "avaricious"; the *Europaeus* is "gentle," "acute," and "inventive," and the *Afer*—called the *Africanus niger* in previous editions—is "crafty," "indolent," and "negligent." To the *Afer* category, Linnaeus adds a perplexing additional note: the women exhibit *mammae* that "lactate profusely," and they are, furthermore, "without shame."[9] Echoing William Davis's much earlier characterization of "Turks" as "sodomites," Linnaeus details the distinguishing characteristics of the women of the *Afer* class to corroborate his assessment of their varietal specificity. Like Davis, Linnaeus offers no evidence for his observation that women of the *Afer* class are "without shame"; on the contrary, shamelessness here is itself evidence, glibly put forward to support his taxonomic division of the human species into different ethnic classes.

Historians of race tend to treat this moment in the *Systema Naturae* as evidence of how early modern and eighteenth-century gender politics implicitly gird processes of racialization, while historians of sexuality, as far as I can tell, have refrained from reading this moment at all. Linnaeus's work as a botanist—and specifically the "sexual system" of botanical taxonomy that he advanced in his *Genera plantarum* (1737) and *Species plantarum* (1753), parts of which were translated into English very quickly—constitutes something of a textual staple in scholarly studies of eighteenth-century sexual practice and organization, but Linnaeus's characterization of women of the *Afer* class as "without shame," a statement that at least gestures toward the question of sexual custom, has languished unconsidered.[10] Although the project of taxonomizing human variety, so central to early modern colonial enterprise and so instrumental in the development of universalizing languages of Enlightenment personhood, is shot through with quiet but ubiquitous narratives about human sexual diversity, eighteenth-century natural history has somehow yet to capture the attention of historians of early sexuality. That absence animates the study that follows.

This book takes as a point of departure the centrality of natural history, and specifically, naturalist investigations into "human variety," to the history of sexuality in the eighteenth-century Atlantic world. Despite the importance of natural history as a major genre of eighteenth-century print, a lively intellectual culture that engaged some of the most prominent scientists of the century, and a hobby of many influential statesmen such as Thomas Jefferson, natural history has received virtually no attention from historians of sexuality. *The Natural History of Sexuality* addresses this problem in two ways. First, I offer close readings of the archive of eighteenth-century transatlantic, primarily North American, natural historical writings, which reveal how persistently this body of literature trains its ethnographic gaze on the question of sexual practices and inclinations. A large number of excellent scholarly studies of natural historical writing have carefully anatomized the contributions of philosophers like Linnaeus, the Comte de Buffon, Adam Smith, and even Thomas Jefferson to contemporary understandings of the significance of "human variety," or racial difference, but no scholars have considered how what I term *the sexual politics of racial difference* changes our understanding of early sexuality.[11] Travel writing, ethnographies, and natural historical texts—the handmaidens of colonialist knowledge production—represent one of the most consistent locations for the representation of sexual behavior, sexual inclinations, and variety in gendered mores in eighteenth-century print. This book seeks to restore this archive to the centrality it merits in studies of early sexuality.

Natural historical texts themselves (monographs, pamphlets, dissertations, essays, and even speeches given at storied institutions such as Philadelphia's American Philosophical Society) account for only a part of the archive at the center of this project, however. Despite the importance of natural historical inquiry to the history of race, the history of science, and, as I argue here, the history of sexuality, many of the texts that were particularly influential within philosophical circles during the eighteenth century did not necessarily see wide distribution or circulation. While scholars have debunked the characterization of natural historical inquiry as a hermetically or exclusively elite intellectual culture, essays such as John Mitchell's *Essay Upon the Causes of the different Colours of People in Different Climates* (1744) or Samuel Stanhope Smith's *Essay on the Causes of the Variety of Complexion and Figure in the Human Species* (1787) simply could not claim the kind of readership that contemporary popular genres such as execution narratives and seduction novels could.[12] But while lengthy natural historical texts—some of them written entirely in Latin, and many bearing expensive bindings that only very wealthy readers could afford—did not themselves circulate widely, the ideas explored within them *did*. Taxonomies of humans, animals, plants, and minerals, detailed descriptions of the "states of society" of cultures around the world, and questions about the origins of human phenotypic difference that were the standard fare of natural historical writing appear pervasively in a wide range of eighteenth-century texts, many of them cheaper and more readily available than works such as Linnaeus's *Systema Naturae* (1735) or Buffon's multivolume *Histoire Naturelle* (1749). *The Natural History of Sexuality* traces the questions, guiding logics, and tropes of representation that characterized naturalist debates about human variety out of natural history and into popular narrative, tracking the way the most widely discussed and contentious questions of natural historical inquiry echo in more widely consumed forms of narrative, such as accounts of crime and punishment and cheaply produced adventure tales.

This study examines four genres of popular narrative—Barbary captivity narratives, execution narratives, cross-dressing narratives, and antivice narratives—to emphasize how all of these genres, which represent some of the most widely read vernacular literature in the eighteenth-century British colonies, explore the sexual politics of racialization and thus must be read alongside their less available and more expensive natural historical forebears. By uniting the archives of popular narrative and natural history and considering how in each body of literature, discussions of racial difference provide the occasion in which sex acts and practices are rendered visible, we are able to appreciate the relationship between these two genres of eighteenth-century print (which are only rarely read in relation to each other), and furthermore, we can begin to chart how sciences of racialization provide an optic through which sexual behavior can be newly seen and anatomized. In that sense, this book offers a new characterization of the relationship between racial difference and sexual diversity in the eighteenth century, in which a particularized practice of ethnographic or scientific *observation* organized around making sense of "human variety" began to scrutinize sexual behavior as something of a racial symptom. In turn, this practice inadvertently rendered sex an observable characteristic of human beings.[13] In the eighteenth century, then, the increasing visibility of sexual behavior arrived epiphenomenally as a product of the science of racial difference.

The second critical insight that this book offers to the study of sexuality in the period before sexology pertains to what these literatures of human variety imported to contemporary understandings of sex and sexual practices. "Race" and "human variety" are concepts that have changed over time, and during the eighteenth century, they were understood in dynamic or at least nonstatic terms. While natural historians from William Falconer to Thomas Jefferson bickered over the etiology and significance of human variety, there was one theory with which *most* natural historians could agree: that human variety was due at least in part to one's environmental circumstances, including exposure (or lack thereof) to the sun's rays, the relative warmth of the climate, the food and drink one consumed, and even the social and political organization of one's society. Indebted to Hippocratic and Galenic medicine, natural historians' beliefs in the environmental nature of human variety derived from early modern medicine, which understood the body as a "porous envelope," eminently vulnerable to the impressions of environmental circumstance.[14] I argue that scholarship in the history of early sexuality has by and large neglected to account for the historical specificity of environmental understandings of the human body in the way it has formulated its animating questions about what sex was, how it was understood, and how it was practiced during the eighteenth century. We have asked questions about the histories of sexual practices, the history of sexual cultures, and even the histories of sexual *identities* without organizing these questions around the kind of environmental logic that defined eighteenth-century understandings of the human body, what motivated its behavior, and how its senses were coordinated. This study restores the environmental body to the center of studies of sex in the eighteenth century, arguing that like other forms of human variety such as racial difference, diversity in human sexual behavior, sociosexual organization, and even sexual proclivity, were explained primarily through an environmental logic in which the climatic, humoral, physical, and social milieu of the body was understood to be a determining force, *at least* as much as individual inclination.

These twinned insights—that the archive of natural historical thought (including some eighteenth-century popular narratives), with its attendant focus on human variety, is central to the emergent analytical visibility of sex in the eighteenth century, and that its environmental logic fundamentally shaped the way sex was understood—stand to profoundly shift the way we study early sexuality. The history of sexuality, and queer studies more generally, has been roundly criticized for its persistent inattention to race; and the history of race, in turn, has not directed much of its energies toward the study of sexuality. In what follows I bring these critical frameworks together, not to offer a flat account of their mutual imbrication but to demonstrate the deeper and richer interpretations of eighteenth-century texts that are possible when we approach the scene of critique with an awareness of the vexed symbiosis that bound sex to racial difference, in that century of the proliferation of the human sciences. Some of the most cherished, defining formations of the study of early sexuality, such as sodomy, assume new shape when considered through this lens.

The subject of numerous book-length studies, sodomy has provided the footing for multiple generations of scholars in their arrivals not only to the study of early sex but to queer studies as well.[15] From John Boswell's 1981 *Christianity, Social Tolerance, and Homosexuality* to Leo Bersani's landmark essay "Is the Rectum a Grave?" (1987), from Jonathan Goldberg's *Sodometries* (1992) to Tim Dean's *Unlimited Intimacy* (2009), sodomy has been adapted to the needs of scholars thinking in variegated ways about the politics and history of illicit sex since the inception of queer studies as a discipline, at least in part because of sodomy's explicit and persistent visibility in even some of the oldest legal, literary, and religious archives. There has been so much scholarship on sodomy, in fact, that on some level it seems difficult to imagine what more could be said. Yet the analytic that this project offers revitalizes even some of the most well-trod ground in early sexuality studies by providing new questions and fresh perspectives. Chapter 2, "The Complexion of Sodomy," turns to Barbary captivity narratives to explore the fears that enslaved Christian sailors express of being forcibly sodomized by their Muslim captors; this was one of the most frequent occasions for the discussion of sodomy in the eighteenth-century (and especially the late-eighteenth-century) Anglophone print record. Noting William Davis's cavalier assessment of the "Turks" he encounters in 1597 as "altogether sodomites," this chapter tracks the longtime Christian orientalist association of Islam with sodomy back to its etymological roots in the biblical city of Sodom.[16] Building on scholarship that has theorized sodomy geographically, I theorize it climatologically, drawing out the way sodomy is etymologically tied to the torrid zones (and all of the "unnatural" sex acts practiced therein); sodomy thus would have resonated within early modern and eighteenth-century climate theory as a behavior associated with darker skin.[17] Reading sodomy as a racialized formation sheds new light on the history of its codification into law and subsequent criminalization, and especially on the characterization of sexual liaisons between men of different classes or social stations as "sodomitical." Coaxing out the racial politics of sodomy demonstrates the investment of antisodomy laws in creating and maintaining racial order and, in turn, indexes the shared fortunes of sexual dissidents and nonwhite peoples in the history of criminalization in North America.

Insisting on the defining power of environmental logic in eighteenth-century understandings of sexual behaviors similarly poses significant challenges to the ways historians of early sexuality have thus far developed the central archives and critical questions of the study of sex before sexology. Despite Michel Foucault's field-defining insight that modern "sexuality" describes a particular, nineteenth-century arrangement of power/knowledge that produced the modern subject—and his implicit suggestion that if modern sexuality is a little over a century old, that there thus must be something else that predated it—we have not questioned our habits of looking for sex in the genitals, in the articulation of desire, or, at the very least, in the human body. Yet a myriad of long-eighteenth-century texts trouble this way of looking, from Linnaeus's characterization of shamelessness as a regional characteristic, to Herman Mann's apologia for sapphism as one of the "new sensations and emotions" provoked by the fervor of the Revolutionary War and indigenous to "Columbia's soil."[18] Furthermore, tracking the environmental framework through which sex was understood in eighteenth-century writing allows us to better see the persistence of environmental or external theories of sexual behavior into the twentieth century, with, for example, the use of behavior modification techniques in both therapeutic and psychiatric settings that sought to change the sexual predilections of both homosexuals and so-called sexual predators from the outside in.[19]

The fifth chapter of this book, for example, turns to cheap, oft-reprinted and extremely popular early-nineteenth-century adventure stories, all of which are set in "Negro Hill" (a mixed-race vice district in Boston) and which promise their readers a peephole into the world of unlicensed taverns, unchecked prostitution, and promiscuous interracial sex. In these narratives, it is not the brothel marms, the drunk sailors, or unrepentant criminals that render "Negro Hill" the city's most dangerous vice district; rather, it is the neighborhood itself, from the music of its dance halls, audible from miles away, to even the white clay that covers the boots of all who pass through it, quietly betraying the wearer's exploits of the previous evening. "Vice," and sexual vice in particular, is represented quite explicitly as a spatial problem in these texts, and when Boston is incorporated as a city in 1822, the new municipal government defines vice in similarly spatial terms; for city officials such as Josiah Quincy, eradicating vicious behavior was as much a question of urban planning as of reforming individuals. Chapter 5 examines how both blackness and sexual behavior acquired a spatial dimension with the development of urban public health infrastructure in the early nineteenth century, in the years just before sexual behavior became a discrete object of scientific study and was captured and organized into *sexology*, the immediate precursor to the regime of modern sexuality. We have not yet developed an analytic for early sexuality that would allow us to adequately theorize what it might look like outside of its imbrication in the subject, or even outside of its implicit relationship to the human. And as I discuss in the epilogue, writing a history of sexuality that is grounded in a less certain relationship to, or even at a distance from, the subject requires a reevaluation of not only how sexual behavior signified or was explained in the eighteenth century, but also, to a certain degree, how we understand it today. This study offers an environmental theory of early sexuality derived from the logics and the archive of natural history, toward the goal of describing a time when sex and sexual behavior were not imagined to exist in a strictly expressive relationship to interiority, a time when sex could be thought at a distance from the subject. This is a tall order, and in this sense, the book neither attempts nor claims the status of an exhaustive or even a particularly comprehensive study of early sex. My aim is to offer an intervention and some provocations and to model the kind of critical reframing of early sexuality that this rich archive demands.

A Historiography of Sexuality for Early America

To offer an environmental theory of early sexuality without first detailing what is designated by *sex* or *sexuality* itself, however, may be to get ahead of ourselves. And because some scholars of early American history, literature, and culture have at times challenged not only the concept of early *sexuality* but even the cohesiveness of the category of "sex" *tout court* in the eighteenth century, it behooves us to take a step back and consider what sex was—before sexuality.

It is difficult for us today, as twenty-first-century readers, to imagine these terms as distinct from each other. To speak of sex acts is to index, if not also to simply presume, a much larger breadth of information about a person, including patterns of behavior, tendencies toward certain acts and away from others, and even social and political identities. Today, "sex" and sex acts frequently stand in metonymical relation to sexuality, serving a predictive function. A sex act, in our time, is never just an act; despite the insistence by radical sexual political movements, queer theorists, and even conservative behavior-modification advocates who insist that sexual orientations are ephemeral and can be changed, it is difficult to wrest sex and sexuality apart.

As scholars of the history of sexuality have carefully argued, however, this close association between sex and sexuality is a specific historical phenomenon, and one that is only about 150 years old. In short, they have argued that iterations of sex did indeed exist before they were captured, organized, and integrated into the modern knowledge regime of "sexuality." Foucault, too, implicitly suggests as much; his account of the emergence of sexuality as one tactic for producing and disciplining the modern subject likewise intimates that sexuality was not *always* the dominant tactic and that sexuality, as a system of coordinated knowledges, did not always exist as such. And while not all historians of sexuality are compelled or convinced by Foucault, most of us more or less agree with his historical timeline. We have proceeded to populate our field of inquiry with articles, collections, and monographs that take as a point of departure the notion that in the era before 1900 or so, sex can and should be studied outside of the context of sexuality. This book thus describes a moment in North America that was not *without* sexuality, but *before* it.

Before sexuality, however, there was still sex. Some scholars of early sexuality have suggested that it may be an anachronism or misnomer to identify eighteenth-century behaviors such as onanism (masturbation), procreative intercourse, sodomy, or bestiality, among others, as *sex,* voicing a concern that it may be impossible to use or think with the term "sex" without invoking or reproducing ways of knowing and categorizing acts and desires that would be anachronistic to the study of the eighteenth century.[20] This study takes as its point of departure the belief that something like what we now call "sex" resonated as an idea of a type of act or realm of experience during this period. While residents of colonial eighteenth-century North America only infrequently used the word "sex," they had a host of other terms available to them (among them *venery, fornication,* and even *love*) to describe the general category of acts, pleasures, proclivities, and prohibitions, as well as a vast vocabulary of other terms to describe specific acts. Far from being a "modern" vocabulary or way of thinking, broad understandings of sex in eighteenth-century colonial North America proliferated, and they derived much of their cultural power from their imbrication in *ancient* sources such as the Bible. Indeed, one of the earliest execution sermons printed and circulated in the British North American colonies contains a long and extremely careful explanation of behaviors that fell under the category of "uncleanness," and these behaviors—all of which fall into modern categories of the sexual—are depicted as, quite literally, related to one another. Samuel Danforth wrote and delivered his 1674 sermon *The Cry of Sodom Enquired Into* upon the sad occasion of the execution of a young member of his congregation, Benjamin Goad, for bestiality.[21] Goad, a seventeen-year-old youth, had been caught in flagrante delicto with a mare in an open field at high noon in his hometown of Roxbury, Massachusetts, and had thus been sentenced to death, in accordance with Calvinist doctrine.

Danforth's sermon voices the stern censure of a community leader who fears that his flock has been secretly led astray.[22] His sermon offers both preparation and warning for his local community, and he expresses a thinly veiled suspicion throughout the lecture that other sinners of Goad's ilk lurked unspoken within his congregation. Danforth cautions his audience of the dangers of continuing their habits of uncleanness after hearing his sermon on the damning nature of these sins, and he includes a careful schematization of the various kinds of "uncleanness" that might lead one to the same unfortunate end as Benjamin Goad's. Enumerating for his community the different categories of "uncleanness"— "fornication" (which, Danforth explains, includes self-pollution, incest, whoredom, and adultery) and "going after strange flesh" (inclusive of sodomy and bestiality)—his exegesis methodically lists a series of acts that he clearly understands to be related to one another by type or kind. While within the context of late-seventeenth-century Puritan New England, Danforth's stern prediction of the inevitable damnation of many of his "unclean" listeners is not particularly noteworthy, the terms in which he casts these sins are. In hell, he argues, they will join all the sinners of their kind who came before them, and, he suggests acerbically, they will form a kind of hellish community made kindred by perversion. In language dripping with sarcasm, he exhorts the unrepentant members of his congregation:

> Be a Comfort to your Lewd Predecessors, who are long since gone down to the lowest Hell, and lie in the nether parts of the bottomless Pit. There are your Voluptuous Fathers, the renowned *Epicures* who lived before the Flood: There is your Libidinous Mother *Jezabel*, with all her Paramours. There are your Filthy Sisters, *Sodom* and her Daughters: There are your wicked Brethren, *Onan,* the unclean Son of *Judah; Hophni* and *Phineas,* the Belial-sons of *Eli; Amnon* and *Absalom,* the Incestuous Sons of *David:* There are your Venerous Kinsmen, the *Nicolaitans,* the *Gnosticks,* the *Menandrians,* and innumerable others. Hasten you after your lecherous Kindred into the stinking lake: sit down with your Brethren and Sisters in the depths of Hell. As you have partaken with them in their sordid Pleasures, partake with them also in their Plagues and Torments. Let thy lustful Body be everlasting Fuell for the unquenchable fire.[23]

Danforth draws this cast of venereous characters from all corners of his wide knowledge of both religious and biblical history; his congregation's "Lewd Predecessors" and "lecherous Kindred" can be found in both the Old and the New Testaments,[24] within large-scale narratives of Judeo-Christian theological history,[25] and are even rendered in figurative language.[26] This eclectic genealogy of Goad's specific sin dates back, for Danforth, as far as the beginning of the world and inculcates the Roxbury teenager into a kinship network that includes "innumerable others" whose "lustful Bod[ies]" tie the earliest of biblical times to Danforth's own moment, in the late seventeenth century.[27] The different historical moments of these unclean sinners collapse onto one another in the eternity of hell, a space where Danforth fears that Goad will likely join his "lecherous Kindred" in an unfortunate moment of overlap between Goad's seventeenth-century crime and every instance of fornication or going after strange flesh that came before it.

There is no question that Danforth identifies acts such as onanism, incest, sodomy, prostitution (as well as more dispositional qualities such as libidinousness, voluptuousness, and generalized venery) as belonging to a related category of acts, pleasures, and inclinations. Indeed, Danforth's ecumenical genealogy of "sordid Pleasures" and "Venerous Kinsmen" *is* in fact a genealogy, in the most straightforward sense of the term: he allegorically arranges these types or genres of sin into a tableau where they are all figured not only as people (and, indeed, many of the sins he describes are eponymously attributed to biblical figures, such as Onan and "Jezabel,") but as *kin*. To scholars who argue that *sex* did not bear any sort of categorical cohesiveness before the nineteenth century, Danforth's sermon issues a strong rebuttal. And while this sermon could certainly not be said to be representative of the way all or even most communities understood the categorical significance of sex acts over the course of the long eighteenth century—an artificial periodizing device that describes roughly 160 years—*The Cry of Sodom Enquired Into* nonetheless gestures to the fact that the idea that specific sins were related by kind existed with at least a minimal circulation at the end of the seventeenth century.

Furthermore, it would perhaps be easier to dismiss Danforth's sermon if it had not, by most scholarly accounts, initiated the North American development of a genre of print called "execution narratives," one of the earliest and most popular forms of Anglophone print to circulate in the colonial Atlantic world.[28] Printed by Marmaduke Johnson in Cambridge, Massachusetts, in 1674, Danforth's sermon became the first of its kind published in colonial North America. Although this sermon was initially published on its own, later execution sermons were frequently accompanied by a myriad of other documents when they arrived in print, ranging from first-person accounts of the commission of the crime, eyewitness accounts of the execution, confessions and last words of the condemned individual, and woodcuts depicting the scene of the execution. Exactly twenty-five years after Danforth's *Cry of Sodom Enquired Into* was printed, Cotton Mather published his *Pillars of Salt: An History of Some Criminals Executed in this Land, for Capital Crimes* (1699). As I discuss at length in chapter 3, Mather's first effort at gallows literature compiled twelve stories of men and women executed for murder, six of which had previously been published and at least one of which was borrowed from a sermon that Cotton's father, Increase Mather, had delivered in 1675, one year after Danforth penned his *Cry of Sodom Enquired Into*. This new form—the compilation—eventually came to define the standard for the execution narrative genre. While it was common for execution sermons to accompany these collections of what Jeannine DeLombard calls "gallows literature,"[29] with advances in print technology by the end of the eighteenth century that allowed printed matter to be produced faster, cheaper, and in larger quantities, execution narratives more and more frequently appeared as sensationalized accounts of criminal deeds, and they included spiritual literature less and less often. While ministers such as Danforth and Cotton Mather saw the print industry as a means of creating a wider distribution for the word of God, readers became print consumers motivated by a variety of interests, only some of which were spiritual.

I tarry on the broad dissemination and circulation of execution narratives, as a genre of writing that flourished both in the British North American colonies and in England itself, because popular print constitutes the archive at the center of this study and has much to tell us about how residents of British colonial North America understood and explained sexual behavior within their communities. As a form of historical writing that claims a basis in truth or reality, "popular narrative" is a capacious term that describes stories that circulated widely and in many forms, including folklore, songs and ballads, novels, plays, poetry, essays, sermons, and "true histories." Importantly, "popular narrative" was not always, or even usually, "popular" in the sense in which we often use the word today; while it is difficult to determine with any accuracy the circulation of the texts at the center of this study, most if not all of the works discussed in these chapters could boast of very few readers by present-day standards, owing to factors including low literacy rates for much of the eighteenth century and the uneven availability of print technology in the North American colonies. Rather, the "popularity" of popular narrative as I discuss it here denotes an array of demotic cultural narratives, practices, and ideas that include, and often made their way into, the print record, although these practices are not limited to that record.[30] And while this book privileges popular narrative that found its way into print, many of the narratives that appear in works that I examine can be traced to folklore and other vernacular sources, and initially they existed primarily in oral cultures. But despite the uneven or perhaps even limited circulation of these texts, they nonetheless provide powerful evidence of what Foucault might have termed "subjugated knowledges" of sexual behavior that have been obscured by the long shadow of the sexual identity paradigm.[31]

Each of the following chapters considers a different genre of popular narrative —natural history, Barbary captivity narratives, execution narratives, cross-dressing narratives, and tales of urban vice—many of which existed in at least a colloquial relationship to real events and people.[32] Reading these genres of popular print together allows us to consider multiple, intersecting technologies for representing sex and sexual behavior within colonial and early national culture. It constitutes my attempt to define an archive of representations of sexual behavior that saw broad vernacular consumption and circulation within the British North American colonies and the early national United States.

The history of sexuality, as a field, tends to favor emergence arguments, and it is tempting to make large-scale claims about how a culture engaged with a series of increasingly proliferate questions about the significance of sexual behavior. This study proceeds from an awareness of the incredible cultural, infrastructural, and in places religious heterogeneity of the British North American colonial world, and even of New England as a region. Like the twenty-first century, the eighteenth was a time of great cultural and epistemological change, and any singular narrative that would attempt to capture the dynamism of the multiple, aggregate, and competing frameworks that worked in both complementary and contradictory ways to describe sexual behaviors during the period would inevitably fail. As a result, I make no claim to the dominance of any one primary mode of understanding or explanation when it comes to sexual behavior. Instead, I trace a particular way of explaining and representing sexual behavior—environmentalism—through a broad range of popular and erudite texts within the archive of eighteenth-century Atlantic writing on human difference, highlighting the quiet ubiquity of their environmentalist logic and imagining how extant historical narratives about the significance of sex acts in the century before sexology might change or thicken if we were to incorporate this way of knowing. This historiographic approach is indebted to Roxann Wheeler's description of the method she brings to her work on the history of race in eighteenth-century England. She writes: "An issue related to accounting for residual proto-racial ideologies of difference is acknowledging the sedimentation of racial ideology. Twentieth-century critics too often interpret the new scientific practices as dominant ones, which has skewed our sense of the extent and immediacy of changes. Many old and new beliefs and practices coexisted and cross-fertilized each other. Rarely was there a complete break with older ways of thinking."[33]

I do not, then, imagine that natural history delivered its environmentalism to the British North American colonies, effecting an immediate sea change in popular thinking about sex. Rather, I suggest that we reorganize our inquiries into the history of sexuality in eighteenth-century British North America such that they center or acknowledge the primacy of environmentalist thinking in contemporary popular and scientific accounts of human difference. I am not asserting that the environmentalist logic with which natural history, in particular, was shot through was the *only* framework that residents of early colonial North America had available for making sense of sexual behaviors. But I argue that the cultural power of this framework—widely acknowledged in studies of the history of race in the eighteenth century—has been vastly understudied regarding what it might have to teach us about the history of sexuality during the same period.

Furthermore, this study negotiates two competing tensions within the history of sexuality, as a subfield. The first is that wry Jamesonian exhortation to "always historicize" and what seems to be the de facto methodological response to his call: archival research, a strict fidelity to primary sources, and letting the eighteenth century (in this case) speak to us "in its own terms," what we might think of as an "originalist" method. The second tension pertains to the politics of the production of historical narrative and the demand for what Lillian Faderman has termed a "usable past."[34] On some level, many scholars of sexuality—and I certainly count myself among them—want to develop histories of the relationship between sex and the sciences of human difference that can be put to work to develop righteous and nuanced accounts of the shared ideological and political territory of sexuality (before and as it came into the purview of science and, subsequently, the state) and racial difference. Each of these historiographic impulses bears its own set of promises and pitfalls. Uncritical archival fidelity runs the risk of reading objects as metaphors for subjects, of assuming the unremitting knowability of the past and thus failing to account for the drive toward a problematic mastery that underlies this scene of interpretation; Anjali Arondekar pithily characterizes this methodological conceit as "the presumption that if a body is found, a subject can be recovered."[35] Yet historical interpretation that clothes the eighteenth century in the vestments of the twenty-first risks the colossal, devastating erasure of the relentless specificity of the everyday realities of people very distant from us, people who were grappling, in their own time and in their own terms, with the both philosophical and ethical question of making sense of human difference.

This book represents an effort optimistically driven by the best parts of each of these two scholarly impulses, and I ask that the reader approach both the primary texts considered herein and the arguments I make about them with a critical generosity, for what these eighteenth-century texts have to tell us may surprise you. They certainly surprised me. When I initially began this project, I had planned to examine the long genealogy of the bind between sexuality and interiority in the specific context of eighteenth-century colonial North America, proceeding from the a priori assumption that such a bind *existed*. As I began to revise the project into a monograph, however, I found myself stalled at every turn: my primary texts were not supporting my thesis, and even when I clumsily searched for primary texts that *would*, I came up empty-handed. Years later, I came to understand that the problem was not due exclusively to my bringing a modern framework for understanding sexual behavior to eighteenth-century representations thereof—it was, in fact, more complicated than that. The problem was that I was failing to see the deeply ecological way that eighteenth-century colonial texts understood human bodies in my analysis of my primary texts, thus limiting my understanding of *sex,* and how sex might be represented, to acts, inclinations, sensations, and pleasures within or between human bodies.

Once I jettisoned these anachronistic understandings and began again, the project came together quickly, in part because its central argument is deeply intuitive. That no scholarship has previously advanced what I am calling an environmental theory of early sexuality is an effect, I believe, of the worst iterations of the twinned historiographic impulses described above: a flat historicism might assume that nothing even akin to sexuality existed before the modern period, and readers seeking a "usable past" in eighteenth-century archives will miss what these archives have to say about the significance of sexual behavior because the way sex is represented therein does not resonate with modern-day commonsense about what sex *is* or looks like. I am at best skeptical that a formation that might be called sexual identity existed in any form during the eighteenth century, yet I simultaneously refuse the idea that sex and sexual behavior was apolitical. Similarly, this project does not seek political or identitarian forebears in eighteenth-century writers, thinkers, and readers who engaged questions about human variety and sexual diversity, but it nonetheless welcomes the similarities, and perhaps even the continuities, between the way "sex" and "human variety"—notions that our current-day epistemes of *sexuality* and *race* evoke with uncomfortable familiarity—were frequently expressed through each other then, and how these figures stand in charged relation to each other now.

Notes on Terminology:
Sex, Sexuality, Race, Environmentalism

One of the unique challenges of studying the eighteenth-century world is not merely that it was organized in such a radically different way from how the world looks today; it also stems from the fact that early Americans made sense of their world through such different knowledge frameworks, and frequently using different vocabularies, than those to which we have access now. Complicating this historiographic conundrum is that a great deal of overlap exists between, for example, eighteenth- and twenty-first-century Anglophone American vocabularies for sex, but the meanings of the same words have changed, sometimes drastically, over time. While certain terms have largely been left behind—one rarely hears discussions of behaviors called "sapphism" anymore—many terms persist, but this persistence cannot be trusted to indicate that, for example, the eighteenth-century use of a term such as "sodomy" indexes the same behaviors, inclinations, or pleasures as "sodomy" gestures to, today. Furthermore, "sodomy," like many other words for sexual behavior that saw a broad and popular usage during the eighteenth century, was not used, defined, or understood in a self-identical fashion even during the same historical periods, in the same specific culture. As Richard Godbeer, Will Fisher, and Jonathan Goldberg have all demonstrated, "sodomy" was a complex term, thick with implication, a term used to describe a sweeping set of acts, ranging from anal sex between men, sex between people of different classes or positions on the social hierarchy, and sex between people of different religious affiliations (especially Christian and non-Christian).[36]

Terminology, then, becomes not only a stylistic but also a historical and historiographic question for any scholar working on the history of race or the history of sexuality in the period before 1900, and it has been the center of many passionate debates. In this book I use both *race* and *sex* to describe facets of eighteenth-century life and forms of human difference that would not have been commonly described in those terms during this period. I do so not in an effort to collapse the eighteenth-century epistemologies surrounding racial difference or sexual specificity that are the subject of this book with those that we employ today, but for the following two reasons: first, *The Natural History of Sexuality*, as both a historical and a methodological project, proceeds from the assumption that residents of eighteenth-century colonial North America relied on radically different logics to characterize and explain human racial difference and human sexual difference during that period; indeed, that is the central argument of the book. While I respect the choices of scholars such as Roxann Wheeler and Andrew Curran to use the terms most commonly employed by eighteenth-century writers and readers to characterize these forms of human difference, I do not believe that using eighteenth-century language significantly aids the project of fully anatomizing or capturing the specificities of the way that people who lived, spoke, read, wrote, and *knew* during this period made sense of "race" and "sex" as forms of difference. Indeed, I do not think that project is possible or, perhaps, even desirable. Furthermore, reaching toward a kind of historical purity or precision around eighteenth-century language actually veils or refuses the reality of the incredible slipperiness and polyvocality of terms like *race, sex,* and many others, in their own time. I use the terms *race* and *sex*—which, as I will demonstrate, were indeed used by residents of eighteenth-century North America, just less frequently and sometimes to different ends than we use them today—because the historical, representational, and cultural disorder that they bring to this project provide an apt echo or illustration of the disorder inherent in the cacophonous chorus of competing voices and ideologies that sought to make sense of these forms of human difference during this period. Finally, despite some scholars' insistence that "early Americans did not recognize the sexual as a discrete category the way people in the present do," I use the term *sex* because, as indicated earlier, I simply disagree with this position.[37] I also use the term *sexuality,* but I use it to designate a particular nineteenth-century sexological and postsexological configuration of sexual knowledge as it was bound to specific historical sciences of interiority. It would be fallacious to assert a rigid break between the presexological and the sexological era, not least because competing scholarly accounts date the emergence of sexology to various points within the sixty-eight-year window between 1837 and 1905.[38] Yet I use these terms, "presexological" and "sexological," as labile indices of a shift in the dominant framework through which sex was understood during the eighteenth and nineteenth centuries. Similarly, I gesture to *early sexuality* (e.g., "an environmental theory of early sexuality") to describe a set of historical and epistemic questions about how sexual behavior was understood before sexology, but I do so with the conviction that *sexuality,* in the Foucauldian sense, was not the dominant vocabulary for making sense of sexual behavior in the eighteenth century, even though there were certainly individuals who imagined sexual behaviors to be expressions of internal disposition or desire.

Second, I use the terms *race* and *sex* because I firmly believe that the debates, ideologies, and structural hierarchies that were organized around these terms and all that they may have indicated during the eighteenth century bear a relationship to those that constellate around these terms today, whether or not any continuity exists between eighteenth-century "human variety" and twenty-first century "race." Then, as now, conversations about human religious, phenotypic, moral, behavioral, and characterological difference were highly politicized, and these ideologies or formations were frequently defined and organized in a hierarchized relationship to one another. Furthermore, legal, infrastructural, and biopolitical architectures of state power saw "human variety" or racial difference, as well as a wide range of sexual acts, as crucial sites for intervention, policing, and control. In many cases, these architectures—I am thinking especially of seduction laws and sodomy laws—persisted long into the twentieth and, in some cases, the twenty-first centuries, forging a distant but nonetheless material tie between the ways early American colonists organized their lives and the way many people in the United States live today. One need look no further than *Loving v. Virginia* (1967), *Bowers v. Hardwick* (1986), or *Lawrence v. Texas* (2003) for evidence of the lasting sociolegal impact of early American disciplinary infrastructures that grew up around sexual behavior.

Thus, in this book, I use the term *race* to describe eighteenth-century understandings of human difference that were captured by the contemporary phrase "human variety," a concept that includes what we would now call "race," but also "ethnicity" and likely also "religion" and even "culture" or comportment more broadly. To be very clear, my use of *race* is specifically *not* limited to forms of anatomical, morphological, or phenotypic difference among humans, but rather takes seriously the voluminous vocabulary and range of questions with which eighteenth-century natural philosophers, ministers, popular writers, and readers approached what appeared to them as the exciting and critical question of human difference. Similarly, I use the term *sex* to denote a broad range of practices (e.g., sodomy, onanism, sex work); bodily experiences and sensations (e.g., drunkenness, passion broadly defined); characterological qualities (e.g., shamelessness, lacking ardor for sex); inclinations (e.g., cross-dressing, rape), including but not limited to genital-based contact between bodies; and "generation," a term the eighteenth-century Anglophone world used to describe what we now call "human sexual reproduction." In this book I look to the realm of the *venereal* in my understanding of eighteenth-century sex; the venereal was a capacious category of human experience that included all forms of appetite (including appetite for sex, food, drink, and money) and that articulated with fleshliness more generally. Finally, I use the terms *environment* and *environmental* to designate the wide range of natural, physical, and social phenomena that were understood by natural historians to exert an effect on the human constitution, including but not limited to climate (heat or cold, exposure to sun, moon, and cosmos), patterns of consumption (food and drink), local flora and fauna, natural landscapes, social landscapes, and human-made architecture.[39] The environmental theory of early sexuality that I develop explicitly builds on scholarship in the history of race that understands the environment as a central, if not defining, term in eighteenth-century theories of the etiology of human difference.[40] While the chapters that follow trace the natural historical interest in the effects of a range of these environmental influences (climate, diet, state of society, exposure to both moral and immoral people and behaviors, the natural world and its flora and fauna, and architecture) on the human body, this study does not prioritize any one of these external influences over the others. Rather, it is by tracking how natural historians as well as vernacular writers imagined the effects of a *broad range* of external influences on the human constitution that we are able to better understand the centrality of the environmental *writ large* as an aggregate and defining factor in eighteenth-century understandings of human behavior.

Paying attention to the kinds of conversations and local knowledges that appended themselves to cultural negotiations of human difference during the eighteenth century—the period in which the British North American colonies declared independence and developed a legal infrastructure that remains with us today—bears the potential to critically historicize all that we think we know about the intricate meaning and impact of racialization and sexuality in our own lives. It also encourages the denaturalization of *sex* and *race* as terms that at times seem both transhistorical and universal; readers may encounter discussions of eighteenth-century sex acts in this book that are not recognizable to them as *sex* and passionate taxonomies of racial differences assumed to exist between peoples who, to a present-day reader, all appear to be *white*. Examining the contours of cultural investigations into the nature of, and the relationship between, sex and race during the eighteenth century thus offers us a generative opportunity to reflect on the contours of these same conversations as they continue to exert dynamic, expansive influence in our lives right now.

The project of producing a history of colonial North American sexuality can never be a neutral or apolitical one; at a time when registries for both "sex offenders" and gay fiancé(e)s are radically on the rise, researching and writing the histories of homophobia, "sex crimes," and sex work needs to be a project embarked upon conscientiously, with a specific view to how these histories inevitably seep into the fabric of North American racial and sexual politics today.

On that note, a final directive on the use of *racialization:* to speak of the history of race, and the ways of knowing that allowed "human variety" or racial difference to be thought, fails to speak with any precision about the process by which human variety was organized into hierarchies and differentially governed. Foucault began to describe this, albeit in a very inchoate way that did not speak to the realities of colonial infrastructure and governance, at the end of his life in his writings on biopolitics; these questions have recently been picked up and revitalized by scholars in postcolonial and critical ethnic studies that add nuance, depth, and historical and geographic precision to Foucault's account of biopower.[41] Following Simone Browne and David Theo Goldberg, I use *racialization* to name the strategic production of human difference in the service of governance: in other words, the production of site-specific modes of knowing human difference that worked in lockstep with settler colonial organization, thus securing the reproduction of particular structures of labor and control.[42] The sexual politics of racialization, then, is a notion that not only indexes that sexual diversity was often expressed within representations of racial difference; it also describes a particular way of presenting *both* specific forms of human variety—blackness, for example—and the sexual specificity that attends them (Linnaeus's "women without shame," or Jefferson's characterization of Native men as despotically abusive to their wives) that justifies colonial intervention (e.g., "the black rapist"). Racialization was both a fact and a strategic process that instrumentalized sex and human difference to ensure a particular organization of racial hierarchy.

Sex and Environmentalism in the Archives

William Cronon's landmark essay on environmental history, "A Place for Stories" (1992), provided one of the first iterations within early American studies of what Monique Allewaert has recently termed the "ecological" status of the human.[43] Cronon's epic call for the decentering of the human in American history, however, had the unanticipated result of not decentering the human in our narratives of early American history so much as *redefining* it. In "A Place for Stories," Cronon writes:

> On the one hand, a fundamental premise of [History] is that human acts occur within a network of relationships, processes, and systems that are as ecological as they are cultural. To such basic historical categories as gender, class, and race, environmental historians would add a theoretical vocabulary in which plants, animals, soils, climates, and other nonhuman entities become the coactors and codeterminants of a history not just of people but of the earth itself. For scholars who share my perspective, the importance of the natural world, its objective effects on people, and the concrete ways people affect it in turn are not at issue; they are the very heart of our intellectual project.[44]

While Cronon is clearly asking his readers to reevaluate what he implies is a hegemonic, disciplinary humanism among historians, this passage also has the effect of calling for a shift in how the human, in even the most human-centered of historical accounts, is understood. While he stopped short of explicitly theorizing the human in ecological terms in this essay, Cronon and other environmental historians ultimately provided some of the critical intellectual and theoretical foundations for later work in eco-criticism and environmental studies that importantly questioned the status of the human as autonomous with regard to the nonhuman world.

Cronon was not, of course, the first to think this way; as a historian of early America, he was undoubtedly aware that he was only the most recent inheritor of a long intellectual tradition of environmentalist thinking. Originating in the ancient world in climate theory and humoral theory, which constituted two of the major schools of thought influencing the practice and theory of health in early America, ecological understandings of the human flourished within the ancient, medieval, early modern, and eighteenth-century European worlds. One's exposure to heat and sun, argued natural philosophers from the Virginia-born John Mitchell to the Scottish William Falconer, affected one's physiology, phenotype, intellectual ability, and character. Eating spicy foods or consuming too much meat could likewise alter the humoral constitution of the body, transforming a healthy, active, passionate person with a "sanguine" complexion into a bilious, melancholic idler. In short, residents of the eighteenth-century Atlantic world relied less than we do today on the notion that the body is an autonomous, closed system or on the idea that one's disposition and morphological qualities are ingrained or permanent. Rather, early American colonists understood their bodies to be porous, deeply susceptible to the influence of the external, nonhuman world, and they understood their particular characterological and even *racial* qualities to be, at least in part, dynamic and an effect of their prolonged exposure to the environments they inhabited.

Environmental influence took a multitude of forms, however; it was not limited to the effects of climate and diet. As Pennsylvanian Samuel Stanhope Smith argued in 1787, one's "state of society" also bore the potential to affect one's disposition, physiology, and capacity: "The state of society comprehends diet, clothing, lodging, manners, habits, face of the country, objects of science, religion, interests, passions and ideas of all kinds, infinite in number and variety. If each of these causes be admitted to make, as undoubtedly they do, a small variation on the human countenance, the different combinations and results of the whole must necessarily be very great; and combined with the effects of climate will be adequate to account for all the varieties we find among mankind."[45] This passage illustrates an important distinction between eighteenth-century definitions of *environment* and eighteenth-century *environmentalist frameworks*. In Smith's statement we see the protean nature of environmental influence; almost every element of one's world has the capacity to shape the way one lives in it. But beyond his enumeration of the possible factors that might make up one's environmental landscape, Smith also expresses his belief in environmental thinking: "combined with the effects of climate," Smith writes, these environmental factors "will be adequate to account for all the varieties we find among mankind." If we are to take Smith and his naturalist peers seriously, environmental factors—and environmental logic more broadly—constitute one of the defining frameworks through which residents of British colonial North America characterized and understood human difference writ large.

I argue that one of the forms of human "variety" that increasingly drew the attention of eighteenth-century natural philosophers was diversity within human sexual morphology and behavioral patterns. While only chapter 1 discusses eighteenth-century natural historical thought at length, chapters 2, 3, 4, and 5 detail how environmentalist thought informed and was shaped by genres of popular print. Chapter 1 charts the emergence of natural history as a critical intellectual movement, detailing its imbrication in the much broader eighteenth-century project of theorizing the origins and function of racial difference; the status of natural history as an early race science explains the cultural strength of the knowledge that this produced. While natural history is rarely if ever considered alongside more conventionally studied genres of popular narrative such as Indian captivity narratives, slave narratives, seduction novels, or cross-dressing narratives, I argue that natural history constituted a form of popular narrative and thought unto itself and that the broad environmentalist logic that natural history both developed and helped to popularize informs and is invoked by all of the popular print media I consider in the chapters that follow. Emphasizing the participation and imbrication of popular narrative in eighteenth-century environmentalist thinking offers new ways of conceiving of how these texts contributed to the development of cultural knowledge about sexual acts and proclivities during this period and shifts the critical conversation surrounding this archive toward a heretofore unexamined facet of the eighteenth-century Anglophone print record.

Popular narrative provides the archival center for this study for many reasons, perhaps the most important being the frequency with which the intersections of questions about human variety and sexual practices are staged in this genre of writing. Genre, as Travis Foster argues, is less about form than it is about *repetition*, and what interests me about the popular narratives that I discuss is less the internal cohesiveness (or lack thereof) of specific types of narratives—Barbary captivity narratives, for example, or antivice narratives—than the sheer repetition of specific types of stories, especially when these stories showcase and at times sensationalize the same questions about human difference that were so central to natural historical writing.[46] This project tracks how sodomy becomes a tropological feature of representation of "Turks," "Arabs," and Islam; how rape is carefully welded to blackness such that crime is discussed in racialized terms (e.g., describing general sinfulness as "Egyptian Lusts"), and blackness is described in terms of sex; how sexual proclivities such as sapphism are posited and taxonomized as indigenous species of the New World; and how sex work and other vicious behaviors get cast as the inevitable result of a too promiscuous culture of interaction between white people and black people.

The Natural History of Sexuality focuses on this body of print not for the way these texts explicitly address transgressive sex acts, as many of them do, but rather for how they echo, contribute to, or expose the logics through which sex acts were explained and understood, situating these explanations in the context of natural historical questions, inquiries, and writings (some of which were being printed by the same printers and print shops at the same time). And like natural historical writing, a form of inquiry that many medievalists and early modernists locate as originating in the medieval period, the genres of popular narratives considered here are for the most part much older than British colonial America; similarly, both natural historical writing and popular narrative rely on a particular claim to veracity or evidentiary use for their power, and they often have roots in stories of real people or events: natural histories are frequently described as "true narratives," and popular print stories about rapists, captives, or cross-dressers are often termed "true histories."

The cross-dressing narratives discussed in chapter 4, for example, have roots in seventeenth-century working-class British balladry and were likely initially songs that were sung and learned among sailors. These stories of women who dressed as men, joined the military, and enjoyed various forms of success, romance, and other gender-bending high jinx were later echoed in performances offered by a real-life female soldier, Hannah Snell, who donned her former military garb and executed drills on stage for London audiences in the 1750s. By the time Deborah Sampson, the historical figure whose fictionalized memoirs provide the focal point for chapter 4, emerged as a public figure in the United States in the first decade of the nineteenth century, her story would have been received within a historical context in which both cross-dressing and social concerns about sapphism were available vocabularies through which her aberrant behavior might be understood.[47] While Sampson was a real historical person whose male military alias, Robert Shurtliff, we can find in Continental Army rosters from the early 1780s, her narrative also joins a long, transatlantic assemblage of stories of cross-dressing individuals. The lion's share of these stories are probably fictional, but some of them are likely based on real events, and all of them share a dubious but possible relationship to "real" histories. Perhaps Sampson's choice was even made possible by the circulation of these popular stories.

Indeed, the widespread availability of many of the stories that popular narratives tell and retell allowed for the dissemination of particular families of ideas throughout Europe and up and down the eastern seaboard of the British North American colonies, and thus one of the unifying features of the texts that constitute the analytic center of this study is the way these texts act as a vehicle for the circulation and diffusion of specific ideas, even as they themselves circulated as objects. Questions of human variety, in particular, would not have been able to find quite the broad cultural traction that they did during the eighteenth century without the realities of colonialism and the burgeoning strength of oceanic and increasingly global markets. Animating inquiries into the significance of all kinds of human diversity—morphological, religious, sartorial, customary—was the experience of contact between cultures, as well as conflict over the control of land, trade routes, goods, and what Marx termed social reproduction.

In this sense, what binds together the archive at the center of this study is not that these texts are from the British colonial world, the Anglophone Atlantic world, or even early America, but rather that they gird and stage a series of features inherent to eighteenth-century settler colonialism. Historians of science have extensively documented the critical importance of natural historical inquiry and knowledge production to colonial exploration and domination;[48] "the search for new specimens" of plants, humans, minerals, and other organic entities that might be put to work or sold to augment imperial wealth was a major justificatory idiom for the colonial project.[49] While popular narrative has not yet been engaged extensively by scholars as a site of readerly engagement with settler colonial practices or knowledges, it would be difficult to deny that moments like the scene in Mann's *Female Review* when Deborah Sampson/Robert Shurtliff kills a Native man while on a scouting mission on the Ohio frontier index the engagement of popular print with the same kind of settler colonial project that natural historical writing supported so explicitly. This study proceeds from the organizing premise that both erudite and popular engagement with questions of human variety and sexual practice during the eighteenth century were inextricably informed by colonial knowledge production and also represented an iteration of it.

Finally, this volume charts the way popular narratives offer a generatively apocryphal account of how sex and sexual behavior became a category of *cultural* or *collective* knowledge in the eighteenth-century Atlantic world. To return to Danforth's *Cry of Sodom Enquired Into,* the circulation of families of ideas in print (pamphlets passed from ministers to parishioners and from community members to friends and family) converts what could have been an isolated, singular iteration of Danforth's individual interpretation of "uncleanness" as a category of sinful behavior into a set of ideas enthusiastically consumed by hundreds of people, perhaps even more, over the next two decades. I am thus concerned here not only with the ways doctors, ministers, and natural philosophers made sense of apparent variations within sexual practices and pleasures across time, among various communities, and in different geographical locations, but also with theories of sexual diversity generated and consumed by a much wider cross-section of society: people who might not have been able to read or write, people who might not have been able to buy or borrow books, and people whose access to narrative might have primarily taken the form of listening, rather than reading—to sermons, to stories, to songs, to advice, or to lectures.

Although literacy rates were quite low for most of the eighteenth century, even some of the poorest and most disenfranchised residents of the British North American colonies had occasional opportunities to engage with popular media. And with literacy rates on the rise even among poor people toward the end of the century, communities of consumers began to drive print markets, increasingly dictating content and production. It would be unrealistic to imagine that we could use the kinds of stories that were printed, the number of editions that these print productions saw, or the number of copies purchased to gauge with any accuracy what the broadest cross-section of the colonial American public thought or knew about sexual behavior—a fuzzy category of eighteenth-century experience that already bears a vexed relationship to archival representation—but popular narrative offers us the best and most generative archive through which to hypothesize possible answers to these questions.

The Sexual Politics of Racialization

That a sexual politics inheres in processes of racialization would be difficult to deny, in the eighteenth century and in our own time.[50] Specific images—the "black rapist," for example, which I follow through a number of execution narratives in chapter 3—have even maintained a certain staying power, remaining legible today as images used to inspire fear and justify the surveillance of people of color. This is neither a new nor an eighteenth-century phenomenon, but characterizing the relationship between race, racialization, and sex remains one of the persistent challenges of studying the mutual emergence of sciences of human variety and sexual difference in British colonial North America. Even more complicated is that the politics of sexuality in our own time are increasingly defined in comparative relation to racial politics, often with an impressive lack of analytic care.

Consider, for example, a late-2012 commercial run by the Human Rights Campaign, one of the largest and wealthiest LGBT political lobbying groups in the United States, to publicize its work in support of gay and lesbian marriage equality. Titled "Dawn of a New Day for Marriage Equality" and narrated by Morgan Freeman, the commercial explicitly invokes Ronald Reagan's famous 1984 "It's Morning in America" presidential campaign advertisement, using montage to present gay and lesbian marriage equality as the most recent struggle of an American civil rights activist tradition.[51] Opening with a sped-up shot of the sun rising over the ocean, the first one-third of the thirty-second ad moves quickly through a series of images we associate with the American patriotic and political landscape: the Statue of Liberty, the US Constitution, black-and-white images of early-twentieth-century women suffragists (one second), and black-and-white images of Martin Luther King Jr. at the March on Washington (more than three seconds).[52] The ad concludes with a fifteen-second montage of gay and lesbian partners and families, coupled with the gentle familiarity of Morgan Freeman's distinctive voice narrating, "America stands at the dawn of a new day. Freedom, justice and human dignity have always guided our journey toward a more perfect union. Now across our country, we are standing together for the right of gay and lesbian Americans to marry the person they love." The advertisement thus draws a strong association between the Human Rights Campaign's vision of gay and lesbian political equality and the most iconic mainstream images of the black civil rights movement. Using a "like race" filmic rhetoric that explains the struggles of sexual minorities through the history of minoritarian racial and ethnic struggle, the ad asks its viewers to understand gay and lesbian marriage rights as inarguable in the same way that its ostensible audience would read civil rights as an uncontroversial political priority.

I refer to this contemporary example of the use of "like race" narratives that link sexuality and race to draw attention to the fact that analogy is one of the most prevalent cultural logics through which the relationship between race and sexuality is currently discussed.[53] Curiously, analogy was also one of the darling explanatory structures of eighteenth-century logic, and this was especially true of eighteenth-century naturalist science.[54] The relationship of analogy, however, was simultaneously *not* a prevalent idiom through which the relationship between human "variety" and sexual behavior was expressed, and the question thus remains of how to describe the way the sexual politics of racialization contributed to rendering sexual practices visible without collapsing sex into racialization or simply characterizing racialization as something of an early modern factory of *sexuality*.

Although it is not an environmentalist account of either racial or sexual difference, let us consider the biblical story of the Curse of Ham, a particularly widespread cultural reference point for making meaning of human variety and one with which both European naturalists and even uneducated residents of eighteenth-century North America would likely have been familiar.[55] The story of the Curse of Ham, from Genesis 9, was commonly used to explain the origins of human racial difference during the eighteenth century. The episode tells the Old Testament story of Noah, his three sons—Shem, Ham, and Japheth—and their families, in the aftermath of the great flood that purged the earth of nearly all of its living inhabitants. As the human survivors of the flood, Noah and his family have inherited the task of repopulating the world, and God famously bids Noah and his sons to "be fruitful, and multiply."[56] After disembarking from the ark, Noah plants a vineyard, the produce of which is made into wine, and as the story goes, Noah

> drank of the wine, and was drunken; and he was uncovered within his tent. And Ham, the father of Canaan, saw the nakedness of his father, and told his two brethren without. And Shem and Japheth took a garment, and laid it upon both their shoulders, and went backward, and covered the nakedness of their father; and their faces were backward, and they saw not their father's nakedness. And Noah awoke from his wine, and knew what his younger son had done unto him. And he said, Cursed be Canaan; a servant of servants shall he be unto his brethren. And he said, Blessed be the Lord God of Shem; and Canaan shall be his servant. God shall enlarge Japheth, and he shall dwell in the tents of Shem; and Canaan shall be his servant.[57]

As Bruce Dain argues, this story was interpreted in a myriad of different ways throughout the early modern period and the long eighteenth century, and although the Bible story does not offer any details about the differences between Shem, Ham, and Japheth, within popular and scientific eighteenth-century thought, it was broadly believed that Shem, Ham, and Japheth were the progenitors of the three races of men: the peoples of Asia, Africa, and Europe, respectively.[58] Perhaps more significantly, this passage from the Old Testament was also frequently used to justify the link between African peoples and providential servitude; because Canaan, as the son of Ham, would have ostensibly been African (and dark-skinned), and because Canaan and all of his descendants were doomed to a life of servitude for Ham's sin, the enslavement of Africans was thus excused as a biblically ordained system of labor. Finally, the Curse of Ham also served as divine evidence for proponents of monogenetic understandings of human variety; humans all descended from the same species (and person—Noah), and human variety could be attributed to God's darkening of Ham as a punitive measure.[59]

Like the popular eighteenth-century theories of human variety that tend to focus on what we now term racial difference, however, the biblical story of the Curse of Ham contained a strong latent narrative about sexual behavior, especially in early modern accounts of the Curse.[60] The King James Bible narrates Ham's sin as simply the act of looking upon his father's nakedness and reporting it to his brothers instead of covering him up. The problematic aspect of Ham's actions lies entirely in his disrespectful attitude toward his father. Yet early modern accounts of this story teem with other implications, and many scholars believe that this is due to a retelling of the story that appeared in Richard Hakluyt's widely circulated *The Principal Navigations, Voiages, Traffiques and Discoueries of the English Nation*, a multivolume collection of primarily travel writings published between 1589 and 1600. In a narrative titled "First Voyage by William Towson," George Best, an explorer searching for the Northwest Passage in 1577, offers a digression on the origins of dark-skinned peoples based on his dramatic revision of the biblical story of the Curse of Ham. In Best's telling, while on board the ark, Noah instructs his sons to "use continencie" and "abstain[] from carnall copulation with their wives." Ham, who was under the impression that the first child born after the flood would "inherite . . . all the dominions of the earth," refused Noah's exhortation to abstain from sex and "used company with his wife, and craftily went about thereby to dis-inherite the off-spring of his two brethren." As punishment for his disobedience, God made him "so blacke and loathsome, that it might remain a spectacle of disobedience to all the world. And of this blacke and cursed Chus [Cham, or Ham] came all these black Moores which are in Africa."[61] In the account that appears in Hakluyt's sixteenth-century *Principal Navigations,* Ham does not earn his curse by disrespectfully looking on his father's naked body while he slept, but by disobeying his father through a form of sexual incontinence; this story, by association, links "black skin to polluted and polluting 'sexual transgression.'"[62] While most eighteenth-century versions of the story of the Curse of Ham are relatively standardized (and do not reflect the version that appears in Hakluyt), many of them still bear traces of various iterations of a sexual narrative. In some versions of the story, as Dain argues, Noah's curse on the Canaanites included a "shamefully elongated" penis alongside the blackening of their skin; and as Dain parenthetically notes, "(sometimes Ham's sin was seen as sodomy or even castration of his father)."[63] Here, too, the curse of racialization bears a sexual politics, even in this early, biblical account.

The sexual dimensions of this major and central story of the origins of human variety call attention to the fact that despite its implicit refutation of an environmental or heliotropic account of racial difference or of human variety, it, too, contains a series of narratives about human sexual specificity that changed over time but nonetheless persisted into the long eighteenth century. What this suggests is that it is not *only* eighteenth-century environmentalist logic that tied emergent epistemologies of racial differentiation to incoherent but popular accounts of sexual diversity among humans. Indeed, the consistency of overlay between conversations about racial difference and piecemeal or isolated narratives about sexual specificity suggests a much older relationship between them. Such forms of narratives about human difference move and grow together, but they were popularized and institutionalized at vastly different rates, at very different historical moments, and in very distinct ways. As a scientific account of human nature, *racial variety* was both an older and a stronger narrative than *sexual diversity* for most of the early modern period; sexual behavior, as an object of sustained attention, really only seems to gain popular traction in the eighteenth century with the arrival of, for example, transatlantic anti-masturbation campaigns.

The challenge that remains pertains to how to characterize this relationship between, on the one hand, the emergence of racial science as a field of knowledge in and of itself, as an epistemology that sedimented and was instrumental in disseminating a particular set of dominant forms or analytics through which human difference might be categorized; and, on the other, the implicit sexual narratives borne by racial science as a form of knowledge, which, though frequently weaker and quieter than the racializing stories in which they were embedded, seem to have accreted and coalesced into a form of knowledge about human variety in and of itself by the end of the eighteenth century. More challenging still is the disappointing paucity of scholarship in eighteenth-century studies that addresses such epistemological questions. It is certain that in the eighteenth century sex was not "like" race; but it does, at times, appear *in* or *through* representations of racial difference. Siobhan Somerville, although discussing the turn of the twentieth century, makes an important and compelling case for why, intellectually, politically, and methodologically, analogies between race and sexuality are at best useless and at worst deeply problematic; the epistemological separation of race and sexuality, she argues, reifies each formation as distinct, nondynamic, and self-identical over time and space. What I describe in this volume, on the other hand, is a relationship between representations of human variety and representations of human sexual diversity that appear in anything other than distinct, nondynamic, or independent terms. In the texts at the analytic center of this study, race and sex are at times hard to tell apart. Indeed, the only thing that could perhaps be said to distinguish these interdependent fields of analysis would be their critical difference in scale. These constellations of circulating forms of knowledge about racial variety and sexual diversity were, during the eighteenth century, "not singular 'events' through which those meanings were simply established once and for all but rather ongoing processes of contestation and accumulation."[64]

The chapters that follow are animated by this concluding question of how to characterize the relationship between human variety and sexual diversity, and they pay close attention to the mutual articulation of these formations throughout a range of different natural historical texts and popular narratives. Sex speaks at different volumes and appears with varying sharpness or clarity in each of these chapters, and what they have to tell us about the way sex was understood in the eighteenth century is uneven, at times even contradictory. The little consensus that they offer pertains less to what sex *was* than to how it was explained and understood, and in these texts, the environmentalism that structures inquiries into the meaning of sexual behavior comes into stark focus. And the cultural backdrop to all of these texts is, of course, racialization in its many iterative scenes, stages, and structures: the gallows, the frontier, the vice district.

Chapter 1, "The Natural History of Sexuality," lingers in eighteenth-century natural history, building an archive for the environmental theory of early sexuality that I illustrate on a much narrower scale in chapters 2, 3, 4, and 5. Chapter 2, "The Complexion of Sodomy," examines the representation—or nonrepresentation—of sodomy in Barbary captivity narratives from the late seventeenth into the late eighteenth century, reading sodomy geographically, through climate theory, to expose its quiet but persistent racial dimension. Chapter 3, "Egyptian Lusts at the Gallows," turns to execution narratives, tracking the figure of habit through ethnographic literature, parenting manuals, and jeremiads, exposing the persistent use of this figure to explain the putative fact of black criminality. Chapter 4, "Botanical Sexuality and the Colonial Landscape," explores the way Herman Mann represents Deborah Sampson's cross-dressing and implied sapphism as a curiosity of the natural environs unique to the New World, relying on taxonomic language to categorize her sexual novelty. Chapter 5, "Negro Hill and the Sexuality of Space," focuses on representations of Boston's "Negro Hill," a district known for prostitution, to follow how municipal definitions of "vice" understand it as a spatial rather than an individualized moral or spiritual problem.

Ultimately, in charting the relationship between sciences of human difference and their expression in sexual diversity—steeped as they are in environmental understandings of the human—this study promises to change not only how we think about early sexuality but also how we think about sexuality *today*. The many eighteenth-century texts considered in this book display ambivalence about whether sexual behavior originates from an individuated human inclination or the susceptibility of the body to the external world, but even that ambivalence itself undermines the subject-centered sense of *sexuality* that characterizes its expressive organization in the modern period. While the genealogy of modern sexuality described here is by no means *singular*—for there are necessarily many origin stories to be told about the emergence of modern sexuality—in suggesting at least a partial disaggregation of the sexual from the human body, this study poses a powerful refusal to the total reorganization of the sexual into the subjective. In the epilogue, "Thinking Sex—without the Subject," I explore the implications of shifting the framework of sexuality away from the exclusive purview of the subject. Examining not how sex became relegated to the subject, but rather how sexuality became housed exclusively in the human body, the book concludes with a nineteenth-century account of how sex became sexuality and how sexuality became human.

CHAPTER ONE

The Natural History of Sexuality

> Men are like plants. The goodness and flavour of the fruit proceeds from the peculiar soil and exposition in which they grow. We are nothing but what we derive from the air we breathe, the climate we inhabit, the government we obey, the system of religion we profess, and the nature of our employment.
> —J. Hector St. John de Crèvecœur, *Letters from an American Farmer,* 1782

Much has been made of J. Hector St. John de Crèvecœur's assertion that "we are nothing but what we derive from the air we breathe, the climate we inhabit, the government we obey, the system of religion we profess, and the nature of our employment," nestled in the middle of the oft-read Letter 3 ("What Is an American?") in his *Letters from an American Farmer.* Readers and critics alike find in Crèvecœur's botanical simile the exceptionalist, patriotic tone characteristic of so many narratives of America: that there is something particular about "Americans" that derives not only from their government and their religion but literally from their land. What animates the present chapter, however, is the less obvious logic that girds Crèvecœur's observation, although it is one with which most scholars of the eighteenth-century Atlantic world will nonetheless be familiar. Here we see Crèvecœur reflecting some of the most basic common sense of eighteenth-century natural science, and of natural history in particular. He professes what would have been taken for granted by his peers: faith in an environmentalist logic that understands human experience, aptitude, morphology, and organization to be intimately shaped, even defined, by its "peculiar" environment.[1]

In this chapter I examine natural history on both sides of the Atlantic for what it might tell us about how sexual behavior was understood in British colonial North America. What we will see is that environmentalism, particularly as it was promulgated by eighteenth-century natural history, offered residents of early America a broad and portable logic for making sense of human sexual behavior. Debates in the history of sexuality have largely focused on the question of whether, in the era before sexology, "sexuality" would have been broadly understood as a category of human experience and, if so, how behaviors that we might call *sex* were practiced and regulated. Stepping to the side of that question, I am concerned here with how residents of the British North American colonies made sense of the sexual practices they participated in and observed, or, differently put, what cultural knowledges governed their approaches to sex. I argue that the environmental thinking that characterized the way the body was understood more generally, suffusing natural history and the emergent human sciences, provided one of the strongest ideological frameworks through which people of that era explained sexual behavior.

Readers encountered this environmental thinking broadly, through a print culture that took many forms. In the following pages, we will explore a range of natural historical writings from the long eighteenth century, many of them initially published in Europe. Most were eventually reprinted in the North American colonies or could be found in the collections of colonial North American libraries. These texts, which include works by Carolus Linnaeus (1735), John Mitchell (1744), Georges-Louis Leclerc, the Comte de Buffon (1749), John Brown (1758), Adam Ferguson (1767), Johann Blumenbach (1775), John Millar (1773), William Falconer (1781), Thomas Jefferson (1781), and Samuel Stanhope Smith (1787), take up the contemporary question—a pressing one within scientific communities at the time—of the origins of what they called "human variety," a term that correlates, though awkwardly, with our own understanding of racial difference.[2] But, as we shall see, an examination of this natural historical archive reveals that the questions about racial difference so central to these texts are embedded with quiet but proliferate narratives about the meaning of sexual behavior as well.[3]

Natural history—a field of study whose development spanned several centuries and engaged thinkers and institutions throughout the entire, interhemispheric Atlantic world—is a frequent site of inquiry for historians of science, and specifically historians of race.[4] This chapter builds on this important body of scholarship to examine a critically understudied element of the environmentalism that it so ably anatomizes. What we will find is that a narrative about human susceptibility to or predilection for specific sexual behaviors inhered in climatic, humoral, and otherwise environmental theories of racial difference.[5] More simply put, this chapter highlights the way eighteenth-century debates about the etiology of racial difference, primarily within the context of natural history, were also telling a story about the etiology of human sexual diversity. Since the environmentalist logic that shaped understandings of what, at the time, was termed "human variety" *also* informed the quiet narratives that these theories told about human sexual behavior, these eighteenth-century writings reveal a vision of early sexuality that assumes a form starkly different from its modern iterations.[6] Unlike the individuating, profoundly embodied, and expressive qualities that characterize modern sexuality, what we glimpse in the ethnographic grooves of eighteenth-century natural history theorizes sexual behavior as highly contingent on one's environment, fundamentally dynamic, and having, at most, few origins in the body at all.

Eighteenth-Century Environmentalism

There is no shortage of scholarship on eighteenth-century environmentalism. That one's environment exerted a deeply influential, if not wholly determinant, force on one's disposition, character, and even morphology was something of a given during the eighteenth century, but most historians of eighteenth-century science locate the roots of this notion in the ancient and medieval humoral theories of Hippocrates and Galen.[7] Within humoral theory, a medical schematic dating back to the classical period that was used to characterize the internal workings of the human body, physical function and disposition were determined by a body's unique mixture of the four humors: blood, bile, black bile, and phlegm. Excess or deficiency of any of the humors could affect or even impede the body's functioning in a number of ways, and each humor was associated with a specific set of qualitative effects and what might be called character types. (An excess of blood, for example, led to a "sanguine" personality, which was associated with being ruddy-faced or bold in nature; an excess of black bile would result in a "melancholic" type, associated with a physically cool body and a calm or quiet affect.) One's humoral composition could, in turn, be influenced or affected by elements external to the body, such as the food one ate or the climate one inhabited.[8] While works by Hippocrates and Galen were not regularly reprinted in the North American colonies (or the early United States), medical texts that relied on Hippocratic or Galenic medicine circulated widely in both Europe and North America.[9] Indeed, the eighteenth-century body was a humoral body; these theories provided the dominant account through which the body was understood. Cotton Mather's 1722 medical treatise *The Angel of Bethesda*, for example, contains specific references to both Hippocratic and Galenic humoral theory.[10]

A basic understanding of humoral theory is critical to understanding the broad logic of environmentalism because it illustrates one of the sharpest differences between eighteenth-century medical science and that of our own time. During the eighteenth century, and in the early modern period more generally, the human form was understood to be fundamentally dynamic, even mutable. Characteristics that the emergence and hegemony of anatomical science in the nineteenth century would eventually define as unchangeable, such as race, were not believed to be so fixed among eighteenth-century, especially early-eighteenth-century, scientists and medical practitioners.[11] Humoralism, as a system of medical knowledge based on the diagnosis of the balance or lack of balance among the humors within the human body, is thus built on a notion of the human form as constantly in flux. The six "nonnatural" influences—"food, immediate environment, exercise, sleep, excretion, and venery"—bore significant potential to affect the balance of humors in the body.[12] Even though the humors were located within the body, humoral theory as a culturally dominant form of medical knowledge understood the body's humors to be constantly interacting with and reacting to specific features of one's physical, natural, climatic, and social environment. Put more simply, the humoral body was "a porous and fragile envelope," profoundly vulnerable to the influence of its surroundings.[13]

Humoral medicine, however, was hardly the extent of eighteenth-century environmentalist thought, although by most scholarly accounts it was one of the earliest and most widely accepted theories to anatomize human susceptibility to environmental conditions. With the burgeoning excitement surrounding natural philosophy (inclusive of medicine, natural history, botany and taxonomy, and more) in the first half of the eighteenth century, came an explosion of environmental theories of human history, development, variety, and degeneration. This body of thought relied heavily on humoral theory but also heavily emphasized the determining power of climate—exemplified in climate theory (frequently termed environmental determinism)—to characterize and etiologize human difference, especially racial difference. As numerous historians of race have demonstrated, the question of origins of human variety or racial differentiation was one of the most pressing scientific debates on both sides of the Atlantic for most of the eighteenth century. Inquiries into how and why different races of humans emerged, as we shall see, became a crucible for larger conversations in theology (monogenism versus polygenism), history (how and when the world began, and stadial theories of human development), and physiology (whether race, especially blackness, resided in the body and, if so, where). Natural historians from both England and continental Europe wrote and lectured passionately on the origins of human variety. The Swedish scientist Carl von Linné (Carolus Linnaeus) published his enormously influential *Systema Naturae*, which, as we have seen, famously organized humans into four types, in 1735; the Comte de Buffon began publishing his multivolume *Histoire Naturelle* in 1749. Later in the century, writers from the Scottish Enlightenment, particularly Hume and Ferguson, eagerly joined the conversation.[14] A wide range of scientific writing on the origins of human variety permeated print culture in eighteenth-century North America, and widely acknowledged theories or debates about the mutability or transformability of one's racial status hinged on popular understandings of the body's vulnerability to environmental influence.[15] The vast popularity and pervasiveness of eighteenth-century racial science served to further disseminate and sediment widely held beliefs about the human body's ability to adapt to or change in specific environmental conditions. The aggregate effect of the broad but uneven circulation of all of these forms of legitimated knowledge—theories of the susceptibility of the humoral body to its environment, racial science that suggested that humans might actually experience a change of skin color in different environments, and so forth—was a large-scale, cultural faith in the power of one's social, natural, physical or architectural, climatic, or gustatory environment to affect one's body, temperament, and character: what was generally termed one's "complexion."[16] Indexing a theory of bodily constitution derived from ancient medicine, eighteenth-century theories of "complexion" described a combination of "temperament and disposition" derived from "the interaction of climate and the bodily humors (blood, bile, phlegm, and choler)."[17]

This is not to say, however, that eighteenth-century environmentalism provided a singular, agreed-upon theory of how the humoral body was affected by its environment. Far from it, in fact. As Roxann Wheeler argues, "climate/humoral theory was, in the Gramscian sense, the common sense of the day and a magnet for contradictory beliefs."[18] Though eighteenth-century cultural common sense supported the consensus that the body was porous, dynamic, and fundamentally vulnerable to environmental influence, there was simultaneously little agreement around precisely which changes the environment—and especially the climatic environment—would effect upon the body or about how those changes actually took place. Climate, however, was in general assumed to wield a powerful influence on the humoral constitution, and references to climatic theories of bodily makeup abounded in eighteenth-century natural histories. William Falconer, in his 1781 *Remarks on the Influence of Climate,* succinctly parses the relationship between climate and other environmental influences in his prologue:

> I must beg leave to take notice of a general mistake, which appears to me to have pervaded the works of every writer upon this subject; which is, the making their positions too universal.
>
> The Effects of Climate, &c. are all of them general, and not particular; and if a considerable majority of the nations, as well as the individuals, that live under a certain climate, are affected in a certain manner, we may pronounce decisively on its influence, even though there may be some exceptions. It must likewise be taken in consideration, that the influence of one of the above causes often corrects the other. Thus a hot climate naturally renders men timid and slothful; but

the necessity induced by a barren country, number of inhabitants, animal diet, and savage way of life, may, any of them, correct this tendency of the climate, and dispose the manners to a different turn.[19]

Falconer, like many other naturalists throughout the eighteenth century, understands the temperament or morphology of human bodies as the comprehensive effect of a number of different environmental influences (what one eats, the climate one inhabits, the nature of one's work) but irreducible to any one of these influences. Yet alongside scientists who were writing earlier in the eighteenth century, Falconer believed climate to be one of the most significant of the external pressures to which the human body is subject. While climate theory and humoral theory were closely linked (and share intellectual origins in Aristotle's "division of the earth into the northern, southern, and temperate zones, which loosely corresponded to people with white, black, and light brown skin"), the effects of specific climates on human populations emerged as something of an independent vector of analysis. By the end of the eighteenth century, climate became one of the most determining factors in studies of the origin of human variety, and it was arguably much more significant to racial makeup than a body's humoral constitution.[20] Especially by that time, natural historical theories of the influence of climate—on the relative degree of heat or cold, dryness or moistness of an environment—proliferated within racial science in particular. Furthermore, in scientific conversations about climate, the term "climate" frequently appears as shorthand for a broader set of environmental factors, the availability of food and the quality of air in particular, such that the multifaceted impact of "climate," as a conglomerate of environmental effects, was conceptually heightened. All of the big-ticket natural historical theories of the origins of human variety, from humoral theory to stadial theory, in some way acknowledge the impact of climate on the constitution of the body. For example, as Samuel Stanhope Smith argues in his *Essay on the Causes of the Variety of Complexion and Figure in the Human Species* (1787), climate was one of the factors defining a region's "state of society," which constituted the origin of human variety, or racial diversity among humans. For Smith, "state of society" "comprehends diet, clothing, lodging, manners, habits, face of the country, object of science, religion, interests, passions and ideas of all kinds, infinite in number and variety. If each of these causes be admitted to make, as undoubtedly they do, a small variation on the human countenance, the different combinations and results of the whole must necessarily be very great; and combined with the effects of climate will be adequate to account for all the varieties we find among mankind."[21]

Climate theory was also especially important to emergent theories that explained human development in geographical terms.[22] Natural historians from all over Europe, the Caribbean, and North America propounded on the influence of heat, cold, dryness, and wetness in the various climatic "zones" of the earth: the temperate, torrid, and frigid zones.[23] Adam Ferguson muses, in his *Essay on the History of Civil Society* (1767), that

> MAN, in the perfection of his natural faculties, is quick and delicate in his sensibility; extensive and various in his imaginations and reflections; attentive, penetrating, and subtile, in what relates to his fellow-creatures; firm and ardent in his purposes; devoted to friendship or to enmity; jealous of his independence and his honour, which he will not relinquish for safety or for profit: under all his corruptions or improvements, he retains his natural sensibility, if not his force; and his commerce is a blessing or a curse, according to the direction his mind has received.
>
> BUT under the extremes of heat or cold, the active range of the human soul appears to be limited; and men are of inferior importance, either as friends, or as enemies. In the one extreme, they are dull and slow, moderate in their desires, regular and pacific in their manner of life; in the other, they are feverish in their passions, weak in their judgments, and addicted by temperament to animal pleasure. In both the heart is mercenary, and makes important concessions for childish bribes: in both the spirit is prepared for servitude: in the one it is subdued by fear of the future; in the other it is not roused even by its sense of the present.[24]

Ferguson carefully delineates the characteristics associated with the different zones of the world: in the torrid zones, he suggests, men "are feverish in their passions, weak in their judgments, and addicted by temperament to animal pleasure," while in the frigid zones, "they are dull and slow, moderate in their desires, regular and pacific in their manner of life."[25] The "perfection" of man's "natural faculties," Ferguson implies, can be achieved only in the temperate zones, a notion that inadvertently offers additional evidence for current-day critiques of enlightenment and the fundamentally racist exclusions attendant to the category of "man."[26] While it is worth noting that not all natural philosophers agreed on the impact of exposure to particular climates—Andrew Curran reminds us that the Abbé de Raynal and the Italian scientist the Abbé Ferdinando Galiani both contested the centrality of climatic influence—climate theory constituted a major contemporary epistemology through which residents of the eighteenth-century Atlantic colonies understood not only human morphology and behavior but also the organization of the world.[27]

These discussions of the possible impact that climate might effect on the human body gained greater currency because they were elapsing during a period of rapid globalization. More than ever before, people born in one climatic zone might well end their lives in another, whether as a result of a chosen journey or a forced migration (emigrating from Europe to colonial cities, for example, or being kidnapped into slavery and transported). As Europe's collection of colonies grew, so too did Europeans' concerns about the effects of inhabiting these colonial climates, many of which were located within the torrid zones (the Caribbean, Central and South America, the central belt of the African continent, India, etc.). The warmer weather and radically different social and economic hierarchies were imagined to pose dangers to both the moral and the corporeal complexions of Europeans living in colonial outposts.

Climate theory thus became a critical epistemology within transatlantic conversations about the possibility of racial transformation in humans.[28] In 1749 the Comte de Buffon advanced the idea that "if a colony of Negroes were transplanted into a northern province, their descendants of the 8th, 10th, or 12th generation, would be much fairer, and perhaps as white as the natives of that climate."[29] Indeed, the possibility that a person might experience a change in skin color, either during his or her lifetime or over the course of generations, indexed the potential of morphological change that might stem from the powerful influence of climate. European powers worried that their countrymen would "degenerate" in colonial locales, becoming darker in skin color and less stringent in their morals. For example, North American–born naturalist Samuel Stanhope Smith, speaking of people of the newly formed United States, argued:

> A certain countenance of paleness and of softness strikes a traveller from Britain the moment he arrives upon our shore. A degree of sallowness is visible to him which, through familiarity, or the want of a general standard of comparison, hardly attracts our observation. This effect is more obvious in the middle, and still more, in the southern, than in the northern states. . . . The change of complexion which has already passed upon these people is not easily imagined by an inhabitant of Britain, and furnishes the clearest evidence to an attentive observer of nature that, if they were thrown, like the native Indians, into a savage state they would be perfectly marked, in time, with the same colour. Not only their complexion, but their whole constitution seems to be changed.[30]

Though North America and its many climates certainly had its defenders, Thomas Jefferson among them, concerns about the possible degeneration of white European colonists abounded among both European and North American scientists; these concerns were largely animated by the problematic climate of British North America, particularly its southern regions.[31] Smith's assessment of the "degree of sallowness" with which the complexion of recent European emigrants to North America had supposedly been imbued as a result of their relocation illustrates how powerful the influence of climate was believed to be during this period; for Smith, such "sallowness" was an early indication of the slow, morphological change that could ultimately turn "an inhabitant of Britain" into a "native Indian."

Like one's humoral constitution, however, climatic influence was only part of the picture, only one of the myriad environmental influences that exerted pressure on the human complexion. Alongside eighteenth-century British and French scientific writers, many of whom, like Robert Boyle and the Comte de Buffon, were widely read in the British North American colonies, work by Scottish Enlightenment thinkers circulated broadly in the colonies as well.[32] Scottish natural philosophers such as Adam Smith, David Hume, Adam Ferguson, and John Millar contributed to circulating theories of racial difference by developing and popularizing stadial theory. More widely known as "four-stages theory" or "conjectural history," stadial theory described a roughly linear trajectory of human development wherein "civilization" was achieved by certain (but not all) societies by moving through four stages of civil organization: hunting and gathering, herding and pasturage, farming and agriculture, and commerce.[33] Stadial theory ultimately introduced a new element to eighteenth-century theories of environmental determinism, by suggesting that it was not merely climate and humoral balance that exerted significant influence on human constitution; quotidian habits and practices, or one's "mode of living," were important factors as well.[34] Eighteenth-century discussions of habits or "modes of living" more generally also assumed pride of place in early modern accounts of racial differentiation, as well as in ongoing understandings of human corporeal dynamism.[35] Samuel Stanhope Smith (quoted earlier) groups this growing number of "modes" together under the term "state of society." For Smith, then, it is one's "state of society," combined with climate, that effects the greatest influence on one's complexion. Modes of living, habits of dress, habits of behavior and socialization, and even inhabitation or the spaces one frequented, were understood by natural historians to bear a critical influence on complexion; I discuss the racial and sexual politics of habit at length in chapter 3.[36] Indeed, one's state of society "came to occupy a paradoxical place in the cultural imaginary": it was "thought *to influence* physical characteristics as an environmental factor itself and simultaneously *to be influenced* by environmental factors (such as climate and geography) in the same way as physical characteristics. Therefore, society could become both the cause of one's degeneration and the resulting proof of it."[37] This circular explanation of degeneration, in particular—the idea that individual degeneration could contribute to the broader degeneration of a society, which could then in turn beget the degeneration of those that live within it—is a hallmark of natural historical thinking and illustrates the competitive tension between internal and environmental theories of human variety. Though environmental factors such as climate or diet were understood to bear the capacity to shape individuals or societies, they still competed, and often circulated in step with, internal explanations of human constitution such as innate depravity or anatomical inferiority (or superiority) that were understood to be able to exert significant influence on the environment in their own right. In referring to manners, habits, and custom, natural historians often simultaneously both invoked rigid, static theories of human difference and also insisted on the possibility of human characterological and morphological dynamism. For this reason it can be challenging to isolate environmental theories of human variety in natural historical writing and thinking. My intent here is thus not to insist that environmental theories of racial difference were the exclusive framework through which the human body and its variations were explained, but rather to insist on the centrality of environmental thinking about the body to the multifaceted emergence of the human sciences over the course of the long eighteenth century. For it is only by training our attention on this often underexamined framework that we are able to see the critical information that it offers about how sexual behavior was understood in the era before sexology.

Thus, natural historical interest in the "state of society" or "modes of living" of North American cultures—both Native and settler colonial—highlights the uneasy tensions between internal or ontological etiologies of human variety, on the one hand, and environmental causes on the other. Maintaining metropolitan manners and ways of life was perceived as critical to the moral and political success of colonial endeavors, and a careful understanding of the state of society of both Native and settler populations in colonial locales, in particular, was both a driving force behind scientific interest and one that critically girded metropolitan structures of control. Many natural historical theories, especially those of Scottish Enlightenment thinkers, asserted the pride of place of the "state of society" over and above other environmental factors that shaped human constitutional development. Indeed, North American natural historians (and natural historians writing *about* the British North American colonies) direct much more attention to the issue of "state of society" than do their European counterparts, perhaps because of the broad circulation of Scottish Enlightenment writings in the North American colonies, or perhaps simply because British North America, as a colonial society for most of the eighteenth century, was the object of the same kinds of broad concerns about cultural degeneration that so defined conversations about life in the British and French Caribbean.[38]

J. Hector St. John de Crèvecœur, a farmer and lay natural historian, dwells at length on what he calls the "manners" of different communities within the United States, by which he means something akin to the way we, today, use the word "culture"; describing the settlers who live at the frontiers of the late eighteenth century, Crèvecœur laments the "ferocious, gloomy, and unsocial" manner of life that they must adopt, as hunters rather than farmers. He writes:

> But to return to our back settlers. I must tell you, that there is something in the proximity of the woods which is very singular. It is with men as it is with the

plants and animals that grow and live in the forests. They are entirely different from those that live in the plains.... By living in or near the woods, their actions are regulated by the wildness of the neighborhood.... They soon become professed hunters. This is the progress.... The chase renders them ferocious, gloomy, and unsocial.... That new mode of life brings along with it a new set of manners, which I cannot easily describe. These new manners, being grafted onto the old stock, produce a strange sort of lawless profligacy, the impressions of which are indelible.[39]

In this passage, Crèvecœur sketches a loosely causal relationship between the state of the "manners" of frontier communities and the natural environment they inhabit. Men who live on the plains and farm develop a starkly different set of manners—and also experience radically different climatic and humoral conditions—than men who live in the woods and hunt.[40] Crèvecœur's reliance on natural historical thinking here is obvious; but also worth noting is how, already in 1782, Crèvecœur is echoing Samuel Stanhope Smith and others by extending and amending theories of climatic determinism to accommodate a plethora of different hypotheses about the sources of environmental influence, with "state of society" or "manners" among them. For Crèvecœur, it is not *just* the proximity to the woods or *just* the natural inducement to hunt (rather than to farm) that contributes to the "degenerated" state of settlers living on the colonial frontier.[41] It is also the food these settlers eat ("eating of wild meat, whatever you may think, tends to alter their temper") and the company they keep ("having no place of worship to resort to, what little society this might afford is denied them"). Colonists living on the frontiers become "half cultivators and half hunters; and the worst of them are those who have degenerated altogether into the hunting state. As old ploughmen and new men of the woods, as Europeans and new-made Indians, they contract the vices of both.... Hunting is but a licentious, idle life, and, if it does not always pervert good dispositions, yet, when it is united with bad luck, it leads to want; want stimulates that propensity to rapacity and injustice, too natural to needy men, which is the fatal gradation."[42]

Manners, for Crèvecœur, are the result of a complicated combination of environmental influences, and they have the capacity to render one vulnerable to what we might think of as moral degeneration: to become "licentious" and "idle" is to be prone to "rapacity" and "injustice."

We must understand Crèvecœur as illustrating the relationship between the social, natural, and climatic features of the world one inhabits and trying to parse how that relationship develops the qualities that we today tend to think of as internal or subjective, such as "rapacity," "injustice," and "licentiousness." While these characteristics are not always, only, or predictably the result of living in specific environments, Crèvecœur implies that one becomes vastly more vulnerable to certain degenerate characteristics in certain environments: namely, colonial spaces. This moment in *Letters from an American Farmer* makes especially clear the relationship between what we might otherwise think of as objective or incidental elements of one's external environment and the likelihood of or vulnerability to developing specific internal characteristics.

Although Crèvecœur likely draws from natural historical thinking, and in particular natural historical theories that explain racial difference, in offering these observations about the relationship between one's physical or natural environment and the development of specific affective or moral characteristics, the qualities that Crèvecœur names as the likely result of a long inhabitation of frontier territories—rapaciousness, licentiousness, injustice, and so forth—are not actually qualities that would have been thought of as wholly or strictly "racial" or as pertaining to human variety during the eighteenth century. While some of these qualities are certainly (and obviously) racialized and would have borne racial implications during the eighteenth century, Crèvecœur's assertion in no way imagines that the influence of the frontier and its natural, climatic, and social environment would have been limited to racial transformation in the *bodies* of the settlers who lived there. The effects of the degeneration he speaks of are broadly characterological. Crèvecœur is describing the potential for changes in complexion, in the eighteenth-century sense of the term: both the physical body and one's "temperament and disposition." While Crèvecœur compares settlers who support themselves by hunting to both ostensibly degenerated "Europeans" and "new-made Indians," he argues that they "contract the vices of both," gesturing to the fact that it is not *only* what Samuel Stanhope Smith calls "a degree of sallowness" that attends their long-term residency on the frontier. Rather, specific moral and characterological dispositions or tendencies are also the wages of long-term exposure to frontier environments.

While these qualities are frequently implied in taxonomies of racial difference—and Crèvecœur explicitly states that there are specific "vices" particular to "Europeans" and "Indians"—one might also contract these vices simply by inhabiting particular environments where the climatic, gustatory, and social conditions are right, without necessarily exhibiting new racial characteristics as well. If these racialized vices are as acquirable as they might be congenital, then, in Crèvecœur's *Letters from an American Farmer,* we also see one of many iterations of what I am calling an environmental theory of early sexuality. While "vice," in the eighteenth century, was an extremely capacious behavioral term, then as now it broadly indexed venereal behaviors such as fornication and masturbation (I discuss vice at length in chapter 5). What Crèvecœur's meditation on the degenerating power of frontier environments reveals, then, is his implicit belief that inhabiting degenerating environments can produce specifically degenerated behaviors, including sexual behaviors. While Crèvecœur's assessment of the state of society and modes of living in frontier environments relies on the same circular logic combining both internal and environmental theories of behavior described above (perhaps the lack of domestic order in frontier communities encourages degeneration into licentiousness, but perhaps some already licentious people seek out frontier communities as sanctuary), the environmental framework through which he imagines vice and licentiousness to be possible results of inhabiting frontier environments comes into stark relief here. Through the racialization of specific temperaments and dispositions—such as viciousness, licentiousness, or rapacity—the frontier, for Crèvecœur, becomes a space where sexual behavior provides the evidence of the racial degeneration that occurs there. The interest in modes of living that courses through natural historical accounts of human variety is, as Crèvecœur demonstrates, especially attentive to sexual behavior, which then becomes a detail adding texture to accounts of different states of society or ways of life. In an effort to illustrate the degenerating effects of specific environments, both highly educated and amateur natural historians enlist putatively observed sexual behavior as evidence of the harmful quality of these environments, inadvertently explaining sexual behavior through an environmentalist framework in the same way they do human variety.

Sex and Environmentalism in the Eighteenth Century

As colonial exploration and domination expanded the purview of eighteenth-century science on both sides of the Atlantic, it is no surprise that race was a perennial topic of curiosity and theorization. But race was far from the only topic. As we shall see, where race was concerned, sex was never far behind. Eighteenth-century environmentalist theories were labile, proliferate, and quite portable, and to imagine that they did not influence contemporary ideas about sexual behavior would be to ignore a major facet of early modern thinking. Popular eighteenth-century books, tracts, and essays on human variety are veritably shot through with narratives, albeit quiet ones, about sexual specificity and sexual variety. In this section I explore a range of the most common and powerful theories of human difference that circulated among eighteenth-century natural historians—among them climate theory, humoral theory, stadial or "four-stages" theory, and theories of complexion and disposition—to identify the way each of these explanatory frameworks linked environmental influences to particular sexual behaviors.

If the human body in the eighteenth century was, as Susan Scott Parrish puts it, a humoral body and "highly vulnerable to outside influence," it was also "responsive to humoral management through the six 'nonnaturals': food, immediate environment, exercise, sleep, excretion, and venery." Let us begin with the last of the six nonnaturals, venery. During the eighteenth century, "venery" was a relatively capacious term that was generally associated with animality in a variety of ways; "venery" was occasionally used to designate hunting or meat, but it was also a category of the human passions.[43] In this period, "venery" was used to describe a wide variety of forms of excess, from gluttony to fornication to drunkenness, some of which did, indeed, belong to the realm of the sexual. That said, in addition to this broad notion of excess, in the context of humoral theory, venery describes something more specific: activities that influenced the flow of fluids in and out of the body. These activities necessarily determined "the right exchange between inside and outside." One of those fluids, which moves from the inside to the outside, is semen.[44] Venery absolutely encompassed a set of behaviors that now fall under the category of "sex acts," and as a result we need to understand sex acts as one of the many possible environmental factors that, in eighteenth-century humoralism, exerted influence on the body's constitution and complexion.

Environmental theories of human variety frequently suggest that sexual or gender particularities accompany other forms of morphological difference; like much of eighteenth-century natural historical thinking, this derives from Hippocratic and Galenic medicine. A 1616 English translation of Juan Huarte's writings (*Examen des ingenios*, 1594) emphasizes that early modern encounters with what we would now term racial and cultural difference relied heavily on Hippocratic and Galenic humoral theories; thus, according to Huarte, when human bodies "be cold and moist more than is requisite, Hippocrates sayth that the men prove eunuchs, or hermaphrodites; and if it be very hot and dry, Aristotle sayth that it makes them curl-pated, crook-legged, and flat-nosed as are the Ethiopians, and if it be moist, the same Galen sayth that they grow long and lithy; and if it be dry, low of stature."[45] A little over a century later, the same ideological association between climate and sexual particularity was expanded upon by Carolus Linnaeus in his *Systema Naturae*.[46] As we will see in chapter 4, in late-eighteenth-century North America, Linnaeus was probably most broadly known for his work on botany, and specifically for his taxonomic system that would provide the foundation for the binomial nomenclature system (e.g., *Homo sapiens*) that we still use today. Like Buffon's *Histoire Naturelle*, which appeared fourteen years later, the *Systema Naturae* attempted a general taxonomy of the various forms of natural life, including a systematic attempt to define the different varieties of humans. *Systema* was an immensely influential project, broadly translated and frequently reprinted, nearly from the moment of its first publication. In the tenth and final edition of the work, Linnaeus elaborated on the categorical differences between the four kinds of humans, *Europaeus albescens*,

Americanus rubescens, Asiaticus fuscus, and *Africanus niger,* offering a description of each racialized type that included a loose concatenation of attributes, among them regional association, humoral makeup, skin color and other morphology, and various dispositional qualities, including "obstinate," "inventive," "merry," and "haughty" (see the figure below):

AMERICANUS   a. reddish, choleric, erect.
Hair black, straight, thick; *Nostrils:* wide; *Face:* harsh,
Beard scanty.
Obstinate, merry, free.
Paints himself with fine red lines.
Regulated by customs.

EUROPAEUS   b. white, sanguine, muscular.
Hair flowing, long. *Eyes* blue.
Gentle, acute, inventive.
Covered with close vestments.
Governed by laws.

ASIATICUS   c. sallow, melancholy, stiff.
Hair black. *Eyes* dark.
Severe, haughty, avaricious.
Covered with loose garments.
Ruled by opinions.

AFER   d. black, phlegmatic, relaxed.
Hair black, frizzled. *Skin* silky. *Nose* flat, *Lips* tumid.
Women without shame. *Mammae* lactate profusely.
Crafty, indolent, negligent.
Anoints himself with grease.
Governed by caprice.[47]

Linnaeus's reliance on humoral theory is obvious: the different varieties of men are characterized by different humoral constitutions, which in turn evidence certain dispositions. In addition, these humoral dispositions contain, in Linnaeus's account, an anatomical dimension; he presents skin color and other morphological features as regular and predictable features of the four types of humans. This categorization made a significant impact on racial thinking in Europe and North America during its time, especially in linking humors to skin color; *Systema Naturae* was one of the most important and widely cited scientific texts, on both sides of the Atlantic, for at least a half century after its publication. Thomas Jefferson, for example, cites Linnaeus frequently in his

MAMMALIA PRIMATES. Homo.

Afer. ♀. niger, phlegmaticus, laxus.
    Pilis atris, contortuplicatis. *Cute* holosericea. *Naso* simo. *Labiis* tumidis. *Feminis* sinus pudoris; *Mammæ* lactantes prolixæ.
    Vafer, segnis, negligens.
    Ungit se pingui.
    Regitur Arbitrio.

Monstro- 1. solo (a) arte (b, c).
    a. *Alpini* parvi, agiles, timidi.
    Patagonici magni, segnes.
    b. *Monorchides* ut minus fertiles: Hottentotti.
    Junceæ puellæ abdomine attenuato: Europææ.
    c. *Macrocephali* capite conico: Chinenses.
    Plagiocephali capite antice compresso: Canadenses.

Habitat inter Tropicos *sponte gratisque*; *per reliquam* Telluris *totam continentem coacte.*
CORPUS erectum, nudum, Pilis raris remotissimis adspersum, subsexpedale.
CAPUT obovatum: *Pileus* pilosus: *Sincipite* obtuso, *Vertice* obtusissimo, *Occipite* gibboso.
Facies nuda: *Frons* planiuscula, quadrata, *Temporibus* compressa, angulis sinu intra Pileum adscendens. *Supercilia* prominula pilis extrorsum imbricatis, jnterstincta. *Glabella* planiuscula. *Palpebra superior* mobilis, *inferior* quieta; utraque *Ciliis* exstantibus subrecurvatis pectinata. *Oculi* rotundi: *Pupilla* orbiculata. *Genæ* gibbæ, molliusculæ, coloratæ. *Malæ* compressiusculæ, uti *Buccæ* laxiores. *Nasus* prominens, labio brevior, compressus, apice altior gibbosiorque: *Naribus* ovalibus, intus

---

Luctus, *Luxuria*, *Ambitio*, *Avaritia*, *Vivendique cupido*; *curat atque etjam post se de futuro.* Plin.
Theologice: Te ultimum finem creationis;
  In Telluris globum, omnipotentis magisterium, introductum; ratione sapiente, secundum sensus concludente, mundi contemplatorem; ut ex opere agnosceres Creatorem omnipotentem, omniscium, immensum & sempiternum DEum.
(*Cætera revelata a Theologis explicanda.*)
Duo namque sunt quæ in cognitionem DEi ducunt, Creatio & Scriptura.
                                               Augustinus.
DEUS itaque Natura cognoscendus, dein Doctrina recognoscendus.
                                                     Tertullian.
Homo solus DEum contemplatur, Naturæ & Revelationis eundem Auctorem: quem te DEum esse jussit, & humana qua parte locatus es in re, disce, Pers. III; 71.

The 1758 edition of Linnaeus's *Systema Naturae* describes humans of the *Afer* taxon as "feminis sinus pudoris" and "mammae lactantes prolixae." Courtesy of Thomas Fisher Rare Book Library, University of Toronto.

1782 *Notes on the State of Virginia*. Perhaps more significant, this simple chart also links humoral constitution and skin color to sexual characteristics, in the case of the variety of human Linnaeus designates *Africanus Niger,* Africans. The women, he advises, are "without shame." Ezra Tawil cautions us to not read Linnaeus's assessment of *Africanus Niger* as one defined primarily by skin color; instead, he insists that we understand "women without shame" (a characterization that would later, as Jennifer Morgan demonstrates, become synonymous with black femininity) in terms of the humoral category of "phlegmatic" and its association, in Linnaeus's formulation, with the continent of Africa and the region or climate of *Aethiops*.[48] Linnaeus very explicitly assumes a relationship between humoral constitution and geographic region and, at least for *Africanus,* sexual disposition ("*feminis sinus pudoris,*" which can be translated as either "women without shame" or "in women: stretched labia"). The incredible popularity of the *Systema Naturae* would have served to disseminate this ideological linkage of climatic region to specific sexual characteristics.

Felicity Nussbaum's 1995 *Torrid Zones* constitutes the only scholarly effort (as far as I am aware) to attempt to sketch the relationship between racial thinking and sexual taxonomies during the eighteenth century, in terms of what she calls "the geography of sexual desire"; her study of British colonial representations of the torrid zones reveals that for residents of the early modern Atlantic world, "the torrid zone exist[ed] in an eternal past, permeated with sexual passion." Citing Scottish Enlightenment philosopher Adam Ferguson, she reminds us that eighteenth-century natural historians understood "that passion is strongest in the locations least known to Europeans." Nussbaum recounts Ferguson's formulation: "Hot climates produce sexual desire, while more temperate climates require greater control and more elaborate ritual. Domesticity, believed to be antithetical to sexual heat, increases the farther one resides from the equator. Warmer climates naturally intensify the amount of sexual activity and consequently produce a larger population that freely indulges its libidinous energy."[49] Ferguson, however, was just the tip of the iceberg. The belief that certain sexual behaviors proliferated in certain climates, or that specific propensities for sex or levels of desire correlated with specific geographic regions, appears pervasively throughout eighteenth-century natural historical thinking. At times, this sexual regionalism takes a slightly more subtle form. Virginia-born John Mitchell's influential *Essay Upon the Causes of the different Colours of People in Different Climates* (1744), for example, never draws any direct relationship between warm climates and specific sexual propensities, but Mitchell does argue that the "yaws," a skin infection common in tropical areas and which Mitchell identifies as a "cutaneous Malady of the Negroes," "laid the first Foundation of the *Lues Venerea* [syphilis]" in the United States.[50]

At other moments, the association between specific geographic or climatic regions and specific forms of sexual behavior is incredibly explicit. We see these links throughout Buffon's *Histoire Naturelle,* a forty-four-volume work of natural history (especially in "Of the Varieties of the Human Species," which appears in his *Natural History of Man*); describing natives of the "Maldiva" Islands, for example, Buffon writes: "These islanders love exercise, and are industrious artists; they are superstitious, and much addicted to venery; though the women carefully conceal their bosoms, they are exceedingly indolent and debauched; they perpetually eat betle and other hot spices. As to the men, they are less vigorous than their spouses would incline."[51] Here Buffon implies a relationship between being "addicted to venery" and two critical environmental factors: the climatic region ("Maldiva" is in the torrid zone) and a spicy diet. Similarly, in Buffon's discussion of the population of Madagascar (almost all of his ethnographies are synthesized from travel narratives), he writes that Malagasy society "abounds in cattle and pasturage. Both men and women are exceedingly debauched; and public prostitution is not followed with dishonor. They love dancing, singing, and similar amusements. . . . Though they have no moveables in their houses, but lie upon mats, they have husbandmen, smiths, carpenters, potters, and even goldsmiths. They eat their meat almost raw, and devour the skins of their oxen, after singeing the hair; they likewise eat the wax with the honey."[52] The discussion of the "debauched" character of both the men and the women of that community is thus sandwiched between a description of Malagasy culture rendered in the idiom of stadial theory ("abounds in cattle and pasturage") and a discussion of their "state of society" (they have no "moveables" in their houses, but they have local industry of various kinds).

Appearing next to his concluding discussion of what the people in these communities eat, Buffon's discussion of sexual behaviors is given without any particular emphasis; the purported sexual proclivities of the Malagasy people are offered as just another ethnographic detail, shoring up Buffon's larger taxonomy (which evokes, at different moments, contemporary stadial theory, humoral theory, and climate theory) of their fundamental difference. Although there is no basis for believing that this account of Malagasy communities bears any relationship to actual life in Madagascar during the eighteenth century, Buffon's epistemologies bring into stark focus the way natural historical discussions of human variety understood diversity within human sexual behavior to be evidence of that variety. What we see, in this ethnographic attention to difference within human sexual practice, is an understanding of sexual behavior as another site through which climatic, humoral, and other forms of environmental influence might be expressed through the human body.

These passages about the Maldivians and the Malagasy people appear— among many, many others like them—amid Buffon's attempt at an exhaustive study of "human variety." His effort is organized primarily through three categories: "colour," "figure and stature," and "the dispositions of different people."⁵³ Scholars of eighteenth-century natural history have discussed at length the contemporary scientific interest in the first two of Buffon's vectors of analysis, skin color and figure and stature (a category that anticipated the emergent human science of anatomy), and the way taxonomies of human variety contributed to then-emergent sciences of racial classification.⁵⁴ Yet Buffon's third category of inquiry, "the dispositions of different people," and all that it may or may not comprehend, has received little scholarly attention.

Linked to one's humoral makeup, complexion, and climatic or otherwise geographic location (think of Crèvecœur's "licentious" frontiersmen), eighteenth-century uses of the term "disposition" are broadly concerned with order, in this case the ordered or disordered state of the body, internally and externally. In this sense, eighteenth-century uses of "disposition" are more or less coterminous with a much more prolific keyword of eighteenth-century science: *complexion*. But "disposition" was also invoked, at times, to discuss more internal features of human experience; like humoral theory, discussions of disposition simultaneously gestured to both internal and external human qualities. And, importantly, disposition, as Buffon uses it, renders racialization an external feature of the body that both predicts and reflects an internal constitution. In disposition, race becomes a descriptive framework that describes both surface and depth. For this reason, Buffon's interest in disposition is particularly generative for what it suggests about the early modern linking of human variety and sexual behaviors, as well as the historiographic relationship between those emergent sciences and emerging analyses of sexual particularity. In Buffon's use of "disposition" as a gloss for racialized characteristics that were often associated with sex—being "debauched," for example—both environmental and more fixed etiologies for sexual behaviors are announced in the same explanation. According to Buffon, Maldivians are addicted to venery and Malagasy peoples are debauched, and in both cases these dispositional characteristics are simultaneously evidence of the profundity of their difference from Europeans and are explained as the result of their climatic and social environments. By materializing debauchery and venery as racial characteristics, Buffon inadvertently offers a second, subsidiary hypothesis about the origins of sexual behaviors, in which these behaviors are themselves dispositions, both environmental and more internalized, or ontological. As we shall see in the four chapters that follow, this tension follows the representation of sexual behavior throughout the eighteenth century. I do not mean to argue in this book that sexual behavior was *never* understood as an internal characteristic. Sex does, at times, appear as expressive, as a reflection of internal desire or experience, or even as a metaphor for one's internal spiritual uprightness or depravity. Rather, I want to insist on the equal importance of environmental theories of human complexion and disposition—steeped as they are in historically specific understandings of the body—to contemporary theories of human morphology, behavior, and society. These environmental theories were equally central and definitive when residents of eighteenth-century North America made sense of their own sexual behaviors and when they considered those that they observed in the world around them.

Like Buffon, William Falconer, a fellow of the Royal Society, carefully delineated the relationship between climate and disposition in his *Remarks on the Influence of Climate*, which was published in 1781 and available in the collections of the Library Company of Philadelphia by 1789. Falconer offers a systematic anatomy of the effects of hot, cold, and temperate climates on various human dispositions, ranging from industriousness and indolence to benevolence and amorousness.[55] In a section unimaginatively titled "Effects of Heat on the Temper and Disposition," Falconer describes the ways heat—and, implicitly, life in the torrid zones—affects the "amorous disposition" of peoples in warm climates:

> To the same sensibility is owing the amorous disposition of the people of hot climates; which disposition again, in its turn, . . . enhances the sensibility that produced it. This, as well as the jealousy that attends love, has always been remarked as a part of the character of those people.
>
> But although the enthusiasm of love be most powerful in such climates, yet this passion is in them far from being of a refined nature in point of sentiment. Beauty, indeed, is highly valued as a possession; but regard, esteem, and attachment, have scarcely any place in the union of the sexes.[56]

Perhaps offering a nod to Buffon's three-part schematic for the categorization of human taxonomy ("colour," "figure and stature," and "dispositions"), Falconer's assessment of human "disposition" is central to his taxonomy of human characteristics and their vulnerability to environmental influence. Combining climate theory and extant hypotheses about human behavior steeped in a sexualized regionalism, Falconer's work is notable for the way it locates the "amorous disposition" as *unique* to the torrid zone. He explains that in warm climes "the passions are naturally very strong, and likewise kept in a perpetual state of irritation from the high degree of sensibility that prevails." The result is not only a higher level of *amorous* passion in residents of the torrid zones, but a higher likelihood of criminal passions as well; the "high degree of sensibility" in warm regions "causes a great multiplication of crimes, by multiplying the objects of temptation." People who live in torrid zones are addled by the two-pronged drawback of the climate: "The passion or inclination is stronger, and the power of restraint less."[57]

Scottish Enlightenment philosopher Adam Ferguson is similarly quite explicit in connecting sexuality and climatic region, but for Ferguson the unrestricted sexual behavior that characterizes life in the torrid zones is intimately linked to a civilizational thesis; unrestrained sexuality points to what he sees as a lack of aptitude for self-governance.[58] In his *Essay on the History of Civil Society* (1767), Ferguson draws something of a map for his readers, walking them from the south, where the passions flourish overabundantly, to the frigid zones, where they wither dangerously:

> The burning ardours, and the torturing jealousies, of the seraglio and the haram, which have reigned so long in Asia and Africa, and which, in the Southern parts of Europe, have scarcely given way to the difference of religion and civil establishments, are found, however, with an abatement of heat in the climate, to be more easily changed, in one latitude, into a temporary passion which ingrosses the mind, without enfeebling it, and which excites to romantic achievements: by a farther progress to the North, it is changed into a spirit of gallantry, which employs the wit and the fancy more than the heart; which prefers intrigue to enjoyment; and substitutes affectation and vanity, where sentiment and desire have failed. As it departs from the sun, the same passion is farther composed into a habit of domestic connection, or frozen into a state of insensibility, under which the sexes at freedom scarcely chuse to unite their society.[59]

Ferguson provides a quite explicit assessment of the way climate influences sexual passion, observing that "the melting desires, or the fiery passions, which in one climate take place between the sexes, are in another changed into a sober consideration, or a patience of mutual disgust."[60] Whereas Buffon distinguishes between the "savages" of North America and those of the African continent based on his observation that "the Torrid Zone is not so hot in America as in Africa,"[61] Ferguson allows for no such distinction. He finds the tempering of sexual passion based on climatic shifts to be true around the globe; it can be "remarked in crossing the Mediterranean, in following the course of the Mississippi, in ascending the mountains of Caucasus, and in passing from the Alps and the Pyrenees to the shores of the Baltic."[62]

Other forms of environmentalist logic also contained implicit narratives linking sexual behavior and civilizationalist teleologies. John Millar, another of the Scottish Enlightenment thinkers, and credited as one of the developers of stadial theory, believed that human societies across the globe progressed through four stages as they grew toward political independence.[63] Although stadial theory does constitute a fairly distinct epistemology through which eighteenth-century thinkers theorized human variety, Millar's *Observations Concerning the Distinction of Ranks in Society* relies heavily and obviously on a climate-based understanding of human vulnerability to the environment.[64] Millar insists on the unique status of the passions "which unite the sexes" (love and the incitement to sexual intercourse) relative to the other human passions ("senses relating to emotional or mental states")[65] and argues that these passions "are most easily affected by the peculiar circumstances in which we are placed, and most liable to be influenced by the power of habit and education. Upon this account they exhibit the most wonderful variety of appearances, and have produced the greatest diversity of manners and customs, in different ages and countries."[66] In this passage, then, we see Millar identifying and insisting on the distinction of the sexual passions as a category of human experience, and one that is particularly vulnerable to climatic conditions.

Stadial theory bore a complex, two-part sexual politics: first, it contained a strong narrative that bound women's "freedom" to advanced stages of civilization and, as I discuss in chapter 2, assumed man's despotic power over women in less "advanced" societies; and second, it contributed to the sexualization of "whole areas of the world,"[67] connecting "homoerotics, the unnatural, and the Other in a sexual geography" that specifically characterized "the torrid zones of Asia Minor as well as Mediterranean Europe."[68] Millar's *Observations* relies on a host of ethnographies and travel narratives to argue that the lack of social and civic hierarchy common in communities in the earliest "stages" of development (and where there is thus a lack of prohibitions on social interaction) allows for widespread sexual licentiousness. "In the most rude and barbarous ages," he writes, "there are no differences of rank to interrupt the free intercourse of the sexes. There are no distinctions among individuals, but those which arise from their age and experience, from their strength, courage, and other personal accomplishments."[69] The way Millar links sexual promiscuity to "the peculiar circumstances in which we are placed" relies less on a sexual regionalism steeped in climate theory than on a civilizationalist narrative that equates civic independence—or the lack thereof—to the governance of one's own sexual body; the inability to achieve the one predicts an inability to achieve the other.

The lack of obstacles preventing "savage" peoples from gratifying their sexual desire in turn slows down their ability to progress up the ladder of civilization, because for Millar it is social hierarchy and the prohibitions it establishes, between certain people or against certain forms of sexual practices, that encourages a degree of rational self-consciousness surrounding sexual behavior. This rational self-consciousness is a prerequisite for self-governance, of the individual body and the body politic alike. Millar laments:

> It cannot be supposed therefore that the passions of sex will ever arise to any considerable height in the breast of a savage. He must have little regard for pleasures which he can purchase at so easy a rate. He meets with no difficulties nor disappointments to enhance the value of his enjoyment, or to rouse and animate him in the pursuit of it. He arrives at the end of his wishes, before they have sufficiently occupied his thoughts, or engaged him in those delightful anticipa-

tions of happiness which the imagination is apt to display in the most flattering colours. He is a stranger to those eager hopes, those anxious fears, which agitate the mind of the lover; and which, by the conflict they produce, have a tendency to enliven his feelings, and to increase the force of his prevailing inclinations.[70]

The "intercourse of the sexes," when properly engineered or shaped by extant social and civic hierarchies (that is, by self-denial) encourages the proper development of passions that later allow for successful self-governance. It is important to note here that Millar does not conceptualize strong sexual or romantic passions as negative or even as experiences exclusive to life in the torrid zones. For civilizational progress to occur, within Millar's stadial narrative, a paucity of sexual passions can become as significant a problem as an excess:

> Among early and rude nations, we can expect but little improvement, with regard to these passions. A savage, who earns his food by hunting and fishing, or by gathering the spontaneous fruits of the earth, is incapable of attaining any considerable refinement in his pleasures. . . . He has no time for cultivating a correspondence with the other sex, nor for attending to those enjoyments which result from it; and his desires being neither cherished by affluence, nor inflamed by indulgence, are allowed to remain in that moderate state which renders them barely sufficient to answer the purposes of nature, in the continuation of the species.[71]

It is notable that excessive sexual moderation or disinterest is frequently associated with Native peoples of North America within eighteenth-century naturalist texts, since parts of North America disputably fall within both the torrid and frigid zones. Nussbaum notes this internal contradiction in Millar's writing as well, arguing that Millar, "although seeming to argue that warm climates heighten sexuality, paradoxically contests environmental determinism to assert that the savage may be less interested in sexual subtleties because he dares not distract himself from grubbing out an existence."[72] This tension—between Millar's ostensible faith in climatic influence and his assessment of people at the putative first stage of human development as problematically, insufficiently passionate—demonstrates the incoherence and contradiction that permeates eighteenth-century natural histories and their attending environmental explanations of sexual behavior. Theories about human vulnerability to environment were wrought, highly contentious, and inclusive of numerous voices and opinions, both authoritative and colloquial, on any given issue.[73]

In an unusual twist, for example, English essayist John Brown relies on the power of the climate to offer an acerbic critique of the sex and gender norms of the *temperate zones,* complaining in *An Estimate of the Manners and the Principles of the Times* that "our effeminate and unmanly Life, working along with our Island-Climate, hath notoriously produced an Increase of *low Spirits* and *nervous Disorders*."⁷⁴ Reprinted in Boston in 1758, his refreshingly dour vision of life in England and other increasingly overcivilized locales describes the "Character of the Manners" of the mid-eighteenth century as "a *vain, luxurious,* and *selfish* EFFEMINACY."⁷⁵ Both women and men, as well as the social and sexual relations *between* the sexes, are, for Brown, poisoned by this widespread effeminacy that results from the toxic combination of the British climate and the greater access to and participation in fashion; he bemoans that "the Sexes have now little other apparent Distinction, beyond that of Person and Dress: Their peculiar and characteristic Manners are confounded and lost: The one Sex having advanced into *Boldness,* as the other have sunk into *Effeminacy.*"⁷⁶ For Brown, this historical change has especially problematic consequences for British youth: "THUS, like *Plants* hastily removed from their first Bed, and exposed to the Inclemencies of an unwholesome Air . . . our rising Youth are checked in their *First Growths;* and either *die away* into ignorance, or, at most, become *Dwarfs* in Knowledge."⁷⁷ The botanical metaphor at the heart of his warning (a trope I return to in chapter 4) and the fact of the reprinting of his book in Boston, where burgeoning anticolonial tensions were beginning to spread, demonstrate that natural history could be enlisted by colonists, too, to speak back to metropolitan power. It was a versatile and quite portable tool.

In North America, one of the most famous works of natural history arrived toward the end of the century, with Thomas Jefferson's oft-discussed *Notes on the State of Virginia*. Jefferson began writing *Notes* sometime around 1780, at the request of François Barbé-Marbois, the secretary of the French legation at Philadelphia.⁷⁸ In 1780 Barbé-Marbois circulated a series of questions to representatives of the thirteen North American colonies about their natural resources and economy; Jefferson complied, penning a series of responses that were eventually printed in 1785. To historians of science, *Notes on the State of Virginia* constitutes a fairly typical effort within the larger eighteenth-century tradition of natural historical writing, in both form and content.⁷⁹ Opening with a discussion of the bounds of the territory of Virginia, Jefferson proceeds to describe the region's natural features (rivers, mountains, etc.), its natural resources (minerals, animals), its natural environment (climate and air), the population (Native, colonial, and enslaved), and the infrastructural conditions of the area (laws, currency, educational institutions, etc.). Throughout much of *Notes,* Jefferson refers to other pieces of naturalist writing, in particular Buffon's *Histoire Naturelle,* which he cites frequently and at length; as Christopher Iannini argues, *Notes* is a document very much in dialogue with long-standing conversations among natural historians around the Atlantic world.⁸⁰

Jefferson's clear invocation of other works of natural history offers a fitting picture of the way ideas advanced by scientists in Europe and the Caribbean found their way to the British North American colonies, where they circulated widely. Furthermore, in *Notes* Jefferson refers promiscuously to each of the different theories of the origins of human variety—climate theory, humoral theory, stadial theory, and theories of complexion and disposition—that I discuss in this chapter, frequently taking issue with other naturalists' disparaging assessments of the peoples, animals, and societies of North America based on its ambiguous situation in the torrid zones. Jefferson is particularly vocal in his ire for Buffon's assessment of the degenerating climate and society of the North American colonies.[81] Importantly, part of Jefferson's defense of North America and its climate includes a refutation of some of the more notably racist ethnographies that appear in Buffon's long essay on human variety in the *Histoire Naturelle*. Jefferson begins by quoting several of Buffon's most damning descriptions:[82]

> Although the savage of the new world might be of about the same size of that of man of our world, that is not enough to make an exception to the general fact of shortening of life throughout that continent; the savage is weak and small in the organs of generation . . . no ardour for his female. . . . Do not look any further for the cause behind the itinerant life of savages or their removal from society: the most precious spark of the fire of nature was denied them; they lack passion for their female, and therefore love for their fellow men: ignorant of the keenest attachment, the most tender of all, their other feelings of these kinds are cold and languishing; they feebly love their fathers and their children; the most intimate society of all, that of the family, has amongst them the weakest of ties; among families, there are none at all; as a result no coming together, no republic, no state of society. Physical love, among them, constitutes the moral of mores; their heart is hard, their society cold, and their empire unforgiving. They regard their women only as domestic servants or beasts of burden whom they task, without management, with the burden of their hunt, and whom they force without pity, without gratitude, to do work that is often greater than their strength: they only have a few children; they lack care; everything stems from this first flaw; they are indifferent because they are not strong, and this indifference for sex is the original sin that withers their nature, that prevents them from flourishing, and that destroys the seeds of life, cutting off the root of society at the same time.[83]

This is one of the several long passages from Buffon's *Histoire Naturelle* that Jefferson reproduces in order to refute, but Jefferson's refutation of this passage is motivated less by any interest in defending Native peoples as a population than by his interest in defending North America against Buffon's accusation that it constituted a degenerating climate. Buffon's assessment of Native Americans as "weak and small," "less strong of body," and less passionate girds his argument about the inevitable degeneration that occurs among animals, human and otherwise, in North American climates, which include both temperate and torrid zones.[84] Jefferson retorts that "human nature is the same on every side of the Atlantic, and will be alike influenced by the same causes";[85] in a London printing of *Notes,* Jefferson explicitly asserts that Europeans, European-Americans, and Native Americans all fall under the Linnaean category of "Homo Sapiens Europaeus."[86]

Importantly, however, this widely cited passage tarries at length on the sexual habits of Native North American peoples, again defining both parts of the body (the "organs of generation") and relative ardor ("passion for his female") as qualities vulnerable to the influence of the specificities of the North American climate, diet, and state of society. More importantly, however, it showcases Jefferson's unique interest in manners, here a metaphor for state of society, or civilization more broadly, as a powerful environmental influence. In *Notes,* Jefferson systematically corrects Buffon's misapprehensions about North America and its peoples but ultimately acquiesces to Buffon's universalizing assessment of Native American gender norms, admitting that "the women are submitted to unjust drudgery." He continues, "This I believe is the case with every barbarous people. With such, force is law. The stronger sex therefore imposes on the weaker. It is civilization alone which replaces women in the enjoyment of their natural equality."[87] For Jefferson, then, it is not anything in the climate or air of North America that encourages degeneration among humans or other animals; rather, it is the lack of civilization in Native societies that renders women subordinate and men weak and dispassionate. Jefferson repeatedly returns to this logic in *Notes,* arguing against climate-based readings of any of North America's drawbacks and relying instead on civilizationalist narratives that draw from stadial theory to explain why certain populations might appear degenerate. Jefferson acquiesces to the popular cultural perception among Europeans, for example, that Native peoples "have no ardour for their female," and explains it by arguing, "It is true they do not indulge those excesses, nor discover that fondness which is customary in Europe, but this is not owing to a defect in nature, but to manners."[88] Repeating this position, he states baldly, "The seeming frigidity of the men therefore is the effect of manners, and not a defect in nature."[89] Here we see the role that sexual behavior plays in eighteenth-century theories of progress (stadial or "four-stages" theories, among others): the passion between the sexes is necessary for both reproduction and advancement, but too little passion between them renders a society despotic and incoherent, and too much degenerates the community. While refusing the common climatic hypotheses about the inevitable degeneration of man in North America—proclaimed both by Buffon and by his contemporary, the Abbé Raynal—Jefferson concedes that differences between Native Americans and Europeans may exist;

but if so, it is a result of "manners" or the "state of society," not an inherent defect in the climate.

When it came to Africans and African Americans in North America, however, Jefferson had no such ambivalence about the possibility of fundamental morphological and characterological differences between varieties of humans. Jefferson reads differences between black people and Europeans as universal and fundamental, and alongside his proposition that slaves be manumitted in Virginia, he was also a well-documented proponent of colonization, the political platform in favor of sending emancipated slaves back to the African continent, and specifically, to what later became the American colony of Liberia. His colonizationist beliefs stem not just from his insistence that Africans and African Americans would suffer as a result of long-standing discrimination in North America:

> To these objections which are political, may be added others which are physical and moral. The first difference which strikes us is that of colour. Whether the black of the negro resides in the reticular membrane between the skin and scarf skin, or in the scarf skin itself; whether it proceeds from the colour of the blood, the colour of the bile, or from that or some other secretion, the difference is fixed in nature, and is as real as if its seat and cause were better known to us. And is this difference of no great importance? Is it not the foundation of a greater or less share of beauty in the two races?[90]

Although Jefferson ardently refuses Buffon's characterization of Native Americans as feeble and dispassionate and urges the universal emancipation of slaves, he reverts to an anatomical explanation of racial difference in his discussion of African Americans; this passage points again to the coexistence of environmentalist and more rigid theories of human variety in works by the same authors, and even in the same texts. Jefferson's understanding of black skin as effectively a symptom of deeper differences that are "fixed in nature" implicitly suggests that this blackness indexes other forms of difference as well, and one of the forms of difference between white people and black people that he makes certain to address is that of their differing levels of "passion" for the female sex. He writes of African Americans: "They are more ardent after their female: but love seems with them to be more an eager desire, than a tender delicate mixture of sentiment and sensation. Their griefs are transient. Those numberless afflictions which render it doubtful whether heaven has given life to us in mercy or in wrath, are less felt, and sooner forgotten with them. In general, their existence appears to participate more of sensation than reflection. . . . It would be unfair to follow them to Africa for this investigation. We will consider them here, on the same stage with the whites."[91] Insisting on considering the effect of the North American climatic and social environment on both black and white peoples ("We will consider them here, on the same stage with the whites"), in *Notes*, Jefferson amends certain older environmentalist theories of human variety to describe a North American climate that neither degenerates whites nor recuperates (or racially transforms) Africans, African Americans, or Native peoples. To develop a natural historical theory of life in North America's varying temperate *and* torrid zones, Jefferson had to rely more heavily on taxonomies of "manners" to develop a cogent theory of human variety. As a result, Jefferson's racial science often seems to anticipate later, nineteenth-century anatomical theories of racial difference.[92] This tension between environmental and "intrinsic" properties illustrates, again, that beliefs about sexual behavior had gradually but widely become embedded in theories of racial difference or human variety. Understanding the coemergence and interrelatedness of these frameworks for eighteenth-century scientific theories of human variety helps us understand the coexistence of both environmental and intrinsic theories of the origins of human sexual behavior as well.

We have glimpsed the wealth of narratives about the origins of human sexual behavior that circulate within eighteenth-century natural historical theories of human variety. Although accounts of sexual variety in North America appear primarily in European natural histories, these texts also circulated in North American personal collections and were later housed in early libraries. And, as Katy Chiles has argued, popular literature of the eighteenth century engaged both explicitly and more subtly with a wide range of natural historical theories of human variety.[93] While many of the scientists penning these inquiries and essays were members of a small, educated elite, European theories of racial differentiation were also presented, debated, and circulated by important early institutions such as the American Philosophical Society. Ideas that were developed in and by relatively inaccessible scientific social and institutional circles frequently saw a much broader and more diffuse dissemination within popular culture as a result of being referred to, explored, or argued within more popular literatures and performances. Furthermore, "scientific" theories—a specious distinction in the eighteenth century, as it remains today—were frequently supported or permeated by forms of basic cultural knowledge that reflected everyday ways of the world. While many if not most natural histories were not always immediately accessible, the multiple forms of environmentalist logic that girded the eighteenth-century human sciences frequently reflected commonsensical ways of thinking about the world that would have been familiar to a very broad segment of the population.

Because sexual practices, differences, and inclinations were so often anatomized by natural historians as part of a larger taxonomy of racial difference—think, for example, of Buffon's *Africanus niger*—writing about human variety had the inadvertent effect of rendering sexual behavior visible in eighteenth-century inquiries into human difference. And because the same environmentalist logic that girded contemporary understandings of racial difference necessarily informed the way these theories represented sexual behavior, in natural historical writing, sex and sexuality bore the same capacity for dynamism, mutability, and transformation as did skin color or other iterations of human variety.

There are important implications here for scholars of the history of sexuality. As we have seen, residents of eighteenth-century North America frequently encountered information about sexuality—whether in its social, cultural, or regional forms—through environmentalist theories of human variation, which understood sexual behaviors to be contingent on a person's natural, physical, climatic, or cultural circumstances (among others). These eighteenth-century sources support two critical and counterintuitive hypotheses about the significance of sexual behaviors during this period.

First, these texts suggest that sex—acts, desires, dispositions—was not entirely or even mostly theorized in privileged relation to the individual. We see sexuality today as the product and property of individuals, who produce specific forms of sexual desires or identities that originate exclusively in their bodies. Christian settlers of early America on the whole believed nearly the opposite: humans were shaped not as individuals but as communities, the result of a broad and unique collection of environmental factors that together made manifest particular social, affective, and sexual forms on their bodies. While there was always an assumed degree of entropy present in the dynamic interface between any singular human body and its environment (such that the resulting behaviors or social forms could never be satisfactorily predicted with any certainty), eighteenth-century natural history demonstrates that the influence of environment was at least as strong as the influence of individual inclination or will when it came to sex.

Second, theorizing sexual behaviors as driven primarily or even significantly by environmental factors rather than human will or subjectivity demands a new set of definitions for sex and sexuality *tout court*. If, as these eighteenth-century texts imply, sexuality is not strictly a human experience but rather the effect of a combination of environmental factors that influence and at times find expression in human behavior, without being limited to or entirely exhausted by these behaviors, then what was sexuality during the eighteenth century, and what does an early history of sexuality look like? If, in the Foucauldian sense, sexuality is a dense and complex grid of locations through which power is organized to render a subject that is both coherent and knowable, what is sex *before* sexuality? Before the subject? And finally, if the archives that have heretofore been central or taken pride of place in studies of the history of sexuality are those that prioritize the subject—first-person accounts (diaries, true narratives, etc.), genital-based descriptions of sexual encounters of all kinds, and court records—to which archives should we be turning to find more accurate or apt narratives of what sexuality was, or was understood to be, during the eighteenth century? In the chapters that follow, I trace sexuality in some of its less recognizable forms through a host of eighteenth-century texts, tarrying on historical moments and cultural productions that were widely perceived as sexual but which represent sexuality and sexual behavior more generally through forms at a distance from the subject or subjective experience. These chapters also highlight the ways even the most recognizable forms of early sexual behavior relied upon eighteenth-century environmental logic, or other ways of understanding sexual behavior drawn from contemporary racial sciences, for their coherence.

This book searches for a way of describing this period in what Peter Coviello calls the "earliness" of sexuality, but in a much earlier "earliness" than the nineteenth-century moment of his inquiries: "the experience of sexuality as something in the crosshairs of a number of forms of knowledge and regulation *but not yet wholly captivated or made coordinate by them*"; the "sense . . . of sexuality as a realm of experience and expression as yet uncodified, not yet battened into place by the discourses in which it increasingly found itself located."[94] Theorizing an eighteenth-century history of sexuality might mean considering the possibility of a realm of experience that is *not* actually a knowledge, in the Foucauldian sense of the term. To be very clear, here, while this book is deeply informed by Foucault's narrative about the emergence of modern sexuality that he details in *The History of Sexuality*, volume 1, it also tells a story that significantly departs from his account, controlling as it nonetheless remains in US American sexuality studies. In this book, I ask what sex was before it was fully targeted by discourse; to be more precise, here I am concerned with how a set of loosely related forms of behavior that trafficked under terms like sodomy, venery, bad habits, bestiality, sapphism, and many more were conceptualized before they could be precisely thought and named within the singular category of "sex." This question is rendered more manifestly complex when we ask what "sex" was before it described a field of power that operates through a roughly predictable or imaginable set of sites (institutions, people, etc.) but while it was simultaneously becoming a central, if not constitutive, component of another discourse, that of racial difference, which at least minimally relied on sex to articulate itself. It is tempting to imagine the eighteenth century as some sort of pre- or proto-discursive moment in the history of sexuality, in which historians of sexuality can just make out the traces of a teleology that would achieve some form of completion in the nineteenth century with the development of sexology. Certainly there is evidence, and a way of organizing that evidence, to support that claim. Under this line of reasoning, "racial difference" becomes yet another location in which the discursive power of sexuality invests, grows thick, and proliferates. And while modern discourses of sexuality are absolutely invested in racism and racial difference, as numerous scholars have shown, there is yet another story to be told about this eighteenth-century earliness of sexuality.

The next chapter takes us to the lives of sailors, based in both England and New England, working to transport commodities between Europe and the British North American colonies. Life aboard ships was difficult work and was attended by no shortage of dangers. One of the most pressing of the risks of the seafaring life throughout the long eighteenth century, as perceived by sailors themselves but also by residents of colonial British communities, was rampant piracy, especially that which was perpetrated by Barbary corsairs, pirates who not only raided ships and occasionally coastal towns to take possession of their goods and vessels, but also took possession of any Christian sailors aboard to be sold on the thriving Islamic slave market that flourished within the Barbary cities of Tunis, Salé, Algiers, and Tripoli. Many Christian sailors who were kidnapped and sold as slaves on the Barbary slave market recorded their experiences in real time and later penned "true narratives" of their enslavement after they were ransomed or redeemed; these texts, both manuscripts and printed editions, grew increasingly popular among North American readers throughout the eighteenth century. Many of these stories reflect the same fear: that of being sodomized by their male captors. The narratives thus provide a radically different perspective on the history of Christian perceptions of sodomy from the way historians of sexuality have tended to theorize it; as I demonstrate, residents of early North America perceived sodomy largely as a racial formation.

CHAPTER TWO

The Complexion of Sodomy

Buggery is a detestable, and abominable sin, amongst Christians not to be named, committed by Carnall knowledge against the ordinance of the Creator, and order of nature, by mankind with mankind, or with brute beast, or by womankind with bruite beast.
—Edward Coke, *The Third Part of the Institutes of the Laws of England*, 1644

These *Turks* are goodly people of person, and of a very fair complexion, but very villains in mind, for they are altogether sodomites, and do all things contrary to a Christian.
—William Davis, *A True Relation of the Travels and Most Miserable Captivity of William Davis*, 1704

Pierre Dan's 1649 *Histoire de Barbarie et de ses Corsaires* (History of Barbary and its corsairs) is a compendious survey of the history, geography, governance, religion, and language of the Barbary states and of their militarized organizations of corsairs (pirates).[1] In terms of form and content, Dan's *Histoire* is not particularly remarkable; it resembles many contemporary natural histories of regions of the African continent penned by European writers, such as John Ogilby's slightly later *Africa: Being an Accurate Description of the Regions of Aegypt, Barbary, Lybia, and Billedugerid* (1670).[2] Unlike contemporary natural histories, however, Dan's *Histoire* includes only one illustration.[3] Expensive editions of seventeenth-century French natural histories often contained multiple copperplate engravings, but Dan's *Histoire* bears only the *Figure Nécessaire: Ou sont representez les divers supplices, dont les Turcs, et ceux de Barbarie, persecutent, et font mourir cruellement les Esclaves Chretiens . . .* (Necessary Figure: In which is represented the various tortures by which the Turks and those of Barbary persecute and cruelly murder Christian Slaves . . .), a carefully numbered table

Figure Necessaire, Ou sont representez divers supplices dont les Turcs, and ceux de Barbarie, persecutent et font mourir cruellement les Esclaves Chrestiens, comme il se voit aux pages suivantes. From Pierre Dan's *Histoire de Barbarie et de ses Corsaires* (Paris: Chez P. Rocolet, 1649). Courtesy of the Thomas Fisher Rare Book Library, University of Toronto.

met entre la page 414. 415. 416.

of twenty-two panels depicting a range of tortures to which Christians enslaved by "Turks" in the Barbary states might be subjected.[4]

Some of these images would have been immediately recognizable to most Christian readers, such as panel 5, where a Christian is being burned at the stake, or panel 16, in which a Christian is being stoned to death. Several other panels portray forms of punishment that were, in seventeenth- and eighteenth-century descriptions of the Barbary states, representative commonplaces. The first panel, for example, depicts death by ganshing, a style of punishment that William Davis anatomizes in his 1597 Barbary captivity narrative (published in 1614); a person executed by ganshing was thrown off a high wall onto sharpened hooks protruding from the wall below.[5] Panel 10 depicts staking, another staple in representations of Barbary tortures (and one described at length in both Davis's narrative and Royall Tyler's 1797 *Algerine Captive*), in which a sharpened post is inserted at the "fundament" (anus) and driven through the body to emerge at the shoulder; and panel 17 depicts discipline by bastinado, or the beating of the soles of the feet, a common complaint of Christians enslaved in the Barbary states.[6]

Many of the images, however, depict excessive, almost artisanal forms of torture or executions that are not commonly (or ever) found in literary or graphic representations of Barbary captivity, forms that, furthermore, might not be identifiable at all were it not for the careful captions in the text of the *Histoire* itself. Panel 19 depicts a Christian sitting on the end of a cannon, with his Turkish captor (we know he is a "Turk" because of the turban he wears and because he is fully dressed) apparently poised to fire the cannon into his backside; panel 11 represents a flaying, complete with a "Turk" triumphantly brandishing a flap of what is ostensibly Christian skin in the background; and panel 9 shows a Christian built into a barrel that is about to be rolled down a hill. From commonly used and represented forms of execution such as burning or stoning, to outlandish and likely fabricated forms such as firing a cannonball into a Christian's "fundament," this engraving provides the only visual representation of any of the information included in Dan's four-part *Histoire de Barbarie*. Indeed, it is these dramatic images—of Christian bodies stripped naked, shaved, dismembered, rent full of holes, and rendered wholly vulnerable to the "brutish Lusts," as one author termed it, of their captors—that typify Christian representations of the experience of Barbary captivity.[7] Whether or not these representations bore any relationship to reality (although there is significant evidence to suggest that, at least in some cases, they did), the circulation of such images in geographies and other natural historical texts, Barbary captivity narratives, novels, poetry, and sermons painted a vivid and often terrifying portrait of life in the Barbary states for readers in both Europe and the North American colonies. Nor was being flayed, trapped in a barrel and rolled away, ganshed, staked, or burned alive the extent of Christian fears of what might befall them if taken captive at sea or during coastal raids by Barbary corsairs. Barbary captivity narratives persistently express fears of being sodomized as well as of falling prey to other forms of what Pat Parker terms "pathic subjection" that were associated with both the Islamic world and efforts to torture Christians into "turning Turk."[8]

I begin with Dan's *Figure Nécessaire* because it neatly stages two of the central concerns of this chapter: the often underexamined relationship between representations of Barbary captivity and natural historical inquiry, and the way Barbary captivity provided one of the most consistent occasions for the explicit discussion of sodomy and other forms of pathic subjection in seventeenth- and eighteenth-century Christian print culture. The lone *Figure Nécessaire* that appears in Dan's *Histoire* points to this overlap; the twenty-two panels depicting the brutal torture and murder of Christians by Muslims are arranged into a table, a visual form that Foucault defines as one of natural history's most important technologies, "a new way of connecting things both to the eye and to discourse."[9] The table is in turn neatly placed in the midst of twenty-two careful textual explanations that caption and explain each of its panels.

This chapter returns to sodomy, one of the most cherished formations for scholars of the history of sexuality, to examine a hitherto understudied feature of its appearance in seventeenth- and eighteenth-century texts: its power as a racializing figure. Much Barbary captivity writing makes heavy use of natural historical and otherwise ethnographic descriptions of the "Mahometan" world, implicitly (and often explicitly) proffering religious and ethnic difference to explain the putatively Islamic penchant for perpetrating physical and sexual harm against Christians. The sodomitical, in these narratives, constitutes a promiscuous tropology that cuts across representations of "Mahometanism" in many different genres and is deployed to assert a fundamental difference between Christians and Muslims. In what follows, I theorize the sodomitical as an expansive assemblage that indexes the act of sodomy while always simultaneously gesturing beyond it to geographies, polities, and even ways of seeing that reveal the limits or exterior of Christian worldviews. Here, the sodomitical is not just a set of acts or a set of meanings imagined to emanate from them; instead, the sodomitical points to ways of life inimical to Christian habits of diet, comportment, governance, and relation. In this sense, the sodomitical is a framework derived from natural history and is thus fundamentally animated by questions of human difference. By focusing on representations of the sodomitical in Barbary captivity narratives, the racialization of both Christianity and Islam comes into stark focus; the sodomitical defines Christianity in the negative, providing a vision of Christianity that turns on the depiction of what Christianity is *not*. While certain Christian religions—Catholicism, for example—were similarly represented as sodomitical in character, a species of religious enthusiasm tending toward lavishness or despotism, my focus here is how both natural historians and religious leaders turned to the figure of sodomy to stage their concerns about ever-increasing contact with the Islamic world.[10]

Importantly, the association of sodomy with "the east" and with Islam is not *entirely* a stereotype deriving from orientalist or proto-orientalist lore; seventeenth- and eighteenth-century Christians would have understood sodomy in relation to its biblical referent, the city of Sodom, in the Land of Canaan, which more or less corresponds to the region that is today termed the Levant.[11] Sodomy bore a distinct geography that corresponded in a labile but persistent way with long-standing European cartographies of human ethnic difference. Thus, if, as William Davis states so baldly in his captivity narrative, "Turks are altogether sodomites, for they do all things contrary to a Christian," and if sodomy was legally understood as the sin "amongst Christians not to be named," then Barbary captivity narratives provide one site where this diametric opposition between Christianity and sodomy is staged and exposed.[12] But if sodomy is truly understood in both highly orientalized terms as a "Turkish" vice, a language so alien to Christianity that it cannot even be spoken by believers, then we must take the explicitly religious (and thus, in eighteenth-century terms, also ethnic) character of representations of sodomy seriously and understand it for what it was, according to these texts. Sodomy conjures a trope as much about race as about sex, a behavior as central to the history of race in the Atlantic world as to the history of sexuality.

In the pages that follow, we explore five Barbary captivity narratives, alongside ethnographic accounts of the enslavement of Christians in the Barbary states, that circulated in both manuscript and print in the Atlantic world between 1653 and 1745. This chapter tracks the persistent invocation and representation of an assemblage of forms of violence that I term the sodomitical, including but not limited to discussions of sodomy, rape, priapic harm, and other forms of pathic subjection. I track the mutual imbrication of natural historical writing and Barbary captivity narratives, paying careful attention to the situatedness of many early Barbary captivity narratives in natural historical anthologies or collections and to the narrative reliance on ethnographic and otherwise natural historical detail in representations of "Turkish" culture in Barbary captivity narratives. The potential payoffs of reading Barbary captivity narratives alongside natural histories of the Barbary coast are manifold, but this chapter is concerned with the way sodomy was understood and explained in eighteenth-century North America. The incredible frequency with which sodomy, or fears of being sodomized, appears in Barbary captivity narratives constitutes an oft-noted but understudied feature of this genre, one that scholarship in the history of sexuality—a field more or less built on landmark studies of sodomy—has largely neglected to address at any length.[13] When we read Barbary captivity narratives' curiously persistent interest in sodomy alongside contemporary natural historical interest in the Barbary region, we are better able to understand how textual invocations of orientalist theories about Islamic depravity, sensuality, and decadence were often themselves steeped in heliotropic or otherwise climate-based theories of human difference.[14] Thus, whereas chapter 1 surveys the quiet sexual environmentalism that lurked latent (or sometimes quite evidently) in natural historical theories of human difference, this chapter offers sustained analysis of one particular iteration of my environmental theory of early sexuality, as it appears in Barbary captivity narratives. Sodomy, as these texts reveal, was understood not only as a problem of polity, discipline, social order, inclination, or innate depravity, but also as a particularly racialized behavior.[15]

This argument constitutes a major departure from almost all scholarly treatments of sodomy in early North America. Those works tend to read the prohibition against sodomy as an indictment of specific types of relationships: cross-class relationships, interracial relationships, relationships between men that appear to preclude or occlude sexual relationships with women, or relationships such as nonmarital or extramarital connections that do not conform to Christian or more broadly European forms of social organization. Is it any wonder, then, that sodomy became defined in English common law—and *continued to be defined* in US laws through the late twentieth century—as the crime "not fit to be named by Christians"? Although contemporary legal language may not be aware of it, these sodomy laws are once again relegating this crime against nature not to homosexuals, but to Muslims.[16]

A final introductory note: this chapter engages in only a very limited, mediated way with Arabic or other non-Anglophone texts from the eighteenth century, owing both to my own lack of appropriate language skills and to the cultural and geographical scope of my research. The representations of both Barbary captivity and the Islamic world that appear in the Anglophone sermons, narratives, and travel writings at the heart of this chapter index popular forms of collective cultural imagination, approbation, longing, and much more. Following Edward Said and Sara Ahmed, I read British and North American textual representations of the Barbary states in phenomenological terms as manifestations of early orientalist discourse, understanding the relationship between Christian readers and the representations of Muslims that appear in these texts to be what Ahmed terms an "energetic" one, one constituted between (and *constituting*) the reading subject and the read or represented *object*.[17] Put differently, while Barbary captivity narratives and other forms of early modern popular print that took Islam as their focus certainly had real-world ramifications and influenced what Christian readers believed they knew, it is critical to remember that many of the most prolific writers on the subject, such as Cotton Mather, had never been taken captive and had never seen Tunis, Tripoli, or Algiers themselves, and that even those writers who *had* such experiences are not necessarily reliable narrators, given the vicissitudes of memory, uneven archival histories, editorial mediation, and what amounts to early Islamophobia. At the same time, I heed the warnings of the many scholars of the global eighteenth century who have cautioned against reading a totalizing or hermetic divide between "East" and "West," "Occident" and "Orient" during the early modern period or imagining that curiosity or interest in the "East" by the "West" was one-sided. Much important scholarship has emphasized that travel writing and ethnography was a deeply global generic phenomenon.[18]

Barbary Captivity and Popular Narrative

Barbary piracy was an enormous problem for both European and North American communities from the early modern period through the early nineteenth century. The Ottoman regencies of Tunis, Tripoli, and Algiers, and the republic of Salé (later Morocco) stretched along the long northern coastline of the African continent and were home to powerful and immensely skilled networks of corsairs who used *raizas*, or raids, to accrue commodities and slaves and to control ocean territory. Corsair networks included the Berber natives of these regencies alongside experienced seafarers from all over Europe, the Mediterranean, and northern Africa. Corsairs were always professed Muslims (for this religious difference is what justified enslaving Christians), but among them were also *renegados*, former Christians who had "turned Turk," sometimes as a result of having initially been taken captive themselves.

It would be difficult to overstate the threat that Barbary piracy posed to European economic prosperity, especially as powers such as England, France, and the Netherlands embarked upon colonial ventures across the Atlantic, shipping commodities back to metropolitan markets hungry for wood, skins, and furs, and products such as coffee, sugar, and spices.[19] Roaming the far reaches of both the Mediterranean and the Atlantic, Barbary pirates would converge upon a ship, taking control (and ownership, if it was not destroyed in the attack) of the vessel and its contents and selling all Christian or otherwise non-Muslim sailors on the Islamic slave market that flourished in Ottoman regencies throughout the seventeenth and eighteenth centuries (or selling them back to their home countries for hefty ransoms). Though they pirated predominantly in the Mediterranean and along the western Atlantic coast, Barbary corsairs engaged in *raizas* and sea-based kidnappings as far north as Ireland and Iceland and as far west as what is now New Brunswick.[20] The sixteenth- and seventeenth-century Islamic slave market—most active in northern Africa, especially in the Barbary states, as well as in parts of the Middle East—was not as extensive or as economically muscular as the simultaneously developing Atlantic slave market, but historians of Barbary piracy estimate that, by 1621, at least 20,000 Christian hostages were being held at any given time in Algiers, and other scholars suggest that as many as 25,000 male and 2,000 female slaves were held in Algiers by the 1630s.[21] While some European powers eventually negotiated treaties with the Barbary powers that allowed their ships to pass safely in exchange for regular and costly tributes, tens of thousands of European and North American sailors were taken captive by Barbary pirates during the early modern period through the early nineteenth century, and a few of those lucky enough to be redeemed went on to write accounts of their experiences.

These accounts, however, appear unevenly in archives, and scholars attempting to study the evolution of Barbary captivity writing in North America find themselves faced with a strange problem. For the seventeenth and early eighteenth centuries, when corsairing was much more prevalent and corsairs more powerful, there are fewer accounts of Barbary captivity available to scholars, especially accounts penned by residents of North America. Yet, by the early national period, and in the years leading up to the Tripolitan Wars (1801–5), when both the Ottoman Empire and corsairing networks were by most accounts weaker than they had been in several decades, an explosion of plays, pamphlets, novels, true narratives, "oriental tales," and poems featuring stories of life in the Barbary states appeared.[22] Because of the immense difference in the number of primary sources available in either physical or digital archives, scholarship in early American studies that has focused on Barbary captivity has tended to cluster around the end of this trajectory of captivity narratives, particularly those from the late eighteenth and early nineteenth centuries. As a result, seventeenth- and early-eighteenth-century Barbary narratives, many of which exist only in manuscript form, have been allowed to languish unconsidered.[23]

Late-eighteenth- and early-nineteenth-century Barbary texts, however, were building on a tradition of Barbary captivity writing that had existed for more than two centuries. From the early modern English stage to the sermons of Cotton Mather, Barbary captivity had long proved itself a fecund occasion for the consideration of encounters with the looming threat of the powerful Ottoman Empire and the Islamic world and worldview it represented. Long before the emergence of the British Empire as a brutal, far-reaching, and administratively powerful regime—a vision we tend to associate with the nineteenth century but is as often projected backward onto its more inchoate colonial moments— England was one of many world powers jockeying for colonial territory during the rapid mercantile globalization of the early modern period. Furthermore, it was by no means the most powerful, wealthy, militarized, or well organized of the many imperial forces that competed for dominance during the sixteenth and seventeenth centuries.[24] The Ottoman Empire, from the medieval period onward, posed a fierce and formidable threat to both the economic prosperity and the territorial sovereignty of western European powers, and despite the declension narrative that scholars of the eighteenth-century Anglophone world tend to generate about waning Ottoman power over time, current scholarship demonstrates that this empire "was emphatically not in serious decline in the seventeenth century, or even, in some respects, for much of the eighteenth century."[25] This should serve as a reminder of the gravity and, frequently, critical concern that characterized Anglo-Atlantic interest in the Ottoman world (which included several of the Barbary regencies). The threat of Ottoman imperialism was of course economic, but it carried with it an additional threat: its defining relationship to Islam. As Samuel Johnson pithily remarked, "There are two objects of curiosity—the Christian world, and the Mahometan world. All the rest may be considered Barbarous."[26] Thus, when Cotton Mather, writing at a distance of more than five thousand miles from the Barbary states, worried in 1698 about the burgeoning power of the Ottoman world, we must read his concern as a genuine reflection of realistic fears about the potential reach of Islam. At the turn of the eighteenth century, it was not a foregone conclusion that the English language, or Christian religions, would maintain their cultural and structural position through the century's end.

Barbary captivity narratives, then, not only registered real experiences but also materialized real fears about the predations of the burgeoning power of the Ottoman Empire. And while there are only two North American Barbary captivity narratives written before 1750 that have received any attention by scholars (the narratives of Joshua Gee [captive 1680–87] and Abraham Browne [captive 1653–68]), the possibility of being kidnapped and ransomed by Barbary corsairs was one with which both North American sailors and their home communities were intimately familiar.[27] In the seventeenth and early eighteenth centuries—long before efforts were in place to develop a national infrastructure to redeem captives—redemption moneys came largely from fund-raising by local Christian churches of all denominations.[28] Even nonliterate or nonseafaring folk in coastal areas would have likely been very familiar with stories of Barbary captivity. Members of Cotton Mather's congregation, for example, would have heard him urge his audience on multiple occasions at the turn of the seventeenth century to pray for the spiritual steadfastness of their peers who remained captive in Barbary.[29] Depictions of Barbary slave markets also consistently appeared in travel writings, ethnographies, and natural histories that were published in England but circulated among North American elites. The story of what might befall a sailor unfortunate enough to be forced into Turkish slavery was, by the end of the eighteenth century, a familiar one; handwritten as well as printed documents telling the stories of kidnapped sailors circulated in the seventeenth- and eighteenth-century Atlantic world, but it is also likely that information about Barbary captivity was shared in an ad hoc fashion among transatlantic communities of sailors, who, in turn, shared these stories with their home congregations.[30] And despite the relatively small body of archival resources available to develop our understanding of what kind of knowledge circulated among residents of the seventeenth- and early-eighteenth-century British North American colonies about the experience of Barbary slavery, the extant narratives exhibit a remarkably consistent set of images, suggesting prior familiarity with the genre rather than uniformity of experience for Christian Barbary slaves. More simply put, by the very early eighteenth century, in both colonial North America and western European seafaring cultures, a clear and entrenched set of images, representations, and narratives surrounding Christian encounters with Barbary pirates was already in existence.

Two of the accounts I discuss here—Abraham Browne's *Booke of remembrance of Gods Provydences toward me* (1653–68) and Joshua Gee's *Narrative of Joshua Gee of Boston, Mass* (1680–87)—were the diaries of men who we know to have been taken captive and redeemed during the second half of the seventeenth century. Diaries both, these narratives are frustratingly partial but nonetheless offer critical glimpses into the experience of captivity. In addition to Browne's and Gee's narratives, I consider Francis Brooks's *Barbarian cruelty*, a salacious account of Barbary slavery initially published in England in 1693 and reprinted in Boston in 1700. Because Gee and Browne both resided in Boston after being redeemed (Gee sailed out of Boston before being taken captive, but Browne was born in Plymouth, England, and later settled in New England), their narratives, alongside Brooks's *Barbarian cruelty*, offer a powerful image of textual representations of Barbary captivity that were circulating in a town of fewer than ten thousand people during a period of just over thirty years. Finally, I turn to two significantly more salacious accounts of Barbary slavery, the anonymous *Description of the nature of slavery among the Moors* (1721), bursting as it is with depictions of "Mahometan" lechery, and William Davis's *Travels and miserable Captivity of William Davis*, a Barbary captivity narrative supposedly penned in 1597 that was included, in 1745, in an anthology of travel and ethnographic writing compiled from the library of the late Earl of Oxford. There is no evidence to suggest that these last two texts did, or did not, circulate in the North American colonies during the early eighteenth century, but the Library Company of Philadelphia lists in its 1789 catalogue a copy of *A Collection of Voyages and Travels*, in which Davis's narrative appears.[31]

When we put this perhaps strange assemblage of texts into conversation with one another, two shared features of these narratives come into stark relief: first, each of these texts, no matter how truncated the narrative, professes an evident fascination with the differences between Christian and "Turk" and a commensurately dedicated attention to the processes by which this distinction is rendered less clear. Offering carefully ethnographic accounts of life in the Barbary states, each text provides detailed information about "Mahometan" styles of food, dress, worship, governance, and social organization and frets over the possible spiritual implications of prolonged exposure to this lifestyle. Second, each narrative depicts the pathic subjection of Christians at the hands of Muslims, at length and in great detail, lingering on iterations of what I am calling the sodomitical—appearing more explicitly in some cases than in others—in their lives in captivity. Read together, these five texts enlist discussions of the sodomitical to emphasize critical questions of human difference, while also folding sodomitical practices, values, and customs into a naturalist vocabulary used to characterize life in the Barbary states more generally, alongside religion, language, geography, and climate. In the aggregate, these texts present the sodomitical less as a stand-alone behavior than as a feature of an idiosyncratic geography that points simultaneously to religion, landscape, and way of life.

Sailor Diarists and Barbary Slaves: Abraham Browne (Captive 1653–1668) and Joshua Gee (Captive 1680–1687)

Joshua Gee's truncated account of his time as a slave in Algiers clearly illustrates his keen awareness of the threat of Barbary piracy. Gee opens his fragmented narrative by recounting the general feeling of foreboding he experienced before going to sea for the first time. Securing his father's permission to ship out as a sailor, he relates: "I had conseved an opinion that it wold be atended with greate evell & distres yet I most prosede. I told som of my frends one of which is living: that I had an aprehenson that I shold be led throw firy trials & carried thro them. . . . he said why will you prosed having suth aprehensions of the voyage I answered I most goe & Can not Avoide it. yes I doe beleve that god will Cary me throw & bring me bak to this place." His general sense of apprehension intensifies as he embarks on his journey, culminating in a dream that comes to him the night before his ship is attacked and he and his fellow sailors are taken captive. Leaving Boston, and London bound, Gee notes: "In owre voige for England we were short of bred. & in a dreame on a day before we weare taken I saw a ship that did suply owr want of bred but at a deare Rate." The "deare Rate" that Gee and his fellow sailors paid for the bread that they so needed was that of their freedom, as the ship was an Algerine vessel that took Gee and the rest of his crew hostage. Importantly, Gee notes that it was not just *any* Algerian ship that approached his own, but the same one that had appeared in his dream the night before: "At site of ye algere ship I aprehended it was the same to aperance I had sene in my dreame & fownde it soe."[32] While this moment of apparent clairvoyance might at first seem just a coincidental detail, Gee uses it to foreground his insistence on his enduring spiritual commitments throughout the rest of the short account, imagining in a way characteristic of Calvinists of his era that his captivity and redemption were ordained for him by God. He understood his captivity not only as a broad test of his faith but as a test of his faith specifically in the face of the exhortation to convert to Islam.

Joshua Gee was born "as early as 1654," although we have no definitive record of his life before his captivity.[33] He lived in Boston, where his father was a fisherman, and eventually began working as a sailor aboard various merchant vessels, conveying loads of tobacco to western European ports. He was first taken captive in 1680, and he remained a slave, working for at least two "masters," for almost seven years. Gee was ransomed on July 14, 1687, after his parents and his younger brother pooled all of their financial assets to secure his release. According to the editor of the sole published edition of Gee's narrative, Albert Carlos Bates, Gee penned his written account of his captivity during the six months that it took him to return to Boston from northern Africa. Since his story was not transcribed and printed until 1943, there is no way to determine to what degree it circulated within his own time, but it is likely that the handwritten account circulated only to a limited extent.[34] The facts of his captivity, however, closely mirrored those that appear in other Barbary captivity narratives that were penned both before and after Gee wrote his, and it is likely that his story was known in his immediate community, especially as his son went on to become a fairly established minister in Boston.

Barbary captivity was a concern not only during voyages that took sailors around or along the Barbary coast, but even on routes that would have been standard for Boston-based laborers, such as between the Boston Harbor and various ports in southern England.[35] Joshua Gee was clearly concerned about this, as his dream indicates; and almost every one of the fifteen pages that make up his narrative describes exactly the kind of torture and violence that probably stoked the fears that had in part animated his ominous dream.[36] A common rhetorical strategy in triumphalist descriptions of Christian resistance to Islamic domination or to the seductive allure of "turning Turk," these descriptions of Christian suffering at the hands of their Barbary captors work toward multiple ends. Gee's representation of the violence to which he was subjected while captive in Algiers demonstrates his understanding of the experience as a test of his faith. At the same time, however, many of the depictions of the violence to which Gee was subject traffic in sodomitical images, as readers vicariously watch Gee and his fellow captives being stripped naked and beaten, their backs, buttocks, and the soles of their feet exposed to both their captors and the imagination of their readers. Here we see the conflation of the "Mahometan" "Turks" and the sodomitical, in which the threat or reality of pathic violence occasions Gee's recommitment to his faith, thus simultaneously defining Gee's Christianity through its imperviousness to the sodomitical predations of Islam.

Consider, for example, this gruesome episode that closes Gee's narrative. Describing a day on which he was put to work constructing a ship for his master, Gee recounts that his master grew increasingly angry at him and "in his grea[t] pashon swore he wold the next daye boare owte my eyes with his knife which he shewed to me: with many evell treatments. I went home with my hart overwhelmed, & spent that night in prayer like that in Isaih 38:13:14. & with that the harts of men are in his hands he Restraneth thear wrath. & . . . the morning I going to my work god soe Canged his hart & his Cownenance: his words weare words of kindness."[37] The master's showing Gee the knife with which he was going to "boare owte [his] eyes" invokes precisely the kind of threat of pathic subjection—the knife being a tool for the penetration of the Christian body—that we see illustrated so graphically in Dan's *Figure Nécessaire*. And here we see Gee's belief that God intervened into this scene of potential brutalization, protecting him from the angry "pashon" (and knife) of his Muslim captor. There is something excessive and incommensurate in Gee's master's threat to "boare owte" his eyes as recompense for what is implied as his mere displeasure. This excessiveness, materializing in a threat to remove Gee's eyes, also bears quietly racial implications that invoke the nominal etymology of the Barbary states themselves: the reaction of Gee's master, like the master himself, is barbaric.[38]

The barbarism of Gee's master calls upon a cultural context that links this scene of subjection to the larger threat of sodomy structuring Barbary captivity narratives. Patricia Parker concurs, in her work on the etymological and cultural overlay shared by references to barbers, barbarity, and the Barbary states within early modern drama. She argues that even the mere reference to "barbarous treatment" of Christians at the hands of Muslims needs to be understood as bearing an indexical relationship to cultural discussions of sodomy.[39] "The network aligning barbering, bodily hair, and barbarous treatment (as well as the multiple implications of cutting and shaving)," she argues, "was in the same period a staple of descriptions of sodomy, the notoriously capacious term that was frequently combined with circumcision as the bodily sign of turning Turk (or Jew)."[40] According to Parker, representations of Muslims cutting Christians would have been read within the wider context of the "pathic subjection" of Christians, a phrase she uses to encompass all forms of submission, including but not limited to sodomitical submission. Crucial to her argument, moreover, is that the "pathic subjection" of Christians to Muslims could not have been understood outside of the European discourse within which Muslims were understood as sodomites; she maintains that "in the same period as such dramatic representations, narratives of Barbary and of bond slaves of the Turk reported the forcible shaving of Christian captives, as well as their barbering in multiple senses, including circumcision, gelding, and sodomizing or pathic subjection."[41]

Indeed, some episodes in Gee's short narrative appear, at first glance, more sodomitical than others. After an awkward truncation in the narrative, for example, Gee describes collaborating with fellow slaves to plan an escape from slavery. Unfortunately, their plan was discovered, and on the next page Gee describes the fate of a French prisoner who ran away. He, "being taken was beaten by our Capen which was a Renegado Greak. till he sesed to Cry groane or ster. & was taken away for ded. but Recovered." Later, during Gee's indenture to his second master, "Covo mestefa," that master realized that "one of owr Company being absent he Coming to the knoleg of it. Cased us all to be laid downe one by one & himself with a Greate Rope abowte a three inch Rope beate us all very sevearly: which I took as a hausell. the slave thus absent was an English man & next day beaten soe Cruely that his flesh espeshally his botoks were forse to be dressed by the dokter his arme soe Lame that had not the use of his Lims for some time after."[42]

Gee lingers on the bodily specifics of the English slave's beating, noting that he was beaten so badly that "espeshally his botoks" needed medical attention. This kind of physiological specificity, lacking elsewhere in Gee's text, is nonetheless characteristic of Barbary captivity narratives; lingering emphasis on the anal and the priapic are a commonplace of this genre, even as other body parts are referred to without detail. As we shall see, the way Gee's extremely short narrative tarries on the "botoks" of a fellow captive participates in a sodomitical discourse even though this scene would never have been defined as such under, for example, British or colonial law. The sodomy here is quite literally "fundamental."

Abraham Browne was an Englishman about twenty-five years older than Joshua Gee.[43] He was born in Plymouth, England, sometime around 1630, to a merchant and shipmaster.[44] On the way back to New England in May of 1655, the company of ships with which Browne was sailing were attacked and eventually taken hostage by "our most crewill Enymes Two Turkes men of war belonging to Sally."[45] As a frequent passenger and laborer aboard ships during the mid-seventeenth century, Browne would have been familiar with the threats posed by Barbary corsairs and also with the experience of Barbary captivity; Browne's own father had also been taken captive by Barbary corsairs on a trip from France to England during the late 1630s, remaining a slave to "Turks" for three years before he was eventually redeemed.

Browne's narrative is similar to Gee's and Davis's in its depiction of the process by which he and his shipmates were taken captive; he and his crewmates were stripped naked and beaten before being taken back to shore and sold in slave markets. Unlike Gee's narrative, however, Browne's is saturated with racialized imagery that echoes natural historical depictions of difference in his day. Browne relies heavily on a racialized rhetoric of animality and otherwise inhumanness to represent his captors in Salé. For example, he describes the "Turkes" descending upon his company of ships as "mor like Ravones beasts then men" (more like ravenous beasts than men), invoking a much older tradition of understanding non-Christians in relation to the bestial, a tradition that, as we shall see, was also used heavily by Francis Brooks in *Barbarian Cruelty* forty years later. What makes his captors "beasts" in Browne's eyes, however, is their excessively humiliating and barbarous treatment of him and his fellow captives. He recounts: "They haveing stript me of all my cloths to an old peace of Lynning drawers I requested them they would have mercy upon me though indeed I was soe faynt by Reason of the Loss of so much blod I would scars stand upon my Leggs, yet all would not prevayle but my drawers they would have of Likewise soe that they Left me nothing but an old Lyning Cap all blody one my heade" (They having stripped me of all of my clothes to an old piece of linen drawers I requested them they would have mercy upon me though indeed I was so faint by reason of the loss of so much blood I would scarce stand upon my legs, yet all would not prevail but my drawers they would have of likewise so that they left me nothing but an old linen cap all bloody on my head). Beating him mercilessly, Browne's captors would not even let him keep his underwear. Naked except for a bloody linen cap, Browne was continually beaten until his "spirets were brought soe Low that [he] did rather desire to dye then to Live." Like Gee, Browne lingers on his "naked posture," alluding to his nudity no fewer than five times in ten pages; and once arrived in Salé, he describes at length his successful attempt to find clothes.

> The next day . . . after our takeing wee were all had upon dex where on us they bestowd some old Ragged cloths only to cover our nakedness for to some that they gave a peare of britches to they gave noe Dublett and to others that they gave a Coate to they gave nether dublett nor britches and to all nether Cap, stockings shurt nor shoose. I had only an old partingale Dublett and one of our men gave me a peare of Canvas drawers which was all the Cloths I had on till I was sold in Sally to my pateron: while wee weare upon that dex that wch did augment my misery, was the seight of those ugly onhumanye cretures wch goe soe disgisd according to there Custome with there heads shaved and their Armes almost naked did teryfie me Exceedingly.[46]

Browne's sullen discomfort at the lack of appropriate clothing, here, pales in comparison to his professed terror at encountering "those ugly onhumanye [inhuman] cretures . . . with there heads shaved. and their Armes almost naked." This passage is difficult to read, insofar as it is not clear who Browne is referring to when he speaks of the men whose heads are shaved. It is likely that he is describing some of the workers aboard one of the men-of-war that took his own ship captive; we can hypothesize that "those ugly onhumanye cretures" may have been slaves, based on their shaved heads.⁴⁷ These slaves could well have been Muslims or at least of northern African or African descent, but it is also possible—and perhaps even likely—that the slaves in question were not "Turkes" but fellow Christians. Browne describes these "cretures" as "disgisd"; the "disgis" in question may be these men's monstrous appearance, with shaved heads and naked arms. In an era when religion (and also race qua religion) could be read on the body through cues in dress and bodily comportment (discussed at length in chapter 1), the "onhumanye cretures" would not necessarily have been identifiable as Christians precisely as a result of being shaved and more or less naked. That Browne reads their nakedness and shaved heads as a "disgis" suggests precisely that they *were* Christians but presenting in the style of "Mahometans." But why did the vision of these men "augment [Browne's] misery" and "teryfie [him] Exceedingly," when Browne does not depict them as bearing any threat?

"Those ugly onhumanye cretures" did in fact represent a very real threat, one voiced consistently in Barbary captivity narratives, that would have indeed terrified any Christian in Browne's position. This threat was twofold. First, Browne may have seen in these "cretures" a vision of what was to come for him during his time as a slave: naked and shorn, lacking all characteristics by which he could be identified as a Christian, he too might become monstrous and "onhumanye." One of the threats that these slaves represented, then, was that of "turning Turk" and the possibility that one could be "turned" without his assent.⁴⁸

The second threat, which likely posed an equally terrifying possibility, hinges on the shaved heads of the slaves. To return to Pat Parker's work on the early modern British stage, she argues that during the early modern period, popular drama often invoked the "characterization of the 'shaven' as a sign of being 'abused against Nature' (the familiar phrase from Romans 1)," a characterization that also bore racial implications as it referred to "the 'beardless' or shaven inhabitants of the Americas and contemporary representations of the 'bearded' Turk."⁴⁹ The racialization and cultural differentiation that are appended to Christian beliefs about being clean-shaven (of both head and face) carry with them the implication of sodomitical behavior, and in this case, forced sodomy. Thus, in noting that the slaves' heads had been shaved, Browne was alerting both contemporary and present-day readers to the threat of being forcibly sodomized by his Turkish captors as part of the process of being forced to convert to Islam.⁵⁰ Reverting once again to the racialized rhetorical distinction between "beasts" and men, Browne understands these potentially sodomized beings as "onhumanye," potentially outside of the realm of the human qua Christian. In this sense, Browne's narrative also actively conflates the "Mahometan" and the sodomitical, both in the content that it represents and in the racialized fears that give shape to even the small pieces of Browne's narrative that are available to us today.

The specificity of details in this scene demonstrates another essential feature of these narratives. Even the truncated pieces that we have of Browne's and Gee's accounts rely on ethnographic discussions of the geography, architecture, religion, language, and style of food and dress of both their Muslim captors and other non-Protestants living in the area (in particular, Jews and "Papists"). While both historians and book historians have long cautioned present-day readers to approach captivity narratives of any kind with a critical curiosity regarding their authenticity or truthfulness, the fact remains that the experience of captivity was a common one in the early modern period through the early nineteenth century, and captivity stories offered information about contact with foreign peoples or cultures that was taken quite seriously in their time.[51] And, in some cases, they should indeed be taken seriously: as Linda Colley notes, captivity narratives, as "a mode of writing rather than a genre," might "form the closest approximation we have for the past to the kind of analyses supplied by anthropologists and ethnographers immersed in alien societies today."[52] Gee, for example, includes a handful of notes on the arrangement of Catholic churches and the preaching of the Mass in his narrative; and Samuel Sewall, an early Massachusetts judge and a friend of Gee's, records in his diary: "Mr. Joshua Gee, sometime Captive in Algeer, tells me June 11, 1694, that the Turks observe an Hebdomadal Revolution as we do; Our first day of the week is their first day of the week; And they call the days by their Order in the Week; One, Two &c. If they have any notable piece of work to doe, they chuse to begin it upon the first day of the Week, bec. God began his Works on that day."[53] The question of how broadly Gee's and Browne's narratives circulated in their day remains unclear, but this evidence that Gee spoke to his friends about "Turkish" customs—and that Sewall was clearly interested enough to jot this information down in his diary—speaks to the mutual imbrication of captivity writing and ethnography. Tracking the circulation of texts is difficult, but tracking interest in Barbary captivity and in the Barbary states more generally is less so.

While this is not an avenue I explore at length in this chapter, seventeenth-century diaries offer perhaps one of the most consistent sources testifying to how stories of Barbary captivity circulated in the seventeenth-century world. A few decades before Samuel Sewall carefully noted the information he had gleaned from Joshua Gee, naval administrator Samuel Pepys got drunk with a number of seafaring men in the Fleece Tavern in London and later recorded:

> Among others Captain Cuttle, and Curtis, and Mootham, and I, went to the Fleece Tavern to drink; and there we spent till four o'clock, telling stories of Algiers, and the manner of the life of slaves there. And truly Captn. Mootham and

Mr. Dawes (who have been both slaves there) did make me full acquainted with their condition there. As, how they eat nothing but bread and water. At their redempcion they pay so much for the water they drink at the public fountaynes, during their being slaves. How they are beat upon the soles of their feet and bellies at the liberty of their padron. How they are all, at night, called into their master's bagnard; and there they lie.[54]

Here, Pepys narrates a moment in which information about Barbary slavery is transmitted not through writing but through the conviviality of tavern life. He was, of course, not impartial regarding the Barbary coast, given his work with the navy and his involvement with the Tangier committee, which would begin a year after he penned this entry. But both Pepys's and Sewall's diaries demonstrate that casual circulation of information about Barbary slavery was frequently bound to a generalized ethnographic curiosity; Pepys's account of "the manner of the life of the slaves there" appears in the quoted passage as itself an ethnographic curiosity describing Algiers more broadly. We should not ignore his elliptical reference to the possibility of sexual subjection. The pointed final detail, of Christian slaves called to lie in their master's "bagnard" at night, invokes a punishment every bit as characteristic of representations of Barbary slavery as the previous invocation of the bastinado.[55]

During the late seventeenth century and into the early eighteenth, Britain and its colonies saw a proliferation of interest in the Barbary states and in the Ottoman Empire more generally.[56] The rise of Ottoman power was a source of both fear and great curiosity in the Christian Atlantic world. Some feared its military power and what they perceived to be the increasing popularity of forms of deism (a term often used to designate any religious movement that was not true religion, from a Calvinist perspective, including Quakerism, Judaism, and Islam). But others found in Islamic republicanism a potentially powerful model for a new relationship between church and state in the wake of the English civil wars.[57] At the same time, by the turn of the eighteenth century, British mercantile and colonial incursion into Northern Africa had long since begun, and shipping and trading sites on the rim of the Mediterranean such as Tangier (possessed by England from 1661 to 1684) "accounted for as much of Britain's trade in terms of value as India and North America did put together."[58] This is all to say that for both Britons as well as residents of British colonial North America, Ottoman power, the Barbary states, and Islam more generally were both foreign and familiar; they constituted a site onto which many fears about religious and state sovereignty were projected but also were a common topic of conversation.

Indeed, moving into the early eighteenth century, representations of the Barbary states, especially those destined for print, become vastly more detailed, relying heavily on ethnography in their narration. The final three representations of Barbary slavery that I discuss in this chapter bear an explicitly different orientation to print; each of them appeared in multiple print editions, and each details the specific features of "Mahometan" life, dress, religion, and diet at great length. Importantly, each of these narratives describes the culture of Barbary slavery alongside other ruminations on life in the Barbary states. The enslavement of Christians appears alongside descriptions of marital traditions and holiday celebrations as just another feature of "Turkish" life. Furthermore, each of these narratives, intended as they were for circulation in print, includes longer, more detailed, and much more explicit discussions of the sodomitical, which are put to work here to aestheticize ethnic and religious difference.

Captive, in Print:
The Natural History of Barbary Captivity

If Gee's and Browne's truncated narratives provide only elliptical accounts of the violence to which the two men were subjected as captives in Algiers and Salé, respectively, Francis Brooks's ninety-four-page *Barbarian Cruelty* (originally published in London in 1693 and reprinted in Boston in 1700) spares no details.[59] Brooks's first-person account is replete with the kind of travel and ethnographic narration that literary scholars now associate with early novels, such as Aphra Behn's *Oroonoko* (1688) and Defoe's *Robinson Crusoe* (1719), which itself contains a brief account of his time captive in Barbary. Brooks prefaces *Barbarian Cruelty* with a dedicatory epistle, first to William and Mary and then to the reader, describing his objective as an effort "to give . . . a clear and particular View of the most remarkable Passages that happened during the unfortunate time of my Confinement among those barbarous Savages."[60] Brooks describes himself as an English sailor, based in Bristol, and claims to have been taken captive by an Algerine ship during August of 1681 while sailing from Marseilles to Tangier. Transported first to Salé and then to Macqueness (Meknes), where he was sold, Brooks's comparatively long captivity narrative moves between first-person narration and something akin to the historical accounts of the political state of Morocco that circulated during the same period.

We know very little about Francis Brooks or about his time captive in Meknes.[61] Brooks's dedicatory epistle to William and Mary serves as a gentle plea for their intercession on the part of those English subjects who still "lie groaning in Slavery"; in this sense, *Barbarian Cruelty* claims the same purpose as Pierre Dan's *Histoire de la Barbarie*, with which this chapter opened. Yet *Barbarian Cruelty* is also a highly sensationalized adventure tale of the author's captivity and escape.[62] Aptly titled, the text contains scene after scene of cruelty enacted by Muslims against Christians in Morocco, but unlike Browne's and Gee's narratives, *Barbarian Cruelty* is neither a didactic text nor one framed by the notion of a test of faith. Beyond occasional parenthetical acknowledgment of the role of God in the survival of many of the men with whom the narrator was held captive (e.g., "Yet through God's Mercy, he was pretty well recovered"), this text is remarkably unconcerned with the spiritual.[63] Instead, its prose seems specifically intended to provoke horror; the long, detailed descriptions of artisanal tortures created by Moroccan slave owners for their Christian slaves, couched in broader ethnographic descriptions of Moroccan life, present a stark departure from the kinds of descriptions that appear in Browne's and Gee's accounts.

Rhetorically, both the eponymous "Barbarians" and the "cruelties" they so enjoy in Brooks's text are presented in language heavily charged with the weight of religious and racial difference. From the very first pages, the narrator conflates Islam with cruelty and Islamic custom with decadence and casts the barbarous treatment of Christians as a fundamental element of both religion and culture. In the dedicatory epistle to the reader, for example, the narrator plaintively explains, "We were confin'd amongst those whose Religion was composed of Cruelty, whose Customs were Extravagant, and whose Usages almost Intolerable."[64] The association of Islamic custom with extravagance cuts across both more value-neutral descriptions of cultural habits and beliefs and more explicitly negative representations of Islamic styles of punishment, providing an early orientalist assessment of the "costume and custom, manners and morals" of "Turks" and "Moors" as peoples.[65] Indeed, as we shall subsequently see in William Davis's narrative, Islamic extravagance can be equally evidenced by the quality of one's bedding and by the pleasure one takes in witnessing an execution. The "cruelty" staged so graphically and relentlessly throughout Brooks's narrative must thus be understood as one iteration of this orientalized excess.

European accounts of the quality of everyday life for those enslaved in Barbary commonly emphasize the special hardships encountered by captives in Morocco. At the turn of the eighteenth century, Morocco was ruled by Moulay Ismail ibn Sharif (often called "Mully Ishmael" in Anglophone Barbary captivity accounts), the second ruler of the Moroccan Alaouite Dynasty, who was rumored by Europeans to be a spectacularly brutal leader and slaveholder.[66] Whether or not this characterization is apt, the figure of the brutal "Mully Ishmael" appears with the persistence of almost a stock figure in both ethnographic accounts of Morocco and Barbary captivity narratives. And it is in Morocco, of course, that Brooks's story is set.

Like Barbary captivity narratives from the later eighteenth century, *Barbarian Cruelty* is notably concerned with specifying the racial diversity of Salé and "Maqueness," but also with transgressions of what the narrator implicitly understands to be the natural ordination of racial hierarchy more generally. Furthermore, this racial difference is not qualified exclusively in terms of religion, the predominant idiom for describing human diversity in that era; Brooks's narrative showcases a vast vocabulary for differences in skin color and tone. Alongside terms traditionally found in seventeenth-century Barbary captivity narratives, such as Moor, Turk, Saracen, Mahometan, Black-a-moor, Negro, Christian, and *renegado,* chromatist terms such as "white," "black," "mollato," and "copper-colour'd" appear frequently in this text and are used to occasion discussions of racial order.[67] This usage may stem in part from European understandings of Mully Ishmael's own racial background; both Brooks's narrative and the anonymous *Description of the nature of slavery among the Moors* (1721), discussed subsequently, specify that this "emperor" "was begotten of a Negro Woman by a white Man . . . and is a *Mollatto* by his Colour."[68] In the figure of Mully Ishmael, we can observe the concatenation of multiple theories of human difference that were employed together to produce the racialization of "Mahometan cruelty," for, bizarrely, we are told that Mully Ishmael's skin actually gets darker as he gets angrier. "When he's in a Passion," Brooks notes, "[Mully Ishmael] looks just as he is, as black as an Infernal Imp; which his Natives take notice of, and can tell when he's angry."[69] Presaging his outbursts of spectacular violence, the temporary darkening of his skin materializes his rage on his body, the blackness of his cuticle serving both an expressive and a predictive function. In the blackness occasioned by his anger, we see the overlay of the Christian metaphysics of blackness and whiteness (a rhetoric explored at length in chapter 3) alongside the more figurative rendering of his "Negro" inheritance emerging visually when the emperor feels inclined toward violence or evil.

In this scene, the tension between internal "Passion" exists in something close to a directly proportionate relationship to the leader's outward blackness, illustrating the complex coexistence of both characterological and environmental theories of racial difference and behavior. When Mully Ishmael becomes angry, he looks "just as he is": "as black as an Infernal Imp." This casual assessment simultaneously suggests a deep, quasi-ontological blackness that reveals itself in the emperor's anger: it is only in his rage that we see Mully Ishmael "just as he is." Yet the leader's racial plasticity is also illustrated here, his skin involuntarily darkening in moments of "Passion" provoked, justifiably or not, by his anger at his subjects and slaves. Predicting his outbursts of excessive violence, this darkening of his skin is shot through with both an environmental and an anatomically proto-hereditary explanation.

Yet, regardless of the etiology of this blackness, this "passion" invokes two critically racialized ideas: first, an orientalized lack of reason that in turn justifies the peculiarly extravagant violence to which he is apparently so predisposed, and second, the diet of the region, which naturalist geographers such as John Ogilby understood to bear the power to make men "fall into the extravagancies of a lustful Passion."[70] While in Brooks's narrative, "Negros" are more or less evenly perpetrators and victims of violence against Christians, the Islamic specificity of Mully Ishmael's darkness is what makes him so dangerous.

Indeed, like the excessive curation of his food, dress, diet, and sleeping habits, there is something sodomitical about these representations of Mully Ishmael's carefully orchestrated tortures.[71] For example, Brooks includes a long narration of a series of events proceeding from the capture of a British ship bound for Barbados by Salé corsairs during 1685. Upon being told that a "Christian Virgin" had been captured during the raid, Mully Ishmael prevails upon the young woman to convert to Islam and marry him. Refusing, the young woman is subjected to a series of tortures, all meticulously described by Brooks, from a scene in which the emperor "caused her to be stript, and whipt by his Eunuchs with small Cords, so long till she lay for dead" to another, in which she is stabbed by the ruler himself with a thousand tiny pin pricks: "Then he prick'd her with such things, as commonly his Women use instead of Pins, being as sharp. Thus this beastly and inhuman Wretch by all ways he could invent, sought to force her to yield, which she resisted so long, till Tortures, and the hazards of her Life forced her to yield, or resign her Body to him, tho her Heart was otherwise inclined." Recalling Parker's theorization of the symmetries between the sodomitical and "pathic subjection" more generally, this scene in which we see a Christian white woman "prick'd" (a priapic term in the eighteenth century as it is today) over and over again with pins enhances Brooks's characterization of the emperor as "beastly and inhuman"; here racialized language joins the chromatist vocabulary discussed earlier, and deployed together, they align the emperor's penchant for particularized torture with his religion, his skin color, and his general despotism. Here, the pricking in question is both stabbing and sexual subjection, as the scene concludes with a narration of the sartorial and social erasure of the woman's Christianity: after she relented, "he had her wash'd, and clothed her in their fashion of Apparel, and lay with her; having his Desire fulfilled, he inhumanly in great haste forc'd her away out of his Presence; and she being with Child, he sent her by his Eunuchs to *Macqueness* . . . to the chief Eunuch, and after that she was delivered of two Children."[72] The ostensible rape of a Christian woman, here, appears as only one abuse in a series that details her forcible transformation from Christian to Muslim.

In *Barbarian Cruelty,* Brooks is notably more concerned with Mully Ishmael's predations toward white, Christian women than he is with the emperor's abuse of men, but the narrative nonetheless contains many graphic depictions of the pathic subjection of Christian men as well. During a subsequent scene in which Mully Ishmael attempts to compel a Christian slave to "turn Turk," Brooks restages the pin-stabbing episode in more graphic, bombastic terms. In response to the unnamed Christian man's refusal to convert, "this inhuman Wretch in great spleen cut him till he fell down, and hack'd and hewed him as if he had been butchering an Ox, and caused the Negro Boys to run his Body fill of Holes with Knives, till his Body was as full of Holes as possible it could be; when he had done so, Bring, saith he in his own Language, four English Dogs to fetch that Dog away."[73] This striking image of the man's body "as full of Holes as possible it could be" presents readers with the image of a Christian form wrested of its physical integrity, "butchered" like an animal, formerly figura but now meat. Alongside subsequent similar images in the narrative of Christians being torn apart—including a scene in which Mully Ishmael tasks his eunuchs to "pull [a] young Child in pieces Limb from Limb"—the vision of sodomitical violence we encounter in *Barbarian Cruelty* is not only in the act of penetration, violence, or sodomy, but in the rendering of the Christian body as eminently vulnerable to such.[74] If in the figure of Mully Ishmael, Brooks's narrative presents us with yet another iteration of William Davis's sodomitical "Turk," then Brooks nuances this representation by revealing its twin in the form of the brutalized Christian body. Indeed, the vision of the sodomitical "Turks" can boast of little impact without an attending image of a penetrated Christian.

Appearing twenty years later, *A Description of the nature of slavery among the Moors, and the cruel sufferings of those that fall into it* was anonymously published by a man claiming to have been restored from his captivity along with almost three hundred others during a series of redemption efforts conducted by Captain Charles Stuart in Morocco between 1720 and 1721.[75] This effort is described at length and in great detail in John Windus's *Journey to Mequinez* (1721), which includes a long list of the names, ranks, and places of origin of the anonymous author's fellow redeemed Englishmen, alongside long ethnographic descriptions of Moroccan life and wildly sensational accounts of the violence to which they were all subjected.[76] Originally taken captive on April 20, 1720, on a voyage between Minorca and Lisbon, our anonymous author pens a curious narrative that combines first-person narrative, third-person anecdotes detailing the brutality to which other captives are subjected, and long ethnographic accounts of Moroccan religion, social customs, and culture. Like Brooks's narrative, this text is explicitly concerned with indicting Moulay Ismail ibn Sharif (who in this text is called "Muley Ismael") for his incredible brutality, but with one critical addition: in this text, the author vilifies Muley Ismael by pointing again and again to his aberrant sexual behaviors, including explicit references to his predilection for young men.[77]

The figure of Muley Ismael that appears in *A Description of the nature of slavery* is a much showier villain than the Mully Ishmael of Brooks's *Barbarian Cruelty;* the narrator describes him as "the most voluptuous Prince in the World."[78] Through detailed, meandering descriptions of the "king's" carefully curated torments, the author describes Muley Ismael's decadent tortures with an even more decadent narrative style, elaborating with a kind of relish upon bizarre and frankly overcomplicated tortures that he devises for his subjects and slaves. The narrator walks readers through a series of grotesque scenes in which women have their breasts cut off by effectively guillotining them between rocks; one of the king's future wives is forced to submerge her feet in hot oil to compel her to "turn Turk";[79] animals are indiscriminately executed;[80] and he tortures his own son, when the son has displeased him, by cutting off one of his feet and one of his hands and cauterizing the wounds in hot oil.[81] Indeed, this vision of Muley Ismael's genius for the development of arcane tortures—the narrator dryly remarks, "the King has lately been so ingenious in new Inventions of Cruelty"—makes the 1649 table of tortures in the *Figure Nécessaire* that opens this chapter seem quotidian and uninventive.[82]

Yet the specifically "voluptuous" quality of many of the tortures that Muley Ismael inflicts on both subjects and slaves speaks to the burgeoning strength of the association, as the seventeenth century gave way to the eighteenth, between aberrant sexual behaviors and Islam. *A Description of the nature of slavery* includes compendious lists of the spectacular tortures, both venereal and nonvenereal, ordered or performed by Muley Ismael upon Christians. The excesses of the emperor's sodomitical tortures are displayed parallel to other scenes of nonvenereal violent torture, suggesting a kind of equivalence between the two. Indeed, in the first five pages of the narrative, we are already introduced to the commingling of physical and sexual violence. Consider this episode, transpiring immediately after the anonymous narrator is taken captive:

> This Cruelty [of their capture] was succeeded by our being put into a Dungeon in an old Castle, wherein we were used with the utmost Inhumanity, till the King's Pleasure was known concerning us; which was, that we, the *English*, should be conducted to *Mequinez*, the Capital City wherein he makes his Residence, there to be dispos'd of to the best Bidders, at the common Market. But first, we were to be brought in Triumph to the Court, there to be survey'd as before; and, as the Humour took that Prince, to be used accordingly, either bastinado'd, or put to the Torture: But as Fortune would have it, who, amidst her very Persecutions, often Times shews some sort of indulgence, one of his Favourite Sultana's [*sic*], an *English* Woman, that, by the Means of insupportable Cruelties, was forc'd to embrace *Mahometanism*, was that very Day brought to Bed of a Son: In Joy for which, he, after making Choice of two young Men, one *James Richards*, and *Henry Negus*, who were very handsome Lads, of about 17 Years of Age, for his own brutish Lusts, caus'd a certain Distribution of mouldy Rice, brackish Sherbet, and dry'd musty Fish, to be given us, and then left us to our respective Fates at the next Day's Market.[83]

In this scene, the horror of being taken captive is immediately replaced with the horror of being sold on the thriving Islamic slave market. Yet even before being brought to the slave market, our author and his peers were to be brought to the Court, surveyed, and either bastinadoed or otherwise tortured at the whims of the prince.[84] Importantly, the author also parenthetically notes that slaves were generally brought to the market and sold, rather than being kept by Muley Ismael, "(unless he took a more than ordinary Liking to particular Persons, such as young Males or Females)," introducing the flexible sexual proclivities of this local prince as an unsurprising parenthetical to his character, a marginal idiosyncrasy or offhand detail. This proclivity is further developed as the narrator celebrates with rather bittersweet resignation his good fortune (as well as that of his fellow captives) that Muley Ismael's wife happened to give birth to a son on the evening when they were all bound to be tortured at court; the bastinado canceled, Muley Ismael elects a different form of entertainment to celebrate the arrival of his child, one every bit as sodomitical as the one initially planned.[85]

This moment constitutes one of the most explicit (and detailed) references to sodomy that I have encountered in Barbary captivity texts from the prenational era. Following Khaled El-Rouayheb, who chronicles how pathic penetration was used widely in the early Ottoman Arab East as a rhetoric or even a tactic of (often militarized) domination, it is important not to attribute any one specific significance to Muley Ismael's pathic use of his slaves.[86] But regardless of what animated Muley Ismael's behavior, Christian witnesses regarded it as sodomitical, in accord with their perceptions of Islamic life and Barbary slavery.

A lingering interest in the priapic also renders this narrative notable, and this interest in the torture of penises, in particular, is one that we shall trace into William Davis's captivity narrative, the last text of our discussion. In the final pages of *A Description of the nature of slavery*, the narrator, with what seems like resignation, offers a summary of the barbarity he purportedly witnessed while enslaved:

> After what has been said, it is needless to speak of the Punishments the poor Captives of *Morocco* are further liable to, since the smallest Faults are so cruelly chastised. It may suffice to let you know, that it is frequent enough to see some impaled, some burn'd alive, and others hung up by the Heels over the *Lime Kilns*. The King often causes them to be Strangled, and sometimes loaded Carts to pass over their Bodies. He has been seen, by Way of Pastime, to make four of his strongest Blacks take a Captive, and toss him into the Air, so as he might fall with his Head foremost, and so repeat it till he died, either by, having his Neck broke, or his Brains beat out. His Brutal Cruelty has prevail'd so far as to cause the Privities of some to be tied with a Packthread, and violently torn off; and it is common enough in *Mequinez*, to see Slaves, an even *Moors*, drag'd at the Tails of Mules, till they die. But what exceeds all this is, that the King has lately been

so ingenious in new Inventions of Cruelty, as to cause Posts to be erected in the Public Market Place, 18 or 20 feet high, with great Iron Hooks, to hang slaves.[87]

This inventory of slaves being impaled, burned alive, strung up, strangled, run over, thrown and broken, beaten, castrated, torn up, dragged, ganshed, and hung would constitute a striking set of images with which to conclude any text from this era, but by concluding with a castration, this litany of atrocities allows an air of the sodomitical to shape the end of the narrative. Having left first-person narration behind after the first one-third of the thirty-eight-page text, the narrator renders this list of cruelties in the disembodied third person (the king "has been seen" to do these things), endowing this passage with the sense that these brutal acts were more or less common knowledge. Yet, compared to other Barbary captivity narratives, or even to geographies and histories of Morocco or the Barbary states, the image of men tying packthread around the "privities" of other men in order to then tear said "privities" off is anything but commonplace. Marrying prohibitions against same-sex erotic contact with Christian indictments of the uneconomical decadence of Islamic violence, this bizarre castration scene would be unimaginable, were it not situated in a story of Barbary slavery.

And indeed, as the eighteenth century progressed, it was the rhetoric of unimaginability that increasingly came to define representations of Barbary tortures, even as depictions of those tortures appeared more graphically in print contexts, and even as they grew more closely associated with the "orient." By 1797, when Updike Underhill, the blundering protagonist of Royall Tyler's *Algerine Captive*, witnesses an execution by staking while captive in Algiers, Underhill at first refuses to "wound the sensibility of my humane fellow citizens, by a minute detail of this fiend like punishment" but then proceeds to describe the execution in rather minute detail.[88] This phobic mode of representing the sodomitical—that is, representing it by claiming to refuse to represent it, what Joseph Allen Boone terms the twinned "history of homoerotic fascination and homophobic aspersion"—starts to appear widely in Barbary captivity accounts by the mid-eighteenth century and develops into a common representational trope by the early national period in the United States, when Barbary narratives experienced an explosion in popularity.[89]

In the final Barbary captivity narrative to be considered in this chapter, *A true relation of the travels and most miserable captivity of William Davis* (1614, reprinted in 1745), we witness the tight bind between the language of unimaginability and the paradoxically unavoidable imaginability of the sodomitical acts, tortures, and even geographies that Davis describes in detail. Imagining the sodomitical through the language of ethnography, sodomitical acts are specifically and at times even literally portrayed as a natural feature of the Algerian landscape. William Davis's captivity narrative, after its first printing in London in 1614, was subsequently anthologized and reprinted multiple times throughout the first half of the eighteenth century in editions of the Earl of Oxford's compilation of travel narratives, *A Collection of Voyages and Travels*. While Davis was an Englishman, we know that his captivity narrative was available in North America by 1789.[90] The narrative begins by describing Davis's departure from England on January 28, 1597, on the ship *Francis of Saltash,* bound for Civitavecchia. He describes a safe voyage and arrival on March 1, 1597, as the *Francis of Saltash* successfully delivered its load of "fish, and herrings, and such-like commodities." A fairly standard description of the different customs and culture of the city follows, and consistent with the writerly conventions we see in contemporary travel narratives and ethnographies, Davis's text dwells on the sexual practices of the peoples that he encounters, especially the sexual practices of women. Over the course of his narrative, Davis works as a galley slave and travels to locations as distant as Brazil and to islands throughout the Mediterranean Sea.[91] The ethnographic character of the account only becomes more pronounced as it continues, although Davis is not the most practiced of narrators; he begins describing "Algier," for example, without ever making it clear to his readers that the *Francis of Saltash* has left Civitavecchia.

Davis's narrative is important in light of its status as a widely and frequently anthologized Barbary captivity narrative that appeared most often in the context of compilations of travel writings that were cited and borrowed by natural history texts. *A Collection of Voyages and Travels,* for example, went through multiple editions in the first half of the eighteenth century, between 1701 and its final reprint in 1747. Davis's narrative appears in the first volume of the two-volume series, which opens with the following introductory note:

> The introduction contains some useful geographical observations on the figure, situation, motion, parts, division, and measurement of the earth: On the longitude and latitude, meridian equator, and different climates: On the origin of mankind, the peopling of the world, the navigation of the ancients, &c. the invention of the mariner's compass; properties of the loadstone; original government, and the several sorts of government both ancient and modern: On trade and commerce, and on the religions and languages throughout the whole world. To Which are added instructions for travelers.[92]

A Collection of Voyages and Travels explicitly aspires to the status of a work of natural history; its promise to detail the different climates, "the origin of mankind," and "the peopling of the world" clearly echoes some of the major works of early racial science discussed in chapter 1. While the tome claims to be a representative study of "geography," it takes care to explain what "geography" entails. The first essay in the first volume, titled "Introductory Discourse Concerning Geography," explains:

> GEOGRAPHY, tho' it literally implies no more than *a description of the whole earth,* as far as it is known to us, is generally intermixed with the *political* and *natural history* of countries, which is more instructing and diverting to the reader than *meer* geography. Consequently its object are the things which are principally considered in every country, *viz.* 1. The elevation of the pole, the distance of the place from the equator and the pole. 2. The obliquity of the diurnal motion of the stars over the horizon of the place. 3. The quantity of the longest and shortest Day. 4. The climate and zone. 5. Heat, cold, and the seasons of the year; also rain, snow, winds, and other meteors. 6. The rising, appearance, and continueance of the stars over the horizon. . . . To which are added, the human qualities of every country and place; *viz.* 1, The stature, shape, complexion, length of life, origin, and diet of the inhabitants. 2. Their commerce, trades, and commodities. 3. Their virtues, vices, learning, genius, and schools. 4. Their customs, marriages, and funerals. 5. Their language. 6. Their civil government. 7. Their religion and ecclesiastical polity. 8. Their towns and places of most note. . . . Without these illustrations, geography would be only a mere skeleton, or at best a body, on which nothing is visible, but a dry skin composed of sinews and bones.[93]

Davis's narrative, then, is not simply intended to offer a glimpse into the life of a Barbary slave, or into the world of the Barbary states; it is put to two tasks in the anthology. First, it introduces the reader to a multifaceted sense of the "geography" of the world of the "Turks"; and second, it details "the human qualities" of the area in which he circulates, including descriptions of "the stature, shape, complexion, length of life, origin, and diet of the inhabitants," "their virtues, vices, learning, genius, and schools," "their customs, marriages, and funerals," "their religion and ecclesiastical polity," and "their towns and places of most note," among other things.

It may seem a small point, but it is critical to note that readers who encountered Davis's narrative would have read it in effectively two registers: first, the story it tells about Barbary captivity would have been a familiar one to most readers in the eighteenth-century Anglophone Atlantic world, and second, the anthology's claim to be a natural history would have been familiar to anyone who might have been able to afford this expensively bound book. Davis's description of Algiers's slave markets, then, would have signified in two ways: it offered an empirical account of life in Algiers; but in noting that slavery persisted in Algiers (as it did in the British North American colonies during the same era), it would have also indicated a specific account of its "state of society." Annexed by the Ottoman Empire in the mid-sixteenth century, by 1600

Algiers had become the epicenter of the North African slave market; as we have seen, it enslaved possibly tens of thousands of Christian seamen. And of course, one of Davis's first descriptions of Algiers addresses its slave markets, emphasizing Algiers's geopolitical centrality in the Christian-slave-trafficking economy. Davis's first few sentences offer his audience an image of a city "very strongly fortified with castles, forts, and platforms, with great store of ordinance planted about it; also there are many gallies belonging to this place, which do much offend the Christians in the taking of their ships, Tartanes and Settees, and other small vessels, making all the Christians that they take slaves and selling of them in their markets like horses: for, according to age and strength they are prized."[94] Slavery, here, appears as a feature of the architectural landscape.

Moving from the landscape to the people who inhabit it, Davis offers a careful ethnography of the "Turks" of Algiers, describing both their bodies and their customs in assiduous detail:

> These *Turks* are goodly people of person, and of a very fair complexion, but very villains in mind, for they are altogether sodomites, and do all things contrary to a Christian; for they never lie in a bed, nor feed upon a Table, yet their feeding and diet is very plentiful, their bedding and apparel is very neat and costly, the manner of their lodging is thus: there is a very fair table in a room, about three yards in breadth, and as many in length, standing in height one yard from the ground, and thereon they use to spread a double quilt very costly, with many curious cushions instead of pillows, whereon he lieth down in his shirt, and linen breeches only, without any other covering, by reason of the heat of the country.[95]

This passage is crucial to my reading of Davis's text, not simply because he identifies the "Turks" as "altogether Sodomites," but because of the accumulation of different ideologies with which both "Turk" and "Sodomite" are endowed, some of which overlap. The first distinction that Davis offers in his description of "Turks" is that they are "goodly people of person," meaning that their bodies are well-proportioned and perhaps tend toward a certain physical beauty, and "of a very fair complexion."[96] Despite their light skin, however, Davis notes that "Turks" are "villains in mind, *for*"—and he forges a relation of causation, here—"they are altogether sodomites." This constitutes a moment of crucial distinction; despite their fair complexion, "Turks" are "villains of mind," owing to their status as "sodomites." Davis justifies his identification of "Turks" as sodomites, however, by explaining that they "do all things contrary to a Christian; for they never lie in a bed, nor feed upon a Table, yet their feeding and diet is very plentiful, their bedding and apparel is very neat and costly."[97] The passage draws a causal relationship between being a "sodomite" and "do[ing] all things contrary to a Christian," importantly defined by how they sleep, eat, and dress.

While the details of this fundamental difference (bed, apparel, diet) are rendered in incidental terms, Davis's careful delineation of the eating and sleeping habits of "Turks" evokes some of the empirical, naturalist idioms for describing human variety discussed in chapter 1. In this passage Davis is in part describing the "habits" and "state of society" of the "Turks" he lives among, but it is important to note that his description begins to change tenor, moving from a more observational dialect into statements that are more qualitative or that offer assessments of the characterological or dispositional qualities of "Turks" (e.g., "fair complexion" but "altogether sodomites"). Furthermore, Davis describes the fundamental difference between Christianity and Islam—religious distinctions that, in the seventeenth and eighteenth centuries, actually described differences in kinds of *humans*—in terms of what he perceives as the excesses of Islamic culture. His description of Turkish sleeping arrangements, in particular, registers Davis's fascination and discomfort with the comparative *lavishness* of the material conditions of everyday life for the slaveholding elite in Algiers. The "very fair" table, the "double quilt very costly," the "curious cushions instead of pillows," and the "Turk's" sleeping "in his shirt, and linen breeches only, without any other covering," combine to create a scene of almost *too much* comfort, of excessive cultivation of sleep. A perhaps perverse degree of *care* has been invested in the culture of rest; the "curious cushions instead of pillows," for example, are not "curious" in the sense that we use the term now, indicating that they are strange, different or unexplainable; the sixteenth-century use of "curious" existed in much closer etymological proximity to the Latin *cura*, meaning "full of care or pains, careful, or assiduous."[98]

Davis's description of the excesses of Turkish "sodomites" importantly reflects long-established cultural assumptions about Muslims that proliferated among Protestants in western Europe, a form of early orientalist discourse that forged an ideological relationship between Islam and sodomy. As Jacob Berman notes, "representations of ethnic figures in Barbary tend to emphasize either Arab degeneracy as a lack of civilization (the dark and naked Bedouins) or oriental decadence as overcivilization (the sumptuous and cruel Turkish ruler)."[99] Because so many Barbary captivity narratives are permeated with moments in which their authors (the Christian captives) come into contact with, or are sold to or by, wealthy Muslims in the Barbary states, these narratives often represent Muslims as wealthy merchants, relying on the presumption of oriental decadence that appears to Protestants as a form of overcivilization that emphasizes the lavish or excessive, a Muslim inclination toward pleasure for pleasure's sake. Formulations of this ilk appear frequently throughout Barbary captivity narratives; in Davis's own narrative, he repeatedly emphasizes that in Algiers "there is great store of gold, and rich merchants," and animals that labor in ceremonies (such as conversions and circumcisions) are frequently described as "being very costly attired."[100] If Davis's assessment of "Turks" as "altogether sodomites" is not due to any sexual practices that he observed during his time in Algiers, it is because Davis understands "Turks" to be "sodomites" in their committed cultivation of a culture of physical excess. Davis's image of the bedroom—which he describes as a private room, in which a Turkish man sleeps alone—depicts a man underclothed but overcomfortable, a sodomy of suggestive decadence. This "homoerotics of orientalism" sees the outlines of aberrant sexual behaviors in lavish sartorial, dietetic, and even sleeping habits, and vice versa, and thus offers a vision of the kind of environmental theory of early sexuality that I am arguing for in this book.[101] While not a wholly external thesis of sexual behavior (Turks, of course, are also sodomites because they are "villains of mind"), it would also be a mistake to ignore the features of the "Turkish" architectural, governmental, and climatic environments that, alongside their religion, encourage practices that align them with the sodomitical.

Within his highly ethnographic assessment of everyday life as a slave in Algiers, like so many other Christian slaves who penned accounts of their captivity, Davis also takes care to record Algerian styles of discipline and punishment. And like the narratives of Browne, Gee, Brooks, and the anonymous author of *A Description of the nature of slavery among the Moors,* Davis's tarries on moments in which this "Turkish" excess becomes, quite literally, embodied, in the narrative's unwavering focus on the naked, subjected, and penetrated male body. The descriptions of the particularities—and perceived excesses—of Islamic styles of punishment and abuse that appear ubiquitously in Barbary captivity narratives, as well as in travel writing that describes Islamic cultures more generally, must be understood as continuous with representations of Islamic culture as lavish or excessive. The long descriptions of the excessive or unnecessary *comfort* in accounts of the Muslim slaveholding class resonate with the descriptions of excessive, unnecessary, and often unwarranted punishments that characterize the experience of Barbary captivity.

An important part of Davis's ethnography of "Turkish" culture is his description of their culture of punishment. His meditations on "Turkish" punishment appear even before he begins to describe his captivity at the hands of the Duke of Florence's sailing fleet in the very beginning of the narrative:

> The manner of their judgment for offending of the law touching death, either for theft or murder is thus: within four hours after he is taken, he is condemned by certain chief soldiers, and presently put to death after this manner, *viz.* ganshed, staked, or beat to death. Their ganshing is after this manner: he sitteth upon a wall, being five fathoms high, within two fathoms of the top of the wall; right under the place where he sits, is a strong iron hook fastened, being very sharp; then is he thrust off the wall upon this hook, with some part of his body,

and there he hangeth for sometimes two or three days before he dieth. Staking to death is thus: a round piece of wood three yards long, and as big as a man's leg, being sharp at one end, is taken and driven in at the fundament of the offender, and out at his shoulder, and so they let him lie till he be dead, which sometimes will not be in a day or two. The manner of beating to death is thus: they take the offender, and lay him down upon his back, being naked, and with two double ropes, two several men, one on the one side, and the other on the other side of him, beat on his belly till he is dead. But these deaths are very seldom used, because they are so fearful to the offenders, yet I have seen them all executed.[102]

In this passage Davis represents the Christian body, in punishment, as becoming excessively and unnaturally exposed. Davis's choice of representative punishments—ganshing, staking, and beating—all require the reader to imagine the body as vulnerable to awkward and unnatural penetration or exposure. The offender who is "ganshed" is impaled, his body dangling on a hook, and left to die; the offender who is staked is, in Christian legal terms, sodomized to death, left to die from the penetration of his "fundament"; and even the offender who is beaten is forced to lie, supine and naked, on the ground before being beaten to death. To describe all of these acts of punition, Davis relies on a rhetoric of exposure; the victims of ganshing and staking ostensibly remain, dying, in public view, sometimes "for a day or two"; and staking, in particular, asks that readers imagine the primal scene in which the pristine "fundament" is first penetrated by "a round piece of wood three yards long, and as big as a man's leg." The simultaneous unbelievability of the brutality of these forms of discipline, working in concert with Davis's very assiduous efforts to represent them, resonates in Davis's persistent return to the language of unimaginability that he invokes in many of his discussions of his own experiences.

For example, a portion of William Davis's eight years as a slave was spent working as a galley slave, which historians of Barbary mercantilism and piracy have unilaterally identified as one of the most painful and strenuous forms of Barbary slavery. Forced to row for hours in the unventilated bellies of war rovers, without food, water, or sleep, while often being mercilessly beaten, Davis describes the pain he experienced as a galley slave in the idiom of unimaginability:

The misery of the galleys doth surpass any man's judgment or imagination, neither would any man think that such torture or torment were used in the world, but only they that feel it. The extremity of misery causeth many a slave to kill themselves, or else seek to kill their officers but we were not suffered to have so much as a knife about us, yea, if we had gotten one by any extraordinary means, and offered any violence to any officer, we should presently have lost our nose and ears, and received a hundred blows to our bare back, and a hundred on our belly with a double rope.[103]

Even though the misery that Davis experienced "doth surpass any man's judgment or imagination," this passage is effective precisely because of Davis's invocation of unimaginability.

The rhetorical unrepresentability of the misery of being a galley slave is evoked as Davis effectively encourages his readers to try to imagine it. The statement that "neither would any man think that such torture or torment were used in the world, but only they that feel it" contradicts itself, especially since readers, most of whom ostensibly had never been galley slaves, finish the narrative with a new understanding that such tortures exist. The unimaginable serves an important function in Barbary captivity narratives, as it provides one of the structuring representational modes through which sodomy is invoked. Indeed, Davis represents sodomy through phobia or interdiction, a negative aesthetics that nonetheless performs positive work in the text. Sodomy, here, is literally the act that shall not be named among Christians; while persistently invoked, it is denied sustained consideration through its interdiction. But the reader is still left with impressions: a "fundament" awaiting staking; Christian bodies, lying naked and supine on the ground as they are being beaten; the biblical story of Sodom and Gomorrah and the sins for which those cities were smitten. Insofar as these images and figures are constantly under brief consideration in this text, but only under the sign of sodomitical interdiction, the reader must read, consider, and reject sodomy, thus participating in a form of reading that I am calling sodomitical.

Jonathan Crewe's insistence on the "mutual implication of the homoerotic and the homophobic" in early modern England can inform our understanding of the relationship of sodomy to Barbary captivity narratives and to Christian representations of Muslims more generally.[104] The effect of evoking the sodomitical puts pressure on representations of the body, and these representations, in turn, often describe bodies that are perforated, penetrated, torn, and unnaturally exposed. Though these references to sodomy refer, etymologically, to a legally defined set of acts (although which acts sodomy refers to differ according to historical period and culture), sodomy tends to serve a much more figurative function in Barbary captivity narratives; I have not encountered a single narrative in which the narrator describes his own experience of sodomy or being sodomized. Sodomy is always discussed in a suggestively lateral manner, and though the presence or frame of sodomitical acts is even more removed within Davis's narrative, it exerts consistent pressure within the text.

Sodomy, as well as the actual city from which the act takes its name, casts a long shadow over even the smallest details in Davis's narrative, often in truly bizarre ways. For example, in describing his travels as a galley slave aboard a ship, Davis offers only the briefest and most passing mention of the time he spends in Palermo; near this city, Davis reports, "there is a town called *Trappany*, in which town there is a monastery, wherein they affirm that the pillar of salt that *Lot's* wife was turned into, coming out of Sodom, is."[105] Nothing further is mentioned about Trappany, the monastery, or the pillar of salt formerly known as Lot's wife; Davis immediately moves on to describe the language and religion (Italian and "Romish," respectively) of the Sicilians. This reference to Lot's wife, and the town from which she was running when she became a pillar of salt, however, further furnishes Davis's narrative with something of a lateral sodomitical frame. It also insists on the geographical inflection of sodomy, as a city situated in the biblical land of Canaan, the land from which, as both natural philosophers and theologians both speculated, brown-skinned peoples (the descendants of Ham) hail. This understanding of the association between the Curse of Ham and the population of the Barbary states was widespread and fairly consistent. To return to the discussion of Pierre Dan's *Histoire de Barbarie et de ses Corsaires* that opens this chapter, the heading of the first chapter of the first book of the four-volume text, an essay on the history and etymology of Corsairs (both the word and the people), is "CORSAIRES de Barbarie heritiers de la malediction, & de la terre de Cham" (Barbary Corsairs, inheritors of the curse and the land of Cham).[106] Beyond the association of the corsairs with Sodom, the city, they are also associated with the Curse of Ham, from the Old Testament story discussed in the introduction and chapter 1, which frequently interpreted racial difference as a punishment for sexual transgression. The loose, associative logic at work here makes it difficult to identify the specific hermeneutics through which the Barbary Corsairs embody the threat of sodomy, but when we think of sodomy—or "Barbary"—as a complex, layered assemblage, the strength of this prevalent Christian cultural association comes into sharp focus.

To conclude my reading of Davis's narrative, I turn to three specific moments in his story, both before he describes being taken captive and during his time as a slave. At several points, Davis lingers on descriptions of the penises of men around him and of the men living in communities that his company visits while he works in the galleys aboard merchant ships. This sustained priapic emphasis in Davis's narrative is not all that unusual; Rudi Bleys notes that European interest in the racialized penis is a common element of travel and ethnographic writing.[107] Travel and captivity narratives in which writers encounter Muslims also often frequently engage the question of circumcision, as circumcision was a cultural practice that differentiated Jews and Muslims from Christians.[108] Yet I suggest that we understand Davis's frequent returns to the penis as a site of inquiry or description as moments that cultivate this associative reading practice, one that relies on an idiom of interdiction to do its representative work.

As Davis recounts his time in Algiers, after identifying "Turks" as "sodomites," he continues to offer an ethnography of bits and pieces of the culture, claiming to base this information upon his own observations. After tarrying on Islamic prayer practices, Davis moves on to circumcisions and details the ceremony during which an adolescent "ride[s] to the place of circumcision, where they cut off the fore-skin of his yard, naming of him *Morat, Shebane,* or *Hosan.*"[109] Shortly thereafter, readers are offered a glimpse of the circumcision ceremony prepared for a Christian who agrees to "turn Turk": "He is put upon a horse with his face toward the tail" and led to the place of circumcision.[110] In a similarly ethnographic mode, Davis describes a culture that he encounters during his trip to Brazil, wherein the men "are altogether naked . . . and taketh a round cane as big as a penny candle, and two inches in length, through which he pulls the foreskin of his yard, tying the skin with a piece of the rind of a tree about the bigness of a small packthread; then making of it fast about his middle, he continueth thus till he have occasion to use him."[111] Furthermore, during Davis's travels to Mogador, he writes, "The *Moors* of this country are very deceitful and treacherous; their clothing is but very naked and thin, for they wear but one flat of thin flannel, wherewith they cover their breasts, backs, and privy parts."[112] Importantly, throughout the rest of the vast array of ethnographic descriptions in his narrative, Davis neither offers any specific description of the "privy parts" of women nor allows for a sustained focus on any other male body part. This narratological priapism is quite marked.

The last example of this lingering interest in the penis requires returning to a passage cited above, in which Davis first invites the reader to imagine scenes in which the body is punctured, penetrated, and unnaturally exposed. One of the subsequent types of violence with which Davis and his fellow galley slaves are threatened—a misery that "doth surpass any man's judgment or imagination"—is punishment for even the threat of mutiny. With a tone of certainty, Davis states that if he or his shipmates had "offered any violence to any officer, we should presently have lost our nose and ears, and received a hundred blows to our bare back, and a hundred on our belly with a double rope, or a bull's pizzle, continuing a slave still."[113] The mention of beating naked galley slaves with a bull's penis (a material used fairly commonly in the construction of whips) further contributes to the priapic tropology in Davis's text, drawing readerly focus not only to specific bodies but to specific body parts. When read in step with Davis's repeated invocation of the imaginability or lack of imaginability of the tortures he sustained while a slave in Algiers, this priapism, too, becomes aligned with the sodomitical by virtue of the way it is represented: unimaginable, but also depicted in surprisingly specific terms.

The narrative's priapic focus prompts the question of how to characterize the relationship between sodomy and the human body. Historians of sexuality have insisted for decades on the promiscuous indexicality of sodomy, and I join them in this assessment; sodomy always signifies beyond itself. Sodomy was never "just" a behavior or an act, even in the rare instances when it could be defined as a limited act or set of acts. Yet specific body parts, especially when appended to specifically racialized bodies, demand a closer consideration of the relationship between sodomy and the physiological. Jonathan Crewe's observation that because of the interdiction of male-on-male venereal contact among Christians, "the widely agitating fact of anal penetration or the fraught, ambivalent, erotic investment of the anus" aligned both the anus and its potential penetration with male same-sex sexual relations.[114] This insight reminds us of the broadly associative logic through which certain body parts passively invoked the sodomitical and through which sodomy accrued meaning more generally. If sodomy is conjured by references to specific body parts, references to specific geographies, and references to specific religions, then it is reasonable to suggest that a gestural or associative relationship coalesced—as understandings of human racial variety hardened, over the course of the eighteenth century, into anatomical and otherwise chromatist understandings of race—between race and sodomy over time, as well.

Conclusion

Michael Warner, in his 1992 essay "New English Sodom," theorized sodomy in geographical terms, arguing that "like the much later coinage 'lesbianism,' 'sodomy' still implies, at however fantasmatic a level, a map of sexual knowledges and exotic origins. No other terms in the language of sexuality have a comparable etymology, as though unlike all other sexual acts—if they even are acts—these two were practiced not by individuals but by cities, islands, or nations. This hidden fantasy about the geography of sex continues to exert some influence, primarily in the assumption that sodomitical and lesbian sex are more germane to public politics than other kinds of sex."[115] Arguing that sodomy provided "a language about polity and discipline," Warner emphasizes sodomy's special referentiality to both specific geographies and specific questions surrounding civic order. This chapter tracks the way references to sodomy that cut across seventeenth-century Barbary captivity narratives do indeed often index concerns about cultural geographies of "polity and discipline," for in the context of biblical teachings, the story of God's smiting of Sodom and Gomorrah is about "national judgment."[116] But if sodomy, as a figure, metonymically gestures to histories of crises in civil and spiritual order, sodomy's geographic etymology simultaneously invokes questions about racial (or racial qua religious) order as well.

If the figure of sodomy was understood in the late seventeenth century to index crises in spiritual and political order, its association with Islam was not only the result of centuries of disparaging proto-orientalist rhetoric. It also pointed to a terrifyingly possible reorganization of world order, in which both Christians and Christianity would be preposterously subjugated to Muslim control. The scenes of pathic subjection that became so characteristic of Christian representations of Barbary might then be read as microcosmically speculative visions of a burgeoning social order that both structured life for Christians within the Barbary states and, more alarmingly, could come to structure life outside of them. As we have seen, Christian cultural or geopolitical dominance was far from taken for granted before the turn of the nineteenth century. And underneath these representations of the brutal subjection of Christians to the whims of Muslim leaders lies a quiet racial narrative (apparent in natural historical writings from Pierre Dan to the Comte de Buffon) that sees in the cruelty of Muslim slaveholders an unfitness for governance more generally.[117] Indeed, natural histories frequently expounded on the "despotic" state of society that they understood to be endemic to northern Africa, partly because of its climatological situation in the torrid zone. So, just as the perception that "Turks" are "altogether sodomites" points to a larger crisis of religious order, the geography of northern Africa and the Levant suggests a racialized disposition toward either too much or too little governance, one that is evidenced biblically in the story of Sodom and Gomorrah.

Furthermore, sodomy, or rather, the threat of being forcibly sodomized, featured heavily in a litany of insults and injuries visited on Christians by Muslims, as part of the process of being forced to "turn Turk." While many Christians claim to have converted to Islam in word only—"turning Turk" could often mitigate the violence of slavery for captives and offer both economic opportunities and the chance to marry, especially for poor Christian men who were unlikely to be redeemed—Barbary captivity narratives reveal even inauthentic conversion to be dangerous, itself a slippery slope to actual conversion. Conversion, in these narratives, is a process depicted almost unilaterally as working from the outside in; both male and female captives who initially convert to Islam to save their own lives are later portrayed as indistinguishable from their Muslim captors. Think, for example, of Abraham Browne's horror at the "seight of those ugly onhumanye cretures wch goe soe disgisd according to there Custome with there heads shaved and their Armes almost naked"; it is impossible to tell whether these men are Christians or *renegados,* and this ambiguity is telling. Or consider, again, the scene in Francis Brooks's *Barbarian Cruelty* in which he describes the Christian virgin taken captive, whom Mully Ishmael tortures by pricking her with pins, ostensibly to force her to relent, convert to Islam, and join his harem; Brooks tells us that this female captive "resisted so long, till Tortures, and the hazards of her Life forced her to yield, or resign her Body to him, tho her Heart was otherwise inclined." Despite the Christian inclination of her heart, Brooks details, this woman ended up a "Moor." Mully Ishmael "had her wash'd, and clothed her in their fashion of Apparel, and lay with her . . . . [and] he sent her by his Eunuchs to *Macqueness*. . . . After that she was delivered of two Children."[118] Steeped in the custom and costume of Mully Ishmael's religion and culture, this formerly Christian woman bears two Muslim children, who are not only Moroccans but ostensibly heirs to the throne. It matters little that this Christian woman's conversion was insincere, for in dress and comportment she looks like a Muslim, and her children are as Muslim as her captor. What she lacks in belief, she makes up for in form. Of course, this mode of representation was also deployed strategically by writers such as Cotton Mather, to cast doubt onto even earnest conversions to Islam, of which we know that there were also many. But this way of understanding the threat of conversion, a constitutive fear informing every Barbary captivity narrative I have read, invokes exactly the kind of environmental theory of religion (and thus race) that I describe in chapter 1, and it is one that also bears a distinct sexual politics.

I emphasize the externality of representations of forcible conversion to Islam to point to the powerful environmental etiology that underlies understandings of sodomy. If a Christian can, voluntarily or by force, become a Muslim, then, following William Davis's logic, that person also becomes a sodomite; sodomites, after all, are simply defined negatively as "do[ing] all things contrary to a Christian." And if a sodomite is something that one can become by virtue of exposure to a foreign religion or a new set of sleeping conditions, or even through being forced to reproduce religion and culture by giving birth to children who will be raised in the Islamic faith, then sodomy, too, bears a set of distinctly environmental qualities that could be embodied by almost anyone. The racialization of the peoples of the Barbary coast, established in part through what Europeans understood to be its biblical and climatic proximity to the actual location of the original city of Sodom, girds the environmental association between sodomy and northern Africa, on the one hand, and sodomy and Islam, on the other. While I don't claim that "sodomy," as a word or an idea, was always or exclusively used to invoke associations of human religious or racial difference, we must acknowledge the persistently racial quality of this term in a broad set of contexts—especially when invoked in natural histories—during the seventeenth and eighteenth centuries. Indeed, this chapter's readings of a series of Barbary captivity narratives demonstrate that sodomy was persistently linked to the Islamic world and was frequently invoked in references to what Christians understood to be the highly unnatural religious hierarchization of Muslims over Christians so fundamental to the infrastructure of Barbary slavery. And while scholars have gestured broadly to the sexual politics of orientalist ethnography, no scholarship has yet explored the possibility that sodomy, even when discovered between white Christians, might have at times been understood as a behavior associated with racial or religious difference. Indeed, this might even explain the persistent criminalization of sodomy, especially in the absence of the concomitant explicit criminalization of other forms of same-sex sexual behavior.[119]

In chapter 3 I turn from sodomy to a differently stigmatized range of sexual behaviors, masturbation and rape among them, to explore emergent eighteenth-century theories of criminality that linked blackness and aberrant sexual behaviors through the figure of habit. As public education, prison reform, and anti-capital-punishment movements all gained traction in the North American colonies, reformers began to set their sights upon identifying how to prevent crime before it happened. Ministers, educators, and authors of parenting and conduct books all found consensus in the idea that habits accrued in childhood (a new episteme of human development largely promulgated by education movements, as Philippe Ariès has argued) largely determined the course of one's later moral, spiritual, and civic comportment.[120] As a result, identifying how and when bad habits began emerged as a central question in eighteenth-century theories of criminality, penned both by philosophers of punishment and reform such as Benjamin Rush and Jeremy Bentham and by ministers charged with delivering execution narratives, the increasingly popular genre of cheap print that some scholars have named as a constitutive precursor to the novel. Habit, a means by which the soul could be disciplined into righteousness or acculturated into evil, offered a theory of criminal behavior at times explicitly drawn from natural historical writings on human difference. In this sense it provided a discursive figure by which blackness was tied to crime, and to sexual crimes in particular.

CHAPTER THREE

"Egyptian Lusts" at the Gallows

On a cold Boston night during the winter of 1721, Joseph Hanno killed his wife, Nanny, as she prepared herself for bed. Complaining later that Nanny had been an *"Insupportable Wretch,"* Hanno hit her several times over the head with an axe and slit her throat before running outside to wake his neighbors, telling them that Nanny had committed suicide.[1] Despite his account of her death, Hanno was suspected of Nanny's murder almost immediately; the coroner's report issued that same evening names him as the guilty party.[2] Hanno was taken to the town prison on Queen Street in Boston, where he was housed for five months as he awaited trial. He was found guilty of murder and sentenced to death by public execution later that year.

Hanno's execution was something of a social event. Per Calvinist custom, a sermon was preached before his death to make meaning of the crime and punishment, holding Hanno up as a warning that might inspire the community to examine their faith and renew their commitments to godly living. Cotton Mather, the venerated local minister, delivered this execution sermon, in which he exhorted his audience to heed the providential lesson of Hanno's crime and the inevitability of divine retribution. The sermon, titled *Tremenda: The dreadful sound with which the wicked are to be thunderstruck,* begs his audience to awaken to the deadly threat that wickedness poses to not only individual souls within his congregation, but to the spiritual well-being of the community as a whole.[3] The sermon is organized around the figure of the "wicked man," and Mather describes for his community in careful, biblical detail how this man might be recognized, in themselves and in other members of his flock. Offering a five-part, categorical anatomy of wickedness, Mather details how the truly wicked man is distinct from a mere sinner (in part because the Calvinist doctrine of innate depravity understood everyone as a sinner): first, the Wicked Man "is a Sinner who *Loves* to be a *Sinner*"; second, he "consents" to sin;[4] third, he "resolves" to sin;[5] fourth, he "*delights*" in sin;[6] and finally, "the Wicked Man" makes a habit of sin, and it is this "*Custom in Sin* [that] is the consummation of *Wickedness;* the finishing stroke in a *Wicked Man*."[7]

In Mather's sermon, as in many execution sermons that came before and after it, casting the condemned as a *habitual* sinner—someone accustomed to sinning—was important for several reasons. Most immediately, Mather's emphasis on "custom in sin" served an explanatory function for a crime that may have seemed shocking and difficult to understand, especially in someone who was likely familiar to members of the community.[8] More importantly, it served a broader didactic function, warning congregants of what might befall those who lacked a vigilant relationship to their own sinful behavior; the implication was that seemingly less serious sins such as an inattention to daily prayer, disobedience to one's parents, or breaking the Sabbath could lay the foundation for more damning behavior later on, such as adultery or murder. For Mather and, as we will see, for many other ministers throughout the eighteenth century, developing bad habits and allowing them to flourish unheeded was a surefire way to lose the grace of God. Mather is so concerned with this "Custom in Sin" because sinful habits are spiritually dulling, rendering one less sensitive to the importance of righteous living, and are apt to further subordinate the faithful to what Mather ominously terms "the *Reigning Power* of *Sin.*"[9] The "Wicked Man," he writes, "*serves divers Lusts.* His Lusts have an absolute Command over him. The Motions of His *Lusts* have an irresistible Force upon him. He can't Resist his *Criminal Propensities.* He can't mortify his *Carnal Appetites.* His unmortified passions for *Vile Things* are too Potent for him. His *Vices* lay Chains upon him; draw him hither and hither to his Confusion. Tho' he knows it will confound him to Sin in such and such things, yet he *must needs go,* when he feels himself driven into it. He is entirely governed by vicious Inclinations."[10]

Bad habits, then, were the gateway to the dangerous loss of both spiritual and bodily governance. Bad habits were not merely isolated behaviors, as Mather himself pointed out: "A *Gracious Man* may *Sin* and may *Repeat* the Sin. But then, he breaks off the *Custom,* with begging & pleading for help against the Sin.... It is the *Custom,* the *Course,* the *Way* of a *Wicked Man,* to do those Things, which he knows are Forbidden by GOD, and Offensive to Him."[11] To make a habit or custom of sin, for Mather, was perhaps more dangerous than the sin itself. To have bad habits was to acquiesce to the despotic governance of vice ("vicious Inclinations").[12]

The racial and sexual rhetoric that inheres in Mather's sermon may not be apparent at first glance, but as we shall see, eighteenth-century concerns about preventing and eradicating vicious habits—voiced by ministers, parents, and teachers—provided consistent occasions for heavily racialized discussions of sexual behavior. And, importantly, these concerns about bad habits drew from a century's worth of natural historical writing that understood habits of all kinds as a critical indicator of one's "state of society" and level of civilization. In Mather's execution sermon, he characterizes Hanno's murderous impulse as a form of sexual domination by eroticizing the despotic relation of enslavement ("His Lusts have an absolute Command over him"). That Hanno was also an African slave brought to North America as a child around 1677 and manumitted in 1707 was not lost on Mather;[13] his assessment of Hanno's vulnerability to the "absolute Command" of his *"Carnal Appetites"* reflects contemporary proto-racial understandings of slavery as both a political and a mental state.[14] As this chapter demonstrates, this articulation (the way habit or customs weave together contemporary theories of both human variety and sexual behavior) appears pervasively in eighteenth-century popular narrative, and in execution narratives in particular.

Tremenda, as a sermon, is also rife with Mather's characteristic literary flourish, and throughout the lecture Mather plays with the simultaneously phenotypic and characterological trope of Hanno's blackness; he refers to Hanno as "the *Black Thing* that you have in Irons here before you," a man bearing "so Doubly and so Deeply *Black* a Character."[15] Quoting the biblical book of Jeremiah, Mather invokes the trope of the "Ethiopian" to explore the relationship between blackness, wickedness, and slavishness to sinful habits. In his discussion of the fifth and final quality of the "Wicked Man"—"Custom in Sin"—he admonishes his congregation that *"Custom in Sin* is the consummation of *Wickedness;* the finishing stroke in a *Wicked Man.* He is, as I may now agreeably express it, an *Ethiopian* in Wickedness. We read of some, Jer. XIII.23. *Who are accustomed to do evil;* and can leave it off no more than the *Ethiopian* can *Change his skin.* They that are *Accustomed to do evil* are *Wicked Men.*"[16]

It is the "*Ethiopian* in Wickedness" that provides the point of departure for this chapter, which turns to execution narratives and other literature about crime to explore how these texts invoke habit as a behavioral and environmental figure central to eighteenth-century understandings of racial difference and sexual behavior. Discussions of the influence of habit appear pervasively in naturalist treatises on the origins of racial difference; here, we will follow this cultural interest in the impact of habitual behaviors out of the realm of natural history and into popular narrative, examining how it appears in gallows literature in particular.[17]

Reading the attention to both good and bad habits that appears in criminal execution narratives reveals habit itself to be an articulating figure, binding contemporary understandings of racial difference and sexual aberrance together in emergent theories of the origin of vice or viciousness. In these texts, racial blackness, in particular, is articulated through criminalized sexual behaviors, and criminalized sexual behaviors are in turn characterized through an aesthetics of spiritual or moral darkness.

The gallows, and gallows literature—broadsides, sermons, confessions and last words, and eyewitness accounts—have long been important sites of inquiry for the study of both early race and early sexuality, in part because the scaffold provided a spectacular point of overlap between the criminalization of racial difference and the criminalization of specific sexual behaviors. Historians of sexuality, however, have tended to approach execution narratives with the methodological assumption that knowledge about sexuality can be gleaned only from texts that are "about" sex; Anjali Arondekar deftly describes this archival fallacy as "the presumption that if a body is found, then a subject can be recovered."[18] Instead of looking for the representation of particular bodies and particular acts, such as bestiality or sodomy, here we will explore the way sexual behavior was thought about and the way environmental logic informed that thinking; as we shall see, this environmental logic persistently tied how people thought about sexual behavior to how they thought about racial difference and vice. By reading for and with habit, we can better understand the sexual politics of racial difference and the complex process by which racialization produces sexual behavior as a discrete formation eligible for analysis.

Reading with habit is surprisingly, and spectacularly, difficult. This is in part because eighteenth-century discussions of habit characterize it in multiple, often contradictory ways. Habit is simultaneously an expression of both good and bad learned behaviors, on the one hand, and of individual inclinations toward goodness or evil, on the other. Good habits can derive from learned practices like prayer, or they can derive, it seems, from some innate spiritual inclination toward goodness. Bad habits can be learned or absorbed at an early age, by following sinful examples set by parents or community members, *or* they might be an expression of individual malevolence or innate depravity. The figure of habit sits at the nexus of environmental influence and individual inclination. Instead of reading habit as a primarily environmental or primarily internal figure, then, habit should be read as a term that points us to eighteenth-century questions about self-governance and about who—or what—exerts sovereignty over individual bodies.

This is why it was such a point of concern for educators, ministers, parents, and statesmen, especially in the beginning of the early national period, and this is why discussions of habit—especially those that appear in execution sermons—constellate so consistently around the bodies of women and people of color. The potential of bad habits to upend the social and civic righteousness of even the most upright of slaves and citizens later led Benjamin Rush, while detailing a dream about becoming the president of the United States in a letter to John

Adams, to bemoan it as "the EMPIRE OF HABIT."[19] Habit was also a term that invoked a long-standing relationship to natural historical thought, but one that was increasingly retrenched by the end of the eighteenth century. Yet in discussions of habit that use it to develop and explore questions about the origins of behavior, including sexual behavior, we nonetheless hear echoes of both natural historical thought and the environmentalist logic that girded it. So, while habit was only sometimes used to describe environmentalist theories of behavior in execution narratives, it always indexes a set of tensions about the origins of behavior, and in this sense it is a term fundamentally bound to the environmentalist logic that this book explores.

Reading with habit demands that we deprioritize the body as the privileged or original location of sexuality—and especially early sexuality—and that when we ask questions about the relationship between sexual behavior and embodiment, we formulate those inquiries with a thick understanding of the environmentalist logic through which the human body was understood during the eighteenth century.[20] Reading with habit also reveals new archives and points to new texts and questions that augment extant theories of how sexual behavior was understood during the eighteenth century. While Joseph Hanno was not arrested or executed for any crime related to what we would now think of as sexual behavior, for his community, his shocking, unprecedented crime nonetheless testified to the governing power of "the Motions of His *Lusts*" and his "unmortified passions for *Vile Things*." Yet far from attributing what Mather calls "*Criminal Propensities*" to a generic innate depravity or even a radically individualized turpitude, ministers throughout the century tended to invoke a Lockean, environmental understanding of crime that saw criminality as the result of the early development of bad *habits*.

Theorizing Habit in the Eighteenth Century

Long before habit became a persistent feature of execution sermons, discussions about the impact of habits had appeared pervasively in natural historical writing. As I explore in chapter 1, natural philosophers such as the Comte de Buffon, John Mitchell, and Samuel Stanhope Smith all hypothesized about the effect of a person's habits on her or his character and even physiology. These habits, which included everything from dress to behavior, from cultural and religious customs to the food and drink one consumed, were believed to exert considerable influence on what some naturalists called "the state of society," a developmental term that described a culture's progress from barbarism or "savagery" to civilizational maturity. Habit was an idea fundamentally tied to scientific debates about human variety, and historians of race often read early modern and eighteenth-century naturalist discussions of habit as a precursor to chromatist or phenotypic theories of human racial difference.[21] Indeed, cultural specificity in dress or custom was itself a means of characterizing human variety during this period, and Valerie Traub reminds us that "if phenotype is a concept based on visible characteristics, then what developed in the early modern period was not a phenotype based on the privileging of skin color, but a strategy of marking differences and similarities through a visual mimesis of nation, religion, lineage, costume, as well as skin color—in a word, *habit*."[22] And, like other theorists of human variety, natural historians imagined the impact of specific habits to have both external (or phenotypic) and internal (spiritual or characterological) effects. In her discussion of the representation of human ethnic variety in early modern maps, Traub highlights precisely this labile tension between the potential external *and* characterological results of specific practices or habits of life; she argues that, "derived from the Latin for 'holding, having, 'havior,' *habit* in this period signified 'the way in which one holds or has oneself . . . a) externally; hence demeanor, outward appearance, fashion of body, mode of clothing oneself, dress, habitation; [and] b) in mind, character, or life; hence, mental constitution, character, disposition, way of acting, comporting oneself.'"[23] Before habit arrived as a focal point in execution sermons as a device to explain the origins of crime, then, the idea that one's bodily comportment might exert an influence on one's spiritual or behavioral state had enjoyed a long tenure in naturalist theories, from antiquity through the early modern period.[24]

To reckon with the way eighteenth-century execution sermons invoked habit, and especially bad childhood habits, to explain later propensities toward crime, we must begin with an understanding of how these ministers and their congregations likely understood the term. Beyond simply an idea about behavior, in the eighteenth century "habit" was a densely significant figure, heavy with implication and at least partially drawn from naturalist vocabularies for anatomizing human racial and ethnic difference. When habit was deployed as a means of making sense of what Mather deemed Hanno's "*Criminal Propensities*," it is reasonable to imagine that congregants would have heard and appreciated the wide range of implications that this term bore. In highlighting the racialized etymological origins of habit in this discussion, I make two arguments. First, even in execution sermons that do not address human difference or people of color, habit was a profoundly racialized term during the eighteenth century. It should come as no surprise that attention to habit appears pervasively in sermons and other print productions that exploit popular interest in criminality; in this sense, the racialization of habit girds both the racialization of crime and the criminalization of people of color, an idea to which we return in chapter 5.

My second argument is thus that because naturalist explorations of habit frequently direct a great deal of attention to the customs, particularly the gendered and sexual customs, of non-European cultures and communities, habit was also an operative term in the discussion of sexual behaviors. This idea is easily observable in sermons like Mather's *Tremenda,* in which sinful or criminal inclinations are represented as *"Carnal Appetites."* During the eighteenth century, in the context of the discussion of criminal behaviors, habit was also a term that connoted sexual practices. In execution sermons as in naturalist philosophy more generally, habit provided a site where theories of human variety were yoked to, and in part *constituted by,* ideas about diversity in human sexual practices, and a site in which socially aberrant sex acts were made visible as objects of inquiry precisely by virtue of their imbrication in racialized bodies. It is no surprise that these popular narratives so often highlighted the crimes of people of color and so often featured crimes related to sex. In them we see how racialization draws its differentiating power from being bound to an association with aberrant sexuality, and sexual behavior, in turn, emerges as an eligible optic for human difference because of its association with racial difference. Naturalist attention to racial difference thus enables the emergence of sex as a discrete category of human experience; human variety, on the one hand, and human sexual diversity, on the other, "articulat[e] themselves in terms of one another."[25]

Nor was eighteenth-century concern with the possible effects of good and bad habits limited to ministers and naturalists. As numerous scholars have demonstrated, the eighteenth century was a time of great change in the understanding of human development; passionate interest in fields such as anatomy and medicine flourished throughout the long eighteenth century on both sides of the Atlantic. It was also the era to which we attribute the emergence of the notion of childhood as a distinct phase of human development and a time when there was growing concern with the education and rearing of children.[26] One of the major sites *outside of* natural philosophy in which habit appeared as a topic of interest and debate was popular literature on the education and upbringing of children. While natural historical texts were only unevenly accessible to broader audiences, child-rearing and early-education tracts, sermons, and books were both written and printed with a larger and less elite group of readers in mind, and discussions of the importance of developing good habits appear consistently in these writings. Indeed, in such texts habit itself seems to belong to the temporal purview of childhood, and didacts from Joseph Belcher in 1698 to John Witherspoon in 1797 urge parents, religious leaders, and other caretakers to take full advantage of the unique impressionability that attends a child's early years. Instilling both spiritual and behavioral habits in young people, it was believed, would render them less vulnerable to the later predations of temptation and keeping bad company.[27]

As Arthur Browne insisted in his 1739 *Religious education of children recommended,* an execution sermon for Penelope Kenny, who had been convicted of killing her bastard child after it was born, "we are naturally inclined to Evil; the neglect therefore of Education lays Children under a kind of Necessity of pursuing what they are by Nature inclined to; Sin and Vice become soon habitual to them; and when once these are grown familiar, there seems to be an End of Liberty and Choice; they are then for the most Part deaf to good Counsel and Instruction: or if they are disposed to hear, they want the Power to follow it, being bound in the Chains of their Sins, and led Captive by Satan, at his Pleasure."[28] Browne echoes Mather's anxiety about the potential for habitual sins to override one's autonomy over body and behavior, especially in children. Even in writings on early education and child-rearing that are less explicitly steeped in the rhetoric of spiritual salvation, habit appears as a question of governance. For example, John Witherspoon, in his 1797 *Series of Letters on Education* (reprinted in 1815 and 1817), reminds his readers that "habits in general may be very early formed in children. An association of ideas is, as it were, the parent of habit. If then, you can accustom your children to perceive that your will must always prevail over theirs, when they are opposed, the thing is done, and they will submit to it without difficulty or regret."[29] For Witherspoon, the habits that are most important to develop in children are not specific behaviors but a comfort with specific relations of governance, or an ability to discern to whose will they ought to submit; whereas Belcher, Mather, and Browne concern themselves with "habits of sin," Witherspoon exhorts parents to cultivate "a habit of submission" in children, which should be "brought on so early, that even memory itself shall not be able to reach back to its beginning."[30] Early habits could also serve something of a predictive function for later behaviors, in the view of Witherspoon and his contemporaries, because they understood childhood to be a time of spiritual and behavioral malleability that would disappear and harden with age. Witherspoon explained it in a 1789 sermon on religious education: "You may bend a young twig and make it receive almost any form; but which has attained its maturity, and taken its ply, you will never bring into another shape than that which it naturally bears. In the same manner those habits which men contract in early life, and are strengthened by time, it is next to impossible to change."[31]

An understanding of the distinct form that discussions of habit take in early education and child-rearing tracts is critical to understanding eighteenth-century gallows literature because, as scholars such as Daniel Cohen have noted, starting in the mid-eighteenth century, ministers drew on environmentalist theories of early childhood development more and more often in their explanations of the origins of criminal behavior.[32] This new way of understanding crime did not replace or even necessarily supersede explanations of criminal behavior long favored by Calvinist clergy—such as innate depravity or the punishment of sin with additional proclivity toward sin—but it added the question of environmental influence and early education to these conversations. This environmental hypothesis grew in popularity throughout the long eighteenth century, and by 1800 ministers regularly referred to early parental influence and spiritual education (or the lack thereof) in their accounts of the fatal trajectories of the condemned. In other words, by the late eighteenth century, "crime became contextual."[33] What emerged was a variety of "habits of sin," including disobedience to one's parents and breaking the Sabbath, that were seen as especially dangerous to the later spiritual rectitude of young people. One of these habits of sin, however, towered above the others and throughout the eighteenth century was seen as endangering children both physiologically and spiritually, promising a later vulnerability to a limitless number of far greater sins: masturbation.

In many ways, masturbation stands in for all that is dangerous about the development of childhood bad habits during the eighteenth century; it is the childhood bad habit par excellence. And while masturbation had never been an unproblematic behavior, spiritually or otherwise—in Samuel Danforth's 1674 execution sermon *The Cry of Sodom Enquired Into* (discussed in the Introduction), for example, he lists "onanism" among the litany of sins related to bestiality, the crime of which the condemned had been convicted—it acquired new force during the early eighteenth century and to an extent maintained that threat through the early twentieth. This renewed interest in the dangers of masturbation was of course not coincidental; it was instead largely the result of the publication of the enormously popular *Onania, or the Heinous Sin of Self-Pollution*. Both the author and the exact first publication date of *Onania* are unknown, but the term "masturbation" first appeared in print sometime around 1708, and Thomas Lacqueur identifies *Onania*'s initial publication date as "in or around 1712," although it was not printed in the British North American colonies until the tenth edition, in 1724.[34] Before its printing in Boston, *Onania* had sold more than fifteen thousand copies, and evidence suggests that at least the 1723 London edition had been circulating in the British North American colonies before it was published in Boston in 1724. Cotton Mather published an antimasturbation tract titled *The Pure Nazarite* in 1723, and John Phillips, the printer of Mather's *Pure Nazarite*, was also the first colonial North American printer of *Onania*.[35]

Masturbation and childhood were ideas defined in relation to each other, and popular interest in each of the concepts influenced the contemporary understanding of the other.[36] As Michael Stolberg details, "masturbation was believed to be particularly prevalent among the young," and children, in turn, bore a special vulnerability that left them predisposed to the solitary vice.[37] He also notes that "among the young as well as among the grown-ups, masturbation was perceived as habit-forming."[38] If, then, during the eighteenth century, childhood was the primary temporal purview of habit, masturbation was *the* habit with which parents, ministers, and other community leaders were concerned. And concerns about masturbatory habits in children only grew louder and more fervent as the eighteenth century progressed; *Onania* matured from a slim volume of less than one hundred pages in its earliest editions in the first two decades of the century into a nineteenth edition, a half century later, that had expanded to more than three hundred pages.[39] Samuel Auguste Tissot, author of one of the most popular medical advice books of the eighteenth-century Atlantic world, published *L'onanisme: Dissertation sur les maladies produites par la masturbation* in French in 1760, and it was translated into English a few years later and reprinted consistently in the British North American colonies (later, the United States), through the first several decades of the nineteenth century.[40] Ministers, politicians, and philosophers from Cotton Mather to Benjamin Rush all addressed onanism—a disease they understood to be both physical and social—in their works.[41]

Associated with youth, masturbation was imagined by these doctors, ministers, and community leaders as especially problematic because of its status as something of a "gateway" sin: yielding to the temptation to masturbate could build habits of "uncleanness" that might later beget sinful behavior of a deeper hue. The 1724 Boston reprint of *Onania*, for example, warns its readers:

> I Have in the foregoing Chapter spoke of some, who gave themselves over to this [masturbation], and yet were Proof against any other gross Sin of Uncleanness, but it is not so with all: Thousands have been guilty of *Adultery*, as well as *Fornication*, who would never have yielded to those Temptations, which overcame them, if they had never been initiated in Lasciviousness, and acquir'd to themselves a Habit of Impurity by SELF-POLLUTION first. In such, not only the grosser Crimes of Uncleanness I just now named, but likewise all others that may be occasioned by them, as *Lying, Forswearing*, perhaps *Murder*, and what not, must be laid to the Charge, and brought in as the Effects and Consequences of their first darling Sin, by which they were infected with a stronger Habit of Impurity, than they could ever have contracted from any other Frailty.[42]

Masturbation, here, does the work of acclimating one to sin; it creates, as Cotton Mather details in *Tremenda*, "custom in sin." The anonymous author of *Onania* draws a clear line between masturbation and later unclean behavior and importantly characterizes this "first darling sin" as having more impact on later behavior than any other problematic childhood habits. The effects of childhood masturbation are not only adultery and fornication, but potentially also "*Lying, Forswearing*, perhaps *Murder*." Cotton Mather, too, saw the spiritual effects of masturbation as extending far beyond the immediate problem of self-pollution, writing in 1723, "Be sure, the Perpetrators of this *Impiety*, bring such a *Destruction* on their *Health* . . . as to render themselves a sort of *Self-Murderers;* They violate the *Sixth Commandment* in their Violation of the *Seventh*."[43]

Eighteenth-century antimasturbation literature reveals the discursive centrality of habit as a framework for understanding problematic behavior. It also points to an incredibly persistent association between habit and sexual behavior. Tracking the proliferation of inquiries into the effects of habits, good and bad, within eighteenth-century print culture demonstrates how wide this figure's field of signification was during the period. Appearing as often in natural history texts as a means of distinguishing among different kinds of humans as it did in child-rearing texts warning parents of the perils of masturbation, habit arrived in gallows literature with a thick set of cultural associations and ideologies attached to it.

In part, the promise and challenge of thinking with habit as a key term in the history of eighteenth-century sexuality is that it is a formation that tries to account for the concatenation of both external and internal influences on the subject simultaneously, in a way that is both explanatory and predictive. In this sense, it is a deeply environmental way of thinking that both draws from and girds the environmental logic through which sexual behavior is explored in the other chapters of this book. Habit is messy, pointing to "an association of ideas," as David Hume imagined it, but refusing to characterize how that association functions or what its effects might be.[44] Habit neither points to a strictly "outside-in" behaviorism that assumes that one's proclivities or inclinations can be entirely shaped by one's practices, nor does it imagine that good or bad habits are the result of unique, interior experiences. Instead, the way discussions of habits—pertaining to clothes, behaviors, customs, even dwellings—appear in eighteenth-century natural history and education texts assumes that a person's actions and inclinations are the result of something like an equilibrium of external influences (practices or behaviors, such as prayer or masturbation) and internal dispositions or inclinations (such as divine election or wickedness). Nonetheless, habit always assumes a relationship to the external world, to the environmental conditions in which a person moves, perceives, and is understood; internal inclination, good or bad, is nothing without the practices and behaviors that give it shape and shape it in turn. As John Witherspoon puts it in *A series of letters on education,* "the form without the spirit is good for nothing; but on the other hand, the spirit without the form, never yet existed."[45]

Habits That Lead to the Gallows

The cultural importance of gallows literature to the development of popular culture cannot be overstated. Scholars such as Daniel Cohen and Daniel Williams credit it with laying the foundations for the popular consumption of literature in North America; others suggest that this genre of cheap, accessible, sensational print created a consumer fondness for proto-psychological forms of characterization that grew more popular and flourished in the novel form.[46] Cotton Mather took to the form for its didactic durability, arguing that "*Sermons Preached,* are like *Showres of Rain,* that Water for the Infant; But *Sermons Printed,* are like *Snow* that lies longer on the Earth."[47] But gallows literature served another purpose as well, providing a major site of cultural engagement with the significance of human differentiation as well as sexual behavior. On the scaffold and in gallows literature—as in the other forms of popular narrative examined in this book—these emergent forms of human difference frequently travel together, and in execution narratives they are often bound to one another in the figure of habit. Habit frequently appears in execution narratives as a racialized trope, a figure used by ministers to grapple with the racialization of crime. This was no coincidence; in the British colonies during the eighteenth century, as still today, execution was a punishment disproportionately reserved for immigrants and people of color, and the executions selected for public consumption in the form of gallows literature were frequently executions of people of color being punished for what we might term "sexual" crimes: adultery, infanticide (often to hide adultery or fornication), and rape.[48]

"Habits of sin" was a favorite item of concern in sermons delivered, and later printed, by Calvinist ministers in the North American colonies, and execution sermons were no exception. Theological discussions of habit appear in sermons from the mid-seventeenth century onward; yet we see a marked rise in attention to the influence of good and bad habits on a person's spiritual and behavioral trajectory in execution narratives in particular in the second half of the eighteenth century. In the slim body of scholarship on this chronological shift, scholars have tended to ascribe this increase in interest to something of a secularization narrative, arguing that accounts of habitual behaviors, and especially illegal behaviors, joined or replaced traditional theological explanations for sinful behavior, such as innate depravity or God's withdrawn grace.[49] If we expand our inquiry, however, we see that this is too simplistic. Attention to the explanatory power of habit did indeed increase in gallows literature over the course of the eighteenth century. However, in print culture more broadly, habit was a pervasive site of curiosity for ministers, philosophers, and other thinkers relatively evenly throughout both the early modern period and the eighteenth century. In other words, the explanatory power and behavioral valence of habit did not arrive with its putative secularization in the second half of the eighteenth century; rather, habit had long been discussed in religious and philosophical contexts.

Furthermore, the appearance of habit in execution sermons seems to be as old as the genre itself, at least in North America. By the end of the seventeenth century, we find admonishments to heed the dangers of bad habits in even the relatively few sermons printed in New England (most of the print material circulating in the colonies at this time was printed in England). And in late-seventeenth-century sermons, such as Josiah Belcher's 1698 *The Worst Enemy Conquered. A Brief Discourse on the Methods and Motives to pursue A Victory Over those Habits of Sin, Which War against the Soul,* the figure of habit is already offering an articulation of the sexual politics of racial difference. In this sermon, Belcher figures sin as an internal threat to the physical and spiritual health of the body, relying on something of a nationalist metaphor to make his point. He warns that "besides the many forreign, a Christian hath many inbred and domestick adversaries; *We have enemies of our own house:* there are not only the temptations of the Devil, the oppositions of the world, which we have to conflict with; but there are also the *lusts of the flesh.*"[50] Acquiescing to temptation, especially when that indulgence is repeated, spurs the same crisis in bodily and spiritual governance that Mather refers to as "custom in sin." For Belcher, habits of sin lead men "into Captivity," "making them *bond slaves for a season,*" and "slaves to [their] lusts."[51] And in *The Worst Enemy Conquered,* as in *Tremenda,* this bondage is explicitly racialized. Belcher repeatedly refers to the ostensibly unbaptized or unconverted soul as "natural man," a biblical phrase that comes from Paul's letter to the Corinthians but one that also both echoes and anticipates civilizationalist rhetoric dividing Christians from the "*Barbarous Nations.*"[52] Belcher further equates the "natural man" with servitude, in his observation that "Sin *rules* and *reigns* in the natural man; he is its slave and vassel," which draws an even tighter association between "natural man" and the race-based system of slavery then emerging.[53]

At the end of the sermon, Belcher returns to his nationalist metaphor, warning of "inbred and domestick adversaries," but sharpens it, adding a new, ethnically specific image. Citing the book of Exodus, Belcher explains: "We read concerning Moses, Exod. 2.11.12 *That when he spied an Egyptian smiting an Hebrew, one of his brethren, he slew the Egyptian.* There are if I may term them so, *Egyptian* Lusts within us, which ever and anon smite our Souls, which are nearer friends to us than a Brother; they are our darlings, our all; when ever therefore we espy these cruel Taskmasters (they would be so) smiting our spiritual part, we must encounter them, and endeavor to slay them."[54] In this rendering of sinful behavior—and not just any sinful behavior, but "Lusts," specifically—as the metaphorical "*Egyptian* Lusts," Belcher unwittingly anticipates the "Ethiopian in Sin" that Mather would describe twenty-three years later. His "Egyptian Lusts" offers a racialized materialization of the "domestick adversar[y]," but in this image it is the Christians who are "under the dominion of Sin,"[55] their spiritual well-being threatened with destruction at the hands of "Egyptian" "Taskmasters." It is no coincidence that Belcher uses "Lust" as a shorthand for the entire, broad horizon of all sin in this sermon, either; for lust, in particular, was the spiritual "domestick adversar[y]" par excellence. Unlike behaviors spurred by bad company or drunkenness, for example, lust tends to index a set of internal inclinations, feelings that are "nearer friends to us than a Brother," "our darlings, our all." And while "lust" could denote a wide range of sins, not limited to behaviors we now think of as sexual, "lust" was a category of experience especially associated with the venereal: fornication, adultery, masturbation, and other forms of what Calvinists termed "uncleanness."[56] That "Lusts" might be figured as "Egyptian" points to another quiet but unsubtle way in which thinking about sex was frequently articulated through an idiom of racial or ethnic specificity during the long eighteenth century, and to the way habit provided a framework in which both sexual trespass and racial difference could be explored.

Only one year after Belcher's *The Worst Enemy Conquered* was printed in Boston, Bartholomew Green, one of the printers responsible for producing Belcher's sermon, published what scholars now know to be one of the first compilations of gallows literature penned in the North American colonies, Cotton Mather's *Pillars of Salt. An History of Some Criminals Executed in this Land, for Capital Crimes.*[57] This 1699 anthology includes the stories of twelve men and women hanged for crimes from infanticide to bestiality, from murder to adultery. Appearing twenty-five years after the first execution sermon published in the North American colonies (Samuel Danforth's 1674 *The Cry of Sodom Enquired Into*), *Pillars of Salt* catered to what even Mather acknowledged to be the increasingly popular appeal of writings about public execution.[58] Unlike most previously published execution sermons, however, Mather's compilation includes several forms of what scholars now identify as gallows literature: narratives of nameless criminals that are simply relayed through a minister's voice;[59] execution narratives offered in interview form, transcribing a conversation between minister and condemned;[60] and first-person accounts of a crime and a subsequent confession, purportedly written or dictated by the condemned person.[61] Mather's *Pillars of Salt* is generally understood to be one of the founding texts of the genre of gallows literature as it developed in the North American colonies, and in his interest in what he terms "habituation" to sin we are able to identify a very early example of the way racialization, sexual behavior, and criminalization are articulated through one another in discussions of habit, a tendency that only increased toward the end of the century.[62]

Pillars of Salt opens with a lengthy prefatory discourse, titled "Reflections on the Dreadful CASE of Sin Punished with Sin," in which Mather seeks to explain why some forms of sin seem to be habit-forming. Mather wonders at the way God's retribution toward the wicked assumes a negative form; he punishes sinners by simply withdrawing his grace and allowing the sinner to descend further into sinful behavior. Turning to the example of "acts of unchastity" at the end of the opening discourse, Mather reminds his readers that "there are Two Things, that frequently occur among us. One thing very Frequent is This; A person that falls into One *Act* of *Unchastity*, if they don't presently with Bleeding Souls fly to the *Blood* of the Lord for Pardon, they are usually left unto another. Yea, and God punishes the *Unclean*, by *giving* them *over*, to proceed from one *kind* of *Unchastity* unto another."[63] "Unchastity" of various kinds becomes a key example for Mather, and just as we see in *The Pure Nazarite* and the anonymous *Onania* two decades later, sins of unchastity are prone to escalation. Thus, Mather warns, "First the young Wretch, confines his *Wantonness* unto himself alone; but he *goes on* to *Fornication,* he *goes on* to *Adultery;* he *goes on*,—to nefandous and stupendous Abominations."[64] The dash in Mather's prose (this piece was not initially delivered as a sermon, as far as we can tell) simultaneously suggests a kind of infinitude and endlessness to the list of unchastities to which the early offender may eventually be inclined and the unspeakability of these "*nefandous* Abominations."[65] The repetition of "unchastity" has a "brutifying" effect, hardening the heart of the practitioner and rendering the person less vulnerable to the ameliorating influence of grace. "These are Sins," Mather continues, "whereof 'tis said, *They take away the Heart:* The Bruitified Wretch that is Habituated thereunto, ha's *no Heart* for Christ, *no Heart* for Penitence, *no Heart* for Piety left unto him. . . . We have seen, that they'l continue their *Whoredomes,* in the very *Prison*."[66] In this impassioned jeremiad, we see the clear trajectory that Mather defines between the practice of myriad of forms of "unchastity"—among them "*Wantonness,*" "*Fornication,*" and "*Adultery,*" but any number of "nefandous" other behaviors as well—that render believers "brutish" by being "habituated" to these practices.

If "habituation" to "unchastity" can render one "brutish," however, Mather also implies that "brutishness" itself suggests an inclination toward unchaste behavior. In the very beginning of the discourse, Mather turns to historical examples, biblical and otherwise, in which "idolators"—"*Barbarous Nations*" who worship the sun, the "Egyptians" who worshipped "a *Bull, a Dog, and a Rat,*" and the "The Learned *Grecians*" and "most Learned Romans" who worshipped "*Feavers*" and "the very *Furies* of Hell"—are punished for their willful ignorance of the true God by being abandoned to sin.[67] Mather offers terse judgment of the many peoples for whom God punishes sin with sin: God "Punished the *Spiritual Uncleannes of Idolatry,* by leaving them to the *Corporal Uncleanness of Debauchery.*"[68] Yet in this passage, Mather's exempla offer a subtle nod to climate theory (and its attending geography-based characterology); as we saw in chapter 2, the designation "Barbarous Nations" referred to a specific geographic region, often including the Levant (and, by implication, Sodom and Gomorrah) as well as Rome and Greece.[69] Mather's repeated references to the sun and to light ("The more *Barbarous Nations* have Worshipped the *Sun,*" and "Truly, there was the *Just Wrath* of God, in this Thing, That the Nations, who had most of *Light*, whereby to *Know God*, and yet *Glorified Him not as God* should become thus *Vain in their Imaginations*") also play with the implicit racialization of the geographic region in which all of these cities were located and the association of barbarity with the torrid zones more generally.

Mather concludes this scene with a delicate figuration of the literalness by which sin is punished with sin. He declares, "They who worshipped God under the Shape, it may be, of *Bruits*, were left by God, unto the *Sun*, which made them worse than *Bruits*."[70] Referring to his characterization of "the *Egyptians*," who "worshipped a *Bull*, a *Dog*, and a *Rat*," and implicitly to his assessment of "the more *Barbarous Nations*" who "have Worshipped the *Sun*," Mather reveals what comes from worshipping "Bruits" instead of the true God: one becomes a "Bruit." But in Mather's illustration, God does not abandon these sinners to their lusts; it is not their sins that make them "Bruits." Instead, God abandons them "unto the *Sun*" (which here also recalls the object that other idolators—the "Barbarous Nations"—worshipped), and it is the sun itself "which made them worse than *Bruits*." Here, Mather offers not one but two implicit theses about how a person might become, as he put it, "Brutified": the first is by being abandoned to sin, but the second explicitly invokes climatological and stadial hierarchies of human difference to suggest the racialization and perhaps degeneration that attends exposure to the sun within the torrid zones. Furthermore, Mather implies a correlation between dark skin and "*Corporal Uncleanness of Debauchery*" or "*Vile Practices*" by arguing that being left "unto the *Sun*" "made them worse than *Bruits*."[71] He deploys climatological imagery to suggest a link between the torrid zones and the idolatry practiced therein: living in these regions meant being abandoned to "farther *Abominations*." Careful attention to the way Mather tarries on "habituation" to "unchastity" brings into sharp focus the sexual politics of racialization in this first North American compilation of execution narratives.

Creatures of Habit: The Cases of Joseph Mountain, Thomas Powers, and Samuel Frost

Whereas Belcher and Mather offer a relatively clear exegesis of the racial and sexual politics of discussions of habit at the turn of the eighteenth century, later execution narratives were rarely quite so explicit, even though in these later narratives, habit appears as an educational and behavioral question with much more frequency. In this section I track discussions of habit through a handful of particularly popular execution narratives that were printed during the last quarter of the eighteenth century: those of Joseph Mountain, Thomas Powers, and Samuel Frost. None of these narratives, nor the sermons that attend them, offers the kind of hermeneutic anatomy of habit that Mather and Belcher offer in their sermons, and the engagement of these later narratives with the cultural significance of racialization, sexual behavior, and crime arrives in a new shape.

While eighteenth-century execution narratives have received less attention from scholars of early America than have other contemporary print forms (such as the novel), the few scholars who have worked on these narratives consistently note both that people from marginalized communities were more likely to be condemned to death in sentencing and that their execution narratives (and attending sermons) were more likely to end up in print.[72] Similarly, what we might now term "sex crimes" (bestiality, rape, adultery, or infanticide—often to hide an illicit sexual liaison) received lingering attention in gallows literature from very early on. Samuel Danforth's *The Cry of Sodom Enquired Into,* cited as the first execution narrative printed in the North American colonies, details the execution of the young Benjamin Goad for bestiality; Cotton Mather's *Pillars of Salt* reproduces twelve execution narratives in the latter half of the text, seven of them concerning crimes of sex.[73] While the early-eighteenth-century press was still more or less under clerical control, by the end of the eighteenth century a significant market for popular print had developed in the early national United States, so market demand may be in part responsible for this trend.[74] The popular press was by no means a secular institutional network at the time when printers were laying the type for Joseph Mountain's narrative in 1790, but scholars of the history of the book generally concur that printers, writers, and ministers alike increasingly responded to market demand in their interventions into the content and design of gallows literature.[75] Thus, while Joseph Mountain's execution narrative contains little spiritual exegesis surrounding the origin or spiritual stakes of his criminal behavior (unlike many other execution narratives, it was published independently of the execution sermon on Mountain, and no extant publications reveal them to have been bound together), the title page itself showily promises the lurid content to be found therein, with its rather bald title, *Sketches of the Life of Joseph Mountain, a Negro . . . For a Rape,* and the subheading that appeared at the bottom of the page: "[The Writer of This History Has Directed That the Money Arising From the Sales Thereof, After Deducting the Expence of Printing, &c. Be Given to the Unhappy Girl, Whose Life Is Rendered Wretched by the Crime of the Malefactor.]"[76]

Joseph Mountain was, for a time, the most infamous highwayman in the United States, and his execution narrative saw at least four editions in the immediate aftermath of his death.[77] Born July 7, 1758, in Philadelphia, Mountain was raised an indentured servant in the house of Samuel Mifflin, a Presbyterian (Samuel's son Thomas became the first governor of Pennsylvania in 1790), and was taught to read and write.[78] Mountain's father was a "molatto" living in Philadelphia, and his mother had formerly been enslaved. Mountain left Philadelphia in 1775 at age seventeen to work aboard a ship, having successfully petitioned his master to allow him to do so. Mountain then sailed to London and promptly deserted his company, and it is in London that his story truly begins. Falling in with two men named Wilson and Hyde, who initiate Mountain into the ways of highway robbery and get him invited into a company of "footpads" (people who rob others on foot) and highwaymen (who commit robbery on horseback), he quickly comes into his own as a self-described "prodigy in vice" and begins an illustrious career as a thief.[79]

In each edition of Mountain's narrative printed in 1790 (the editions vary in length from roughly a dozen to about thirty-two pages), the reader follows Mountain as he robs ostensibly hundreds of people and, furthermore, robs them with style. Mountain is very much a highwayman of the British print tradition; he is a folk hero, and his story is an adventure tale. Daniel Williams argues that the narrative's popularity derived from precisely its generic novelty: "Readers had never before encountered a character like Mountain. He was the first English highwayman shaped for an American audience, and he was the first to so clearly embrace his outsider status."[80] In a particularly memorable episode, while riding through Warrington searching for potential "object[s] of plunder," Mountain came across a gentleman and asked him for the time. The gentleman "drew his watch, and told [Mountain] the hour." Mountain recounts the short conversation that ensues: "I observed, 'You have a very fine watch.' He answered, 'Fine enough.' 'Sir, 'tis too fine for you—you know my profession—deliver.'" The gentleman, remarkably, refuses and attempts to ride away, but Mountain grabs the bridle of his horse, brandishing a pistol, and promises him, "Deliver—or I'll cool your porridge." In another scene, Mountain glibly mocks an unfortunate Quaker while he robs him outside of Manchester, England, poking fun at his plain speech and refusal to take oaths. Mountain recalls, "It was nearly 9 o'clock when I met him. I enquired if he was not afraid to ride alone. He answered, No. I asked him his religion; he replied 'I am a Friend.' I observed, 'You are the very man I was looking for—you must deliver your money.' He seemed very unwilling, and said, 'Thou art very hard with me.' I replied, 'You must not *thou* me.' He then gave me his plain gold watch, 6 guineas, and four bank notes of 20l. each." Mountain gleefully recounts his own actions after robbing the man: "To compleat the iniquity, and exhibit the extent of my villainy I then took a prayer-book from my pocket, and ordered him to swear upon the solemnity of God's word, that he would make no discovery in twelve hours." The Quaker declines to take the oath for religious reasons, but Mountain clearly delights in embarrassing his victims. After robbing one Thomas Reevs, he comments, "To insult him in his distress, after committing the fact, I pulled off my hat, made a low bow, [and] wished him goodnight."[81] Mountain does not invoke the language of habit to describe his exploits, but nonetheless he resorts to an at times almost listlike, repetitive narration of his robberies. The entire account is rendered in the key of this narratological bravado, and more than two hundred years later, it is not difficult to imagine why this execution narrative was so popular.

There is, of course, no way of knowing whether, or to what extent, *Sketches of the Life of Joseph Mountain* reflects the experiences or voice of Joseph Mountain, the actual person who was convicted of rape and executed in New Haven. While Mountain's narrative stands out among contemporary gallows literature in part because of its distinct writerly voice, criminal accounts attributed to the condemned were generally produced within the context of, as Caleb Smith puts it, "coercion, collaboration, and ventriloquism."[82] Whether or not the narrative bears any relationship to the life of the actual Joseph Mountain, this publication was enormously popular, and accounts of both Mountain's early life of crime and the rape for which he was ultimately executed were reprinted in newspapers as far away as New York City.[83]

American slavery looms large in Mountain's narrative and clemency petition, both of which are rendered in the first person. Few execution narratives address or even gesture to the realities of enslavement in the first person, but this narrative is quite explicit about the effect of slavery on the understanding of racial difference—and alliance—in the United States. While in England, Mountain marries Nancy Allingame, an eighteen-year-old white woman who possesses about five hundred pounds and a house in Islington; he exhausts her financial resources before leaving her six months later. In some editions of Mountain's narrative, the statement pertaining to his marriage is followed by a note: "The reader will please to recollect, that Negroes are considered in a different point of light in England, from what they are in America. The blacks have far greater connection with the whites, owing to the idea which prevails in that country, that there are no slaves."[84] Indeed, Mountain is keenly aware of how differently he is perceived in England (and not only because of the practice of slavery), and this may in fact be what leads to his fatal crime and eventual execution.[85]

Upon returning to the United States in 1790, after an absence of ostensibly more than twenty years, Mountain arrived in Boston and immediately set out for New York, stopping in Hartford on the way. In Hartford, he stole five dollars "from the cabin of a sloop lying in the Connecticut river" and was

> immediately apprehended, carried before George Pitkin; Esq. and adjudged to be whipped ten stripes. The sentence was executed forthwith, and I dismissed. This was the first time I was ever arraigned before any court. No event in my antecedent life produced such mortification as this; that a highway-man of the first eminence, who had robbed in most of the capital cities in Europe, who had attacked

gentlemen of the first distinction with success; who had escaped King's-bench prison and Old-Bailey, that he should be punished for such a petty offence, in such an obscure part of the country, was truly humiliating.[86]

Mountain's first experience of capture and conviction was, in legal terms, relatively uneventful; he was caught, arraigned, and whipped ten stripes. Neither jailed, fined, nor held, he was, at least according to the narrative, dismissed almost immediately, such that he arrived in New Haven the following evening. Despite the lack of legal ramifications, however, the incident appeared to injure Mountain's pride. Indeed, his stated embarrassment at his arrest in East Hartford smacks of something like cosmopolitan snobbery: not only does he get caught, but he bemoans getting caught "in such an obscure part of the country." In England, Mountain's criminal prowess had made him almost famous—he recalls returning to the headquarters of his fraternity of highwaymen after a particularly successful trip, fondly reminiscing: "I never shall forget with what joy I was received. The house rang with the praises of Mountain"[87]—but in small-town rural Connecticut, he is arrested, whipped, and humiliated.

This episode directly precedes Mountain's account of the crime for which he is eventually executed, the rape of a young white woman.[88] According to the narrative, the rape actually occurs the day after he is dismissed from his arrest and whipping in East Hartford. Importantly, Mountain insists that he did *not* in fact rape the young woman, although he admits that he taunted her and humiliated her in much the same way he had mocked the Quaker he robbed. He recounts:

> That I abused her in a most brutal and savage manner—... her tender years and pitiable shricks were unavailing—and ... no exertion was wanting to ruin her, I frankly confess. It was a most cruel attack upon an innocent girl, whose years, whose entreaties must have softened an heart not callous to every tender feeling. When her cries had brought to her assistance some neighbouring people, I still continued my barbarity, by insulting her in her distress, boasting of the fact, and glorying in my iniquity. Upon reflection, I am often surprised that I did not attempt my escape; opportunity to effect it frequently presented before I was apprehended. Yet, by some unaccountable fatality, I loitered unconcerned, as tho' my conduct would bear the strictest scrutiny.[89]

The overlap with Mather is instructive. While Mountain's description of this "most cruel attack" is made with something like bravado—he does appear to be bragging of his own callousness—this utterance also invokes some of the rhetoric in Mather's *Tremenda* and *Pillars of Salt*, surrounding the way "custom in sin" hardens the sinner to the consequences of his or her actions and deafens the person to the intervention of God's grace or the community.

But perhaps more intriguing is the question that arises in twenty-first-century readers, graced with the hindsight that comes with a historical awareness of the disproportionate persecution and prosecution of black men (especially when it comes to cases of sexual assault against white women): how could Mountain ever assume that he would *not* be suspected for the crime, even as early as 1790, and even in rural Connecticut? This scene should be viewed, I believe, as a moment of cultural mistranslation, where Mountain presumes he will be perceived one way—as an innocent bystander—only to find himself perceived in a radically different and explicitly racialized light. The surveillance and criminalization of people of color was not a new or rare thing in colonial New England (see chapter 5), and a sensationalized association between blackness and cupidity had long been established in natural history, among other eighteenth-century sciences. Mountain's narrative thus leaves us with a strange question: was it possible that Mountain, twenty years absent from the United States, having come into his adulthood in England, a society with a markedly distinct culture of racist etiquette, was somehow not aware that he would immediately be perceived as the culprit in the rape at hand? Differently put, was he not aware of the cultural habits of racism—the "association of ideas," as Witherspoon might have it—that rendered the story of rape a story that frequently governed any relation between black men and white women? The association of ideas that tied blackness to sexually aberrant behavior was not limited to print and was not an incidental, abstract, or merely theoretical way of understanding the relationship between race and sexuality. Again, habit was both explanatory and predictive, and as a logic that sutured specific, usually criminalized sexual acts to racialization, as a way of thinking, habit had a direct effect on how people of color in particular were understood under the law and penalized. Mountain was remarkably ignorant of how he would be perceived; from the witnesses who happened upon him to the judge who sentenced him, Mountain looked like a rapist, regardless of the facts of the case.

If Mountain did not cast his own criminal history—or the rape for which he was convicted—in the language of habit, the public response to his arrest and conviction certainly did. For James Dana, the minister who penned his execution sermon, Mountain's "habits of sin and excess" were precisely what rendered the death penalty a necessity in the case. Dana, a fifty-five-year-old minister of the First Church in New Haven, composed Mountain's execution sermon, *The Intent of Capital Punishment,* against the backdrop of a major anti-capital-punishment movement that had been agitating against the death penalty for years. The sermon stands out as a relatively rare instance of ministerial apologia for what was otherwise rapidly becoming an unpopular cause.[90] *The Intent of Capital Punishment* also directly responded to Mountain's own insistence on capital punishment as the prerogative of God alone, which Mountain articulated in a clemency petition he submitted to the Connecticut Assembly on August 15, 1790, five days before his death. In it, Mountain questions the right of the state to sentence a man to death and offers to be sold back into slavery or castrated as a commutation of his sentence.[91] Dana's sermon draws heavily on William Blackstone's definitive writings of two decades earlier on the definition of capital crimes and justifies execution as an unfortunate necessity to ensure the health of the state: "Those who are so depraved in their moral character that they can neither be cured nor endured, are in the civil state what a mortified member is in the natural body. As such a member must be amputated for the preservation of the body, so persons of such depravity must be cut off for the preservation of the state."[92] For Dana, Mountain represents precisely the kind of "mortified member" that endangers the health of the community, and he offers Mountain's long history of criminal offenses as the most compelling evidence for this claim.

Like many ministers before him, Dana emphasizes Mountain's early criminal acts as a narrative technique to contextualize the rape of which he is accused, to illustrate the spiritual hardening that necessitates his execution, and to warn young people in the community against making the same mistakes. Dana expounds: "He began his career of wickedness at or before *seventeen*. This is the most critical and dangerous stage of life. . . . See in this instance of what moment it is to form early habits of virtue—to hear counsel, and turn at reproof—to shun the paths of vice. . . . Small sins lead to great ones."[93] In the notion that "small sins lead to great ones," we again hear Mather, as well as the anonymous author of *Onania,* in their warnings against childhood masturbation; it bespeaks an understanding of sin as progressive and as at least partly a behavioral phenomenon. Dana exhorts his congregation to build "habits of virtue" early in their lives as a preventive measure, one much less arduous than "the much more difficult business of reforming and renouncing habits of sin and excess."[94] Furthermore, it is not clear that Dana actually believes that "habits of sin and excess" can be effectively reformed or renounced, as he warns his congregants, "When sinners have assumed the boldness to leap over the boundaries of laws human and divine, they gradually lose all tenderness of conscience, and proceed from evil to evil."[95]

The solution, then, is obvious: Dana exhorts his community to abstain from evil deeds in the first place, but this exhortation assumes what by now we recognize as a familiar form: quoting the book of Jeremiah, Dana admonishes: "Resist the first solicitation, as he, 'How can I do this great evil, and sin against God?' Whenever tempted, allow time for this reflection . . . 'Can the Ethiopian change his skin, or the leopard his spots? Then may those who have been accustomed to do evil, learn to do well.' "[96] Unlike Mather, who cites this verse from Jeremiah in *Tremenda* and recasts the Ethiopian as the "Ethiopian in Sin,"

Dana offers no exegesis of the passage; it is intended to be self-explanatory. It cynically suggests not only that habitual behavior makes it impossible to "do well" but that habitual sin renders a person irredeemable, unable to even *learn* to do well later. This reasoning, of course, justifies "cut[ting] off" Mountain.

But like Mather's "Ethiopian in Sin," Dana's "Ethiopian" simultaneously offers a figurative rendering of the habitual sinner and refers directly to the "Ethiopian" (a generic term that was used to refer to people with dark skin through the end of the eighteenth century) who was about to face the gallows: Joseph Mountain. While this passage of Jeremiah does occasionally appear in other execution sermons about white men or women, usually as the introduction to a discussion of criminal habits, it is certainly no coincidence that Mather and, later, Dana use this verse in relation to the consequences of habitual sin in black men. And we see, in their use of this verse, how blackness itself becomes consonant with being "accustomed to do evil" and begins to exhibit some of the racist characterological ideologies with which it would become so tightly bound in the nineteenth century.[97] Indeed, blackness here appears as the very monument to habit, denoting a lack of self-governance and a lack of capacity for improvement or sovereignty. Cotton Mather himself actually implies as much more than eighty years beforehand in his *Negro Christianized* (1706), a sermon exhorting his congregation to educate their black servants; he muses: "It may seem, unto as little purpose, to *Teach,* as to *wash an Aethopian.* But the greater their *Stupidity,* the greater must be our *Application.*"[98] Ironically, numerous scholars of the history of race have demonstrated how skin color, and blackness in particular, was understood to be "transformable" during the eighteenth century; despite Mather's unoptimistic assessment of the possibility that blackness could be washed off, the mutability of race was a live question for many of Mather's contemporaries.[99] And as we have seen, habits of living or one's state of society were thought to wield an enormous influence on one's racial status.[100] In execution sermons, however, habit was not a dynamic figure so much as a predictive or diagnostic one. "Habits of virtue" might prevent the development of habits of vice, but habits of vice, especially when established during the impressionable childhood years, were nearly impossible to change.

Published only six years after Mountain's narrative, Thomas Powers's execution narrative capitalizes on the increasingly powerful association between blackness, sexual incontinence, and vicious habits.[101] The title page of Powers's execution narrative, *The Narrative and Confession of Thomas Powers, a Negro . . . for a Rape,* like the title page of Mountain's, sensationally advertises that the accused is black and that he is accused of rape. Powers's narrative, however, does not offer readers a charming if remorseless folk hero whose exploits resemble some of the most popular picaresque fiction of the day. Instead, his narrative tells of a "naturally vicious" boy raised by good parents, educated, but corrupted in his youth and unable to submit to even the most basic practices of sexual continence or self-restraint. Powers recounts in his narrative raping a Lebanon, New Hampshire, woman without apparent premeditation or consideration and then returning home to the family for whom he worked to play checkers with their children.

If Mountain was made evil by "custom in sin," Powers, at least according to his narrative, was born that way. In his first-person account, Powers recounts: "[My father,] being a very pious man, endeavored to instruct me in my several duties, to God, to my parents, and to all mankind; far as my young and tender mind was capable of receiving any virtuous impressions. But I was naturally too much inclined to vice, to profit by his precepts or example; for I was very apt to pilfer and tell lies, if I thought there was any occasion." He uses the idiom of the "natural" to explain his illegal behavior at three points throughout the ten pages of his short narrative, narrating that he was "naturally too much inclined to vice," "naturally vicious," and "naturally inclined to be light-fingered."[102] Unwittingly invoking the double valence of the term "natural" (the unconverted or unbaptized "natural man" of Belcher's sermon and the indigenous and putatively uncivilized "natural man" or "state of nature" seen in Rousseau, Locke, and in other naturalist philosophy), Powers's narrative implicitly links blackness to innate viciousness.[103] Although the sermon delivered by Noah Worcester on the day of Powers's execution returns again and again to the figure of habit in his warnings to the young people of his congregation, Powers's narrative suggests that viciousness is a component of blackness, a narrative technique discussed at greater length in chapter 5.[104]

What is interesting about Powers's narrative is that the events he describes, without their narrative frame, differ very little from those in Mountain's. He also starts stealing at a young age and furthermore describes being introduced to sex at age nine by a "young Negro woman" (ostensibly another servant in the house of Isaac Johnson, the man to whom his father had bound him out). The events in question would square nicely with the kind of narrative strategy surrounding the development of "custom in sin" or "habits of vice" that is so prevalent in execution narratives throughout most of the eighteenth century. In Powers's narrative, however, the trajectory takes a different shape. It is not vicious behaviors or the early development of bad habits that renders Powers vicious; the narrative argues that his viciousness is innate, not habitually acquired. In this narrative we again see the vexed and multivalent status of habit as an explanatory tool, as well as its confusing relationship to both the internal and the external. Powers's bad habits are represented as stubborn reflections of his innate badness; even his pious father's attempts to teach him otherwise necessarily fail. Furthermore, Powers's viciousness appears to be specifically tied to sex. For example, although he admits that he was "naturally too much inclined to vice, to profit by his [father's] precepts or example" by the age of only two or three, by Powers's own account the actual enactment of his own viciousness does not seem to begin until he is nine. He states: "When I was nine years old, I was put out to live with Isaac Johnson ... and very early began the practice of villainy and debauchery. It was here I began my career in the gratification of that corrupt and lawless passion which has now brought me to the threshold of eternity, before my years were half numbered." Although Powers offers an assessment of himself as naturally inclined toward vice, the primary form that his vice takes throughout his narrative is "that corrupt and lawless passion"—lust—the practice of which he first learns from the older servant woman who lived and worked in the same house: "Being one Sunday at home from meeting, with nobody but a young Negro woman, who lived in the house, she, enticing me to her bed where she was sitting, soon taught me the practice of that awful sin, which now costs me my life."[105]

When he becomes an adult, Powers's sins simply grow more numerous and more egregious, culminating in the rape of a woman in his community. And the link between blackness and viciousness only grows more apparent as the narrative continues. By the time Powers moves with his master, Isaac Johnson, from Norwich, Connecticut, to Lebanon, New Hampshire, in 1793, he admits "committing a number of crimes, which black as I am, I should blush to repeat."[106] In this statement, Powers's blushing suggests that he is ashamed of his behavior even *despite* his natural viciousness. While he likely is playing with the double meaning of blackness (a spiritual blackness and the blackness of skin), in this statement we see an explicit acknowledgement of blackness as a physiological state that predicts a moral one, as both an explanatory and a predictive figure.

This innate viciousness is only confirmed by Powers's own reported lack of thought before the rape to which he admits in his narrative and by his lack of concern in the wake of committing it.[107] The incident began when he came across an acquaintance while en route to a wrestling match in early December 1795. Powers nonchalantly reports, "After a little querying with myself, and finding nothing to oppose, but rather the devil to assist me, I determined to make an attempt on her virgin chastity." He succeeded in his design, "then left her, and went to the place proposed." He concludes this anecdote by noting, "But the evening being far spent, I returned to my master's house and sat down, as usual, to play chequers with the children."[108]

This chilling final detail weds Powers's alleged viciousness to thoughtlessness or irrationality, his irrationality to ungovernability or lack of restraint, and the lack of restraint or incontinence to blackness itself. As Daniel Williams neatly argues, "throughout the narrative, incontinence and disobedience were linked; not only was the former a form of the latter, but sexual excess, as a lack of restraint, also represented an inability to restrain the self in all areas of social participation, especially within the family."[109] Powers's supposed innate badness is belied, however, at the end of his narrative when, after being convicted and sentenced to death, he asks permission to send an apology letter to the woman he raped. And the supposed naturalness of his evil disposition is further undermined by his report, "Upon hearing my sentence, it had no impression on my mind—for my heart was hardened beyond description."[110] That his heart was hardened recalls the logic of the progressive nature of habitual sin evoked by Cotton Mather, James Dana, the anonymous author of *Onania*, and other writers considered in this chapter; a heart that was "hardened" was once softer. But in Powers's narrative, we see a crucial difference. This hardened heart, paired with Powers's own account of his natural viciousness, suggests that his hard heart is a feature of his innate badness rather than a disposition he acquired through bad habits. Here the fixedness of Powers's viciousness offers a bizarre reflection—or anticipation—of the kind of fixity that would characterize later biological and anatomical understandings of blackness in the nineteenth century, theories of racial difference that saw blackness as both a physiological and characterological state.

Perhaps nowhere is the persistent association between cultural discussions of habit, sexual incontinence, and racialization more clear than in execution narratives that feature white people. While these narratives were certainly an important vehicle for cementing the association between non-European peoples and crime (in part because the majority of execution sermons and narratives that were selected for publication featured the stories of crimes committed by immigrants and people of color), habit provided a broadly portable framework for explaining the path that led many men and women to the gallows. The lives and vicious courses of white people were also sometimes explained as the result of the early development of bad habits, and in these sermons the racialization of habit appears especially stark. In the sermon that Aaron Bancroft delivered at the execution of Samuel Frost, a white man convicted of killing both his father and, later, the ward of his own estate, Bancroft refers to Frost as a "savage," fully governed by the habits of sin that eventually led to Frost's crime and execution.

Unlike Joseph Mountain's and Thomas Powers's execution narratives, Samuel Frost's was published and distributed only as a broadside, likely on the day of his execution. Printed in advance so that it could be sold and distributed at the execution, the narrative is a poster-size document (53 × 44 cm) containing a woodcut illustration of the gallows, Frost's confession in the first person, a third-person account of Frost's life and crime, and a verse titled "Poem, on the Execution of Samuel Frost."[111] The broadside would have augmented Bancroft's execution sermon, given on the day of Frost's death. Importantly, Bancroft's

The Confession and *DYING WORDS* of SAMUEL FROST,
Who is to be Executed this Day, October 31, 1793, for the
Horrid Crime of MURDER.

I WAS born at Princeton, in the County of Worcester, and Commonwealth of Massachusetts, on the 14th day of January, 1765. My father's name was John Frost, he had four sons the eldest of whom died when he was nineteen years old—two of my brothers are yet living. My mother is dead; I always regarded her, and ever thought my father had no affection for her, and that he used her ill; this induced me to kill him, which deed I own with a handspike, and then beat his brains out. I was apprehended and committed to goal in Worcester, and tried before the Supreme Court, and was acquitted contrarily to my expectations.

My mother died when I was about fourteen years old, and I always supposed her death was occasioned by the bad treatment she received from my father. He was very churlish, and was void of all affection for his family.

After I was acquitted, and got released from goal, I went and lived with Mr. George Parks; afterwards with Mr. John Gleason, then with Mr. Phinehas Gregory; after living with them, I went and lived with Mr. Ezekiel Sawin, then went a second time to Mr. Parkis, after this I lived with Mr. Jesse Fisher, and then with Mr. Rice. All these people lived in Princeton. Whilst I lived with Mr. Rice, I arrived at twenty one years of age; I left Mr. Rice, and went to live with a relation, Mr. Benjamin Wilson, where I tarried about two or three weeks, and then went to Capt. Elisha Allen's, where I lodged one night. From Capt. Allen's I went and lived a short time with Mr. Solomon Parker, and then with several other persons, but was only a little time with each—then I went to live with Mr. Ephraim Osgood; I left him and went to the house where I formerly lived with my father, and tarried there about five months, in the year 1786; but receiving an affront, being told I do something against my inclination, I went off without taking leave, and took to the woods—I wandered about for three days and nights; In the day time I kept in the woods, and at nights went to farms and lodged in barns, unperceived. In the woods I got and eat berries, and gathered apples in orchards, and on them I lived during the three days and nights.

Being tired of living in the woods, I went to Fitzwilliam, to the house of Benjamin Wilson, my relation, beforementioned, who had removed from Princeton to that place. I arrived there on the 22d day of August, 1786, at breakfast time.

I was very hungry and eat heartily.—I tarried here until the 5th day of September, the same year.

Soon after this time a number of people were going to Worcester, to stop the sitting of the Court—they asked me to go with them, and treated me. I wanted to see what they would do, and went with them as far as Holden—at this place, stopping at Davis's tavern, I went out to pick some apples to eat; after which I laid down on the ground and went to sleep—when I awoke, I thought I was doing wrong to go with those people to stop the Court, and would not go with them any farther. I left them and went to Princeton—there I met with more people going to Worcester for the like purpose—they gave me some drink. I left them, and went to Mr. Stephen Merrick's where I lodged that night. I then worked for several persons for a few days; and on the 11th day of September, 1786, I went and lived with Capt. Elisha Allen's—who took me because it was the desire of a number of people. It was town meeting day, and I went to the meeting.

I lived with Mr. Allen some time—when choosing to go away—I went off at a time when I was sent to pasture with the cows—I was gone almost three days, living as I could—I spent three coppers whilst I was gone, and went almost as far as Boston—then I returned to Capt. Allen's again. I went off several times afterwards, and was absent sometimes a longer and sometimes a shorter time, but did not say any thing by going away but flogging when I returned. Considering myself as a slave, I have thought I had as well die as live as I did.—I had a small estate and wanted to work on that, but I could not.—Mr. Allen had the care of my estate, and I supposed was paid for my living with him out of it. I thought several times I would kill Mr. Allen, and then thought I would not.— At length, on the 16th day of July, 1793, I effected it thus—Capt. Allen was going to set out cabbage plants, and ordered me to go with him, and to get a hoe; he went to the spot for planting, and I after the hoe; when I returned with the hoe, I found him stooping to fix a plant—I then thought it would be a good time to put my design in execution, and accordingly went up to him and gave him a blow with the hoe on his head, and repeated it. When I was about giving the third blow, he said, "Forbear Sam, you have done enough." I made no reply, but continued repeating the blows until I supposed he was dead. I had beaten his head so as it had made a large hole in the ground and his brains came out. After I had finished him,

I went into the woods not far distant, where I remained four days, living on berries, &c. Whilst I was in the woods, I heard the voice of the people who I supposed were after me; I heard them call me—but they did not find me; at length I was tired of being in the woods, as it was hard fare—came and loitered about a house where I had formerly lived, and was discovered by some of the family, who took me in to the house and gave me some victuals, and secured me. I was then brought to Worcester gaol again, and now shall certainly be hanged. I declare that I always have had a great aversion to stealing and telling lies, and think them to be great crimes.—I always meant to tell truth, and never stole, excepting taking a few apples from orchards may be called theft.

The foregoing account was taken from the mouth of Samuel Frost, in prison.

ACCOUNT of SAMUEL FROST.

SAMUEL FROST was certainly an extraordinary character—his mind was evidently not formed altogether like those of other persons.— He thought it no great crime to kill such as he supposed treated him very ill—and did not appear to have a just conception of the heinous crime of Murder.—His education appears to have been very indifferent, but his natural capacity in many respects seemed to be equal to persons in general, whose minds have not been cultivated. He had high notions of honesty, and appeared much offended when his honesty was suspected.—He appears also to have been a person who regarded truth; and he valued himself upon his probity and sincerity.—One striking proof of his dislike to falsehood appeared when he was induced before the Supreme Court—he plead guilty—he was told that he might plead not guilty, was urged so to do, and was remanded to prison, in order that he might consider of the plea he had made, and retract it, &c. yet notwithstanding on his being again brought into Court, he persisted in pleading guilty. He was sensible of favours granted him, and expressed his gratitude for them—Yet, he had a most savage heart which nothing could meliorate, and he would talk with the same calmness and composure of the horrid Murders he had committed, as though the persons who fell a sacrifice to his fury, had been of the brutal creation. He could read and write, and often was found, whilst in prison, reading in the Bible—yet he shewed no signs of contrition especially for the unnatural murder of his father. Notwithstanding he read the Bible, he was not fond of conversing with the Clergy, and in general of the many who

visited him, few of them could get him to talk with them. He went two sundays to meeting after his sentence, but more through persuasion than inclination; and though much urged refused to go again.—On the first Sabbath that he attended divine service he appeared to be offended with the minister, because he mentioned the murder of his father. Frost said he did not like to be twitted of that—that it was an old matter, and was settled long since.—On being asked by the High Sheriff if he wished to have a sermon preached on the day of his Execution, he answered that he did not care any thing about it—he said he might not be brought back to prison, but be carried from the meeting house to the place of execution.

He had been known to say, that his killing Capt. Allen was rather more than he ought to have done—and that he would, if it was not done, stay seven years first. He was a most dangerous person to society.—On being asked if he had his liberty, if he would kill any other person, he answered there was one more he believed he should. He told some persons who visited him one day, that he believed his father and Allen had a very tough time of it—Being asked why he thought so, he said he had been beating his head against the walls of the prison, in order to know how they felt whilst he was killing them.

He appeared to have a very confused idea of a future state—supposed he should go to Purgatory—and said he believed the Devils wore large black wigs—and many other such chimerical expressions of folly and absurdity he uttered respecting a future existence.—He did not express pleasure that the time of his execution was fixed at so distant a period—he wished, he said, to have it over a fortnight sooner. On the whole, so a man, he was savage—void of all the finer feelings of the soul, and destitute of the tender affections of filial love and gratitude—He appears to have been a being cast in a different mould from those of mankind in general, and to be the connecting grade between the human and brutal creation.

The above will enable the reader to form some idea of his rational faculties.

He was about five feet four inches high, rather slenderly built, and very strong. He had a peculiar way of tossing or twitching his head, and his countenance was very unpleasant.

Printed and sold at Mr. Thomas's Printing-office, in Worcester. Price 6d. Also, A Poem on the Occasion. Price 9d.

sermon does not appear on the broadside, as the broadside was printed in advance of the execution.

Frost's execution was something of an event, in its moment. And for twenty-first-century readers, it is easy to see why; his case is a bizarre one, even by our standards. While the crime for which Frost was to be executed was the murder of Captain Elisha Allen, a resident of Frost's hometown of Princeton, Massachusetts, who we believe controlled Frost's estate, the crime was more sensational because it was not, in fact, Frost's first murder. Ten years earlier, at age eighteen, Frost had killed his own father, and in much the same way that he killed Captain Allen. He "beat his [father's] brains out" with a handspike in September of 1783 while they were digging a ditch together; in July 1793 he beat Captain Allen to death in a cabbage patch with a garden hoe, until "his brains came out." Acquitted of the murder of his father by reason of insanity, Frost had spent the next ten years staying with various members of the Princeton community, working in their homes and on their lands before returning to his familial home under the care of Captain Allen. While all accounts of Frost's life and crime insist that he had sound "rational faculties," the facts of the case—he had earlier been found not guilty by reason of insanity, and his family's estate had been assigned by his community to an executor (Allen) rather than being left in Frost's hands—suggest that Frost's legal competence was definitively in question.[112]

Spectators flocked from miles around to witness Frost's hanging. Contemporary newspapers estimate that two thousand people attended the execution, speaking to the public interest in, and knowledge of, the case. And, as was standard for executions, a local minister was selected to give a sermon, so on October 31, 1793, Aaron Bancroft delivered his sermon, *The importance of a religious education illustrated and enforced,* just before Frost was, in the expression of the day, "turned off."

In his sermon, Bancroft locates the root of the accused's profligacy not in his innate depravity or inner evil, but rather in his lack of early religious instruction, which led to the development of "habits of sin."[113] Reminding listeners that "it is easier to *prevent* than to *cure* habits of vice," he exhorts his audience to attend to the religious training of its youth:

> Early impressions are deep; they are influential. We are all sensible of the influence of principles, that were implanted in our minds before reason had attained its strength. We are all sensible with what difficulty we conquer the influence of opinions and persuasions that, through false methods of education, were impressed upon our infant minds. . . . What salutary effects, then, may not be expected from early impressions of piety, and right moral principles of action, which will receive increasing strength and support from the decisions of a sound

judgment, the dictates of an enlightened conscious, and from every habit of just reflection?[114]

For Bancroft, childhood is a period of critical vulnerability to the impressions of the outside world. By developing proper "habits of just reflection" through religious education, children are armed with the tools necessary to stave off the spiritual or moral contamination that might attend exposure to bad examples or otherwise negative "opinions and persuasions." Forgoing such education could be fatal. "Look among those," Bancroft intones, "whose first impressions were made amidst scenes of vice, and who received their education under examples of sin, whose first essays of reason and activity were amidst scenes of corruption and profligacy, and you will find degrees of youthful depravity, at which you must be astonished; and the recovery of youth, of this character, from the habits of sin, is extremely difficult and altogether improbable."[115]

Bancroft needed no more convincing evidence for this point than Samuel Frost, the man sitting before the audience preparing for his own death later that day. Unlike Mountain or Powers, Frost was, in twenty-first century parlance, a white man, and he was, we suspect, propertied; Captain Allen had control of his estate, suggesting that Frost did in fact have an estate. And unlike the misdeeds of which Mountain and Powers were accused, Frost's crimes—the murder of Captain Allen and the patricide ten years earlier—did not involve sex.

Yet over the course of Bancroft's sermon—and throughout the "Account of Samuel Frost" that appears on the broadsides sold at his execution—Frost is depicted in heavily racialized terms that align him with "brutal creation" and the lower passions more generally. Differently put, the way Bancroft and the author of the "Account of Samuel Frost" understand him to be controlled by his bad habits means that Frost is represented in the racialized and sexualized rhetoric of habit that we have been exploring in this chapter. Frost, a white man, becomes a "savage," ruled by his "passions and inferior desires."[116] In this sense he is represented on a continuum of racialized criminalization that aligns him with the bonded Mountain and the enslaved Powers.

When Bancroft's sermon turns from its stated topic—the importance of early religious instruction—to address Frost himself, Bancroft is unambiguous about what he believes are the origins of Frost's "dissocial, his cruel and malignant disposition": "For an answer to our question we are led to the early period of his existence."[117] Between Frost's lack of early instruction and the negative example of his "cruel and barbarous" father, "all the rough, malignant, and revengeful passions, acquired strength, and obtained an habitual control of [Frost's] mind."[118] Depicting Frost as controlled by his passions, Bancroft's sermon continues to deliver an explicitly civilizationalist rhetoric that aligns Frost with animals and unconverted humans; in other words, Bancroft represents Frost in the same kind of language that natural historians like Thomas Jefferson deployed in descriptions of Native Americans, as atavistic forebears of modern, civilized man. Addressing Frost directly, Bancroft admonishes him, "You have committed injuries in society, for which you can make no reparation. You have been guilty of a crime, which a distinguished heathen nation esteemed so unnatural, *as to render a name for it unnecessary.*"[119] Bancroft is referring here specifically to patricide,[120] in framing Frost's crime as not only a sin but as a civilizationally backward crime considered unnatural even by "heathens." To emphasize the point, Bancroft continues, defining murder as "a crime capital in all civilized governments," thus pointing to Frost's implicit lack of civilization.[121] Following Erin Forbes, we see in Bancroft's sermon a portrait of Frost as a failure of liberal personhood, too bound by his lower desires to agree to be bound by the law, unable to embody the self-sovereignty or sensibility necessary to be a citizen and remain in the community.[122] Painting him as an atavistic relic of the new nation's indigenous "past," Bancroft offers a dire assessment of Frost's civic and spiritual maladjustment: "Yet in a country, where all men enjoy those advantages for moral and religious improvement . . . we behold him a savage, possessed of the most malignant and revengeful passions."[123]

The civilizationalist rhetoric that Bancroft uses to describe Frost is echoed even more explicitly in the anonymous "Account of Samuel Frost" that appears after Frost's putatively first-person account on the broadside. The writer tarries at length on Frost's mental state in the "Account," opening with "Samuel Frost was certainly an extraordinary character—his mind was evidently not formed like those of other persons." Importantly, Frost's "extraordinary" mind is here, too, depicted as existing in a natural state; the writer muses, "His education appears to have been very indifferent, but his natural capacity in many respects seems to be equal to persons in general, whose minds have not been cultivated." Both Bancroft and the writer of the "Account" insist that Frost is mentally able but not cultivated, a slippery if familiar distinction that echoes ethnographic writing about Native American peoples. The question of Frost's soundness of mind is in fact left to the reader; the "Account" concludes by simply stating, "The above will enable the reader to form some idea of his rational faculties." The idiom of liberal personhood girds the "Account," and measured against the humanist metric of the rational, Frost again and again comes up short.

His shortcomings, according to both Bancroft and the unnamed author of the "Account," are due to his lack of training in the "habits of reflection" that Bancroft names as so critical to right upbringing. This is not speculation, for we see the anomalies inherent in Frost's pattern of reflection reverberating through his entire first-person narrative in the first and second columns of the broadside. His decision to murder Captain Allen is itself presented as an incidental decision; the narrative muses, "I thought several times I would kill him, and then thought I would not." Ultimately, Frost's murderous resentment toward Captain Allen stems from a cause that would have been only questionably reasonable at the time; Frost does not like working for him and does not like being held to standard labor expectations, choosing instead to disappear for periods of time to live in the forest and subsist on fruit and vegetation. He grumbles, "I . . . did not get any thing by going away but a flogging when I returned. Considering myself a slave, I have thought I had as well die as live as I did." Again, this moment implicitly points to a lack of reasonable reflection on Frost's part. He admits that he is getting paid for his labor on his own land (controlled as it was by Captain Allen), and he clearly knows what a slave is and that he is not a slave. If Frost had not murdered Captain Allen, we could perhaps read his statement as hyperbole or a figure of speech, but there is something grim about this statement, as it appears in Frost's explanation of the murder; instead of continuing to live in a way in which he would "as well die as live as I did," he killed Captain Allen.

The "Account" even includes a rather devastating vision of Frost's attempt at reflection as he both emotionally and physically imagines what both his father and Captain Allen had felt as he beat their brains out: "He told some persons who visited him one day, that he believed his father and Allen had a very tough time of it—Being asked why he thought so, he said he had been beating his head against the walls of the prison, in order to know how they felt whilst he was killing them." Unable to sympathize in a speculative or reflective way with his father and Allen, we see Frost sympathizing nonetheless (Forbes terms this Frost's "radical empathy")[124] by beating his own head against the walls of his prison cell, trying desperately to emulate the feelings his victims had experienced. Here we see yet another vision of Frost's subordination to his habits; unable to stop beating people's brains out, he continues, even while incarcerated, to attempt to beat a head in, even when that head is his own. Thus, across these several representations of Frost's reflectiveness—or lack thereof—we find accounts of his thinking that illustrate his "natural," uncultivated state.

Frost's bad habits—or simply his lack of good ones—here demonstrate how a potential citizen, an ostensibly propertied white man, is able to join the racialized ranks of natural man. Indeed, the specter of degeneration is everywhere in this narrative, glossing what twenty-first-century readers might otherwise speculate to be Frost's disabled or neurologically diverse status. The "Account" offers this terse assessment: "On the whole, as a man, he was savage—void of all the finer feelings of the soul, and destitute of the tender affections of filial love and gratitude—He appears to have been a being cast in a different mould from those of mankind in general, and to be the connecting grade between the human and brutal creation."[125] Plainly invoking a myriad of natural historical theories about the origins of modern man, and perhaps anticipating debates over polygenesis that gained traction among Atlantic scientists in the early through mid nineteenth century, the author of the "Account" identifies Frost as somewhere between human and brute, a rhetorical position frequently reserved for non-European and non-Christian peoples. His subordination to his habits, that slippery slope that makes one more likely to "commit the most atrocious sins with unblushing impudence," furthermore makes Frost party to the kind of bodily incontinence that, we are told, led Powers and Mountain to commit rapes.[126]

Frost's execution narrative, the story of a white man, seems at first glance to have little to do with sex. Yet the way Bancroft's narrative depicts him as quite literally a *creature* of habit nonetheless frames him within a vocabulary linked to racialization (a form of difference modified through its connection to aberrant sexual behaviors) and to physical incontinence, which was often materialized on the body, in these texts, through racialization. The social problems of nonwhiteness, socially unacceptable forms of sexual behavior, and crime sediment in eighteenth-century interest in the power of habit, and this is why habit is an absolutely critical site for further inquiry into the history of sexuality during this period.

Conclusion

Tracking debates over habit in eighteenth-century gallows literature reveals the deeply behavioral and environmental cast of contemporary theories about the origin of criminality. Ministers and other community leaders understood childhood as a critical time for the development of good habits that would spiritually and behaviorally inoculate young people against later exposure to vice. At the same time, for those young people who unwittingly developed habits of sin early in life, it was often difficult if not impossible to change course. The cultivation of good habits such as Sabbath-keeping, obedience to parents and masters, and daily prayer offered an ecumenical form of bodily and spiritual civilization that transformed children from their natural, innately depraved state into adults who could exercise self-restraint, govern themselves, and be thoughtful participants in their duties to their communities and to their faith.

Habit, of course, was a portable framework that girded not only contemporary theories of criminality but also contemporary theories of reform and rehabilitation. Prison-reform movements of the late eighteenth and early nineteenth centuries found in habit a mode of engineering spiritual and moral improvement. The shift from the prison to the penitentiary, in particular, marked a movement that theorized reform in largely behavioral terms, understanding habits of work and prayer as critical tools in the project of bodily and spiritual transformation. The physical and social environment, too, played an important role in the restoration of the soul, and regulating the environment and behaviors of criminals was one of the highest priorities for those hoping to rehabilitate them.[127] While they were not identical conversations, the way ministers, reformers, and other community leaders understood the importance of developing good habits traveled alongside and frequently echoed theories of habit that were then transpiring among naturalists. Conversations about the civilizing powers of diet, dress, labor, and specific forms of government (especially democracy) that appeared prolifically in travel writing and natural philosophy mirrored those happening in religious communities, and particularly in religious literatures of crime, execution, and reform. What these conversations shared was a profound belief in the determining power of the environment in the constitution of the human body and the conviction that both bodies and souls could be disciplined, at least in part, from the outside in by elective participation in modes of living designed for civilization, improvement, and reform.

In both execution sermons and natural philosophy, discussions of habit were often occasioned by inquiries into the origins or significance of racial difference ("human variety") and into the origins or significance of sexual behavior. These inquiries do not usually arrive to the fore as inquiries of the same order; in much of the gallows literature and many of the natural history texts I examine, racial difference constitutes something of a primary or independent form of difference, whereas sexual behavior often appears as a second-order object of inquiry. It comes into relief only as it is related to racial difference, as a characteristic or object of curiosity evidencing the profundity of racial difference, the more elementary and comprehensive form of human variation. In execution narratives, however, we see sexual behavior emerge as the bad habit par excellence; as we saw in Mather's *Tremenda* and Belcher's *The Worst Enemy Conquered*, "Lust" becomes the shorthand for the discussion of habits, sinful habits in particular. One's relationship to "lusts" of all kinds thus becomes a benchmark of a person's or a community's capacity for self-government, and this capacity in turn is read as a barometer of personal or collective capacity for civilization. The faithful govern their lusts; the natural man is governed by them. Sexual behaviors, racial difference, and self-governance thus provide the triumvirate of ideologies out of which habit emerges as a central and conjoining idiom in the developing sciences of natural history and criminal justice. And this is why the history of eighteenth-century sexuality cannot be understood without at least a lingering attention to contemporary discussions of habit. In these conversations, we see residents of the eighteenth-century Atlantic world puzzling over whether sexual behavior is an innate inclination or the product of environmental influence; whether sexual behavior responds to disciplinary efforts; whether sexual behavior is "natural" or epiphenomenal, a characteristic of racial difference or an experience of an entirely separate order; and whether aberrant sexual acts such as bestiality or rape, in particular, belong to a class of their own or are categorically related to a much wider range of criminal behaviors. Gallows literature, in its enormous marketability, offered a site through which the reading public engaged these questions on a broad scale. And in it, the sexual politics of racial difference come into stark, undeniable focus. What execution narratives teach us about the history of early sexuality is that from an early moment, sexual behavior was always at least partially understood as an effect of the determining influence of one's environment, either directly—for example, through keeping bad company—or indirectly, as an effect of processes of racialization (exposure to the sun, for example; recall Mather's warning "They who worshopped God under the Shape, it may be, of *Bruits,* were left by God, unto the *Sun,* which made them worse than *Bruits*").[128]

But what do these inquiries into the meaning of habit teach us about execution narratives? By directing our attention to the centrality of discussions of habit in eighteenth-century print culture more broadly, we are able to more carefully contextualize gallows literature as a print phenomenon that engaged extensively with some of the most pressing questions of the time surrounding the content and significance of racial difference. Scholars of the genre have written careful, compendious studies of these narratives as they proliferated—and changed—across the eighteenth century and into the nineteenth, and they have universally noted that gallows literature is one of the most common and popular print media for the representation of blackness. Gallows literature might even be said to be about blackness, *tout court.* Yet historians of race have not engaged at length with the genre. I join Jeannine DeLombard in insisting on the centrality of these narratives not only to the making of African American literature but also to the making of public notions of racial difference more generally.[129] Similarly, historians of early sexuality have looked extensively to gallows literature to study the histories of acts, practices, or legal frameworks but have neglected to examine these narratives for what they might tell us about the ways sexual behavior was understood or explained. When we follow discussions of habit as they appear in gallows literature, we read these narratives differently and are able to see them not just as harbingers of the emergence of popular fiction as both a market and a new form of literature, but also as early articulations of the questions that informed the development of modern categories of race and sex.

CHAPTER FOUR

Botanical Sexuality and the Colonial Landscape

In 1800 printers in both New York and London released a "re-published" edition of *The Unsex'd Females,* a popular political tract by the London curate Richard Polwhele. First published anonymously in 1798, Polwhele's extremely popular poem was a vitriolic attack on Jacobin gender politics as a whole, but it reserved the largest part of its ire for one of the most notable "unsex'd females" of the decade, Mary Wollstonecraft. Historians and literary critics of Jacobin and anti-Jacobin politics alike have explored at great length the relationship between *The Unsex'd Females* and the essay to which it, in large part, responds, Mary Wollstonecraft's *Vindication of the Rights of Woman* (1792). Often overlooked, however, is that both the form and the content of Polwhele's poetic diatribe also parodically address another controversial and quite popular piece of literature from the 1790s: Erasmus Darwin's *Loves of the Plants* (1789), an epic poem based on Linnaean taxonomy that described the sex lives of plants in pornographic detail.

To Polwhele, Wollstonecraft's radical feminist politics and Darwin's lavishly copulating flora both reflected the same social ill: the violation of "Nature's law" of feminine modesty.[1] Though Wollstonecraft's work became culturally associated with sexual lasciviousness because of popular knowledge of her many romantic affairs, Polwhele's anxiety about the sexual implications of the rise of popular interest in botany was derived from the science itself. In the wake of the 1735 publication of Linnaeus's *Systema Naturae,* which introduced a "sexual system" of taxonomy that identified plants by the visibility or invisibility of their reproductive organs, botany emerged as a sexual science, and cultural practices of plant exchange, dissection, and illustration exploded among the middle-class and the wealthy.[2] According to Thomas Hallock, botany became a major cultural idiom defining public discussions of sex in the latter half of the eighteenth century; he argues that "from the mid-eighteenth century through the 1790s, plant study intersected with issues of sexual propriety," a statement with which Richard Polwhele would certainly have agreed.³ Keeping this in mind, Polwhele's critique of radical feminism takes new shape when we consider his complaint that female modesty was imperiled by Wollstonecraft's call for women's education. The American edition of *The Unsex'd Females* is permeated with assertions of his haughty disapproval of women practicing botanical science in the name of education. Describing a botany class, he writes that young female students

> With bliss botanic as their bosoms heave,
> Still pluck forbidden fruit, with mother Eve,
> For puberty in sighing florets pant,
> Or point the prostitution of a plant;
> Dissect its organ of unhallow'd lust,
> And fondly gaze the titillating dust.⁴

This series of lusty images—heaving bosoms, "forbidden fruit," "pant[ing]," "prostitution," and the plant's "organ of unhallow'd lust"—renders Polwhele's tract perhaps more pornographic than even *Loves of the Plants,* the poem it satirizes. In the American edition, Polwhele footnotes this passage, explaining: "Botany has lately become a fashionable amusement with the ladies. But how the study of the sexual system of plants can accord with female modesty, I am not able to comprehend." In a somewhat voyeuristic validation of this claim, he also notes, "I have, several times, seen boys and girls botanizing together."⁵ Polwhele's disgust at coeducational "botanizing" clearly demonstrates his belief that the implicit sexuality of botanical taxonomy spurred an explicit cultural awareness of botany not only as a science of sex but also as an outlet for human sexual practice.⁶

This chapter considers the widespread popularity of the "sexual science" of popular botany (and the Linnaean taxonomic system upon which it was built) at the end of the eighteenth century to demonstrate that transatlantic Anglophone culture drew from a wide array of resources, including the natural environment itself, to identify and name various *types* of human sexual characters. Whereas previous chapters have considered how natural historical interest in racial difference rendered sexual behaviors visible by citing specific sexual practices as symptomatic of religious or other cultural differences, this chapter looks at the way the "sexual science" of botany—itself an outgrowth of natural history—was used to naturalize the presence of British settlement in North America in the era of its transition from colony to nation. Focusing on a little-known "unsex'd female" of the 1790s, this chapter considers the historical problem posed by Deborah Sampson, the real-life, cross-dressing, "lesbian-like" protagonist of Herman Mann's *Female Review, or Memoirs of an American Young*

Lady (1797). Mann's fictionalized memoir of Sampson, who was named the "Official Heroine of Massachusetts" in 1983, tells the story of her decision to dress as a man, join the Continental Army, and fight in the Revolutionary War, living as a man called Robert Shurtliff for more than a year.[7] Mann's relatively obscure novel describes both the military and the romantic exploits of his unconventional hero(ine), at times going into explicit yet confusing detail concerning the nature of Sampson's relationship to her "sister sex."[8]

The portrait of Deborah Sampson that Herman Mann paints in *The Female Review,* as well as public depictions of Sampson's life as a whole, is haunted by sapphism. A term used throughout the 1790s, *sapphism* joined "lesbian" as a geographically specific term (like "sodomy") that referred to the possibility of erotic or eroticized relations between women.[9] The text is permeated by representations of Sampson's relationship with other women that clearly appear socially aberrant—and thus demanding apology, explanation, laughter, or defense —to Mann, as well as to the readers whom he understands to be his audience. Just as important, although also often overlooked, these moments are almost entirely described within a botanical idiom. In *The Female Review,* Mann's depictions of Deborah Sampson clearly belie his suspicion that there is something fundamentally *different* about his protagonist, and Mann charts this difference in naturalist terms; he describes her as "singular," "rare," and "miraculous," yet always still "natural."[10]

As we will see in what follows, the colonial landscape is critical to this representational trope. Mann enlists older, natural historical interest in "wonders" or "exceptions" of nature to describe Sampson, while simultaneously contextualizing Sampson's behavior as a logical response to the urgency of her historical and political situation: the Revolutionary War. Like contemporaries such as Benjamin Rush, Mann was interested in what the political and structural changes that attended the transition from colony to republic had wrought in people who had lived through it. Whereas Rush later identified a series of "diseases of the mind" specific to the experience of revolutionary upheaval (*anarchia,* for example, a "form of insanity" provoked by an "excess of the passion for liberty"), in *The Female Review* Mann glosses Sampson's sapphic behavior as an expression of "sensations and emotions different from whatever [she] had before experienced," a reaction to political events "both before and after the first engagement at Lexington."[11] In this sense, Mann's depiction of Sampson portrays her as something of a "new breed" of person, a novelty of "Columbia's Soil." Sampson's sapphism is thus recast and naturalized as a feature of a comparatively new colonial landscape (the "New World") and of the United States' nascent republicanism more generally.[12] Sampson's strange gendered and sexual behavior thus indigenizes her in this text; her sapphism is, bizarrely, what identifies her as an authentic product of her political and geographical circumstances. It is what makes her "An *American* Young Lady." This mode of representation critically relies on two ideas. First, Mann's assessment of Sampson's sapphism as a feature of the altogether new "sensations and emotions" inherent in the postrevolutionary United States evokes natural historical tropes that saw sexual perversity as innate to foreign lands and colonial locales (think, for example, of William Davis's assertion that "Turks are altogether sodomites" [see chapter 2]). The novelty of Sampson's sapphism reflects the "newness" of the land and the Republic.

At the same time, this trope relies on a settler colonial logic that asserted the newfound indigeneity of those who were formerly British settlers and now were "Americans," naturalizing them as the true and original possessors of North American lands. Critically, in *The Female Review* Sampson's prowess as a soldier is proved not only on the battlefields of the Revolutionary War, but on the Ohio frontiers as well, where her successful murder of a Native man also testifies to the success of her efforts to both live and fight authentically as a man. Sampson's sapphism and the broader logic of settlement thus appear in the novel as mutually imbricated; indeed, Sampson's sapphism strangely indigenizes her whiteness.

This chapter thus focuses on the sympathies between natural historical science—the "sexual science" of botany, in particular—and settler colonialism. Whereas previous chapters have identified how we can draw quiet or implicit narratives about the origin or meaning of sexual behaviors out of natural historical studies of non-European or nonwhite peoples, this chapter is primarily concerned with sexual practices associated with whiteness or with settler communities. Late-eighteenth-century botanical science provided a consistent forum for the discussion of sexual diversity, largely because of the popularization of botany that attended the publication of Erasmus Darwin's *The Botanic Garden* in 1789. Since botanical taxonomy is built upon a specialized structure of the categorization of knowledge organized around the sexual morphology and reproduction of plants, popular botany offered a new set of tools that made it possible to imagine human sexual behavior as a categorical form of human diversity.[13] It is clear that Linnaeus's work effected a widespread shift in cultural perceptions of the meaning of race, but we tend to overlook a corollary shift: the popularization of Linnaean taxonomy changed American cultural understandings of sexual behavior.[14] In the aftermath of the publication of Darwin's *Loves of the Plants,* sexual behavior newly appeared as an object of scientific inquiry, and different forms of sexual behavior were understood to bear a fundamental or essential relationship to the nature or character of the plant.[15]

But what of a case like Sampson's, in which her sapphism was explained and classified through the frame of newness—indigenous to the New World, native to the new Republic?

Sampson's aberrant gender and sexual behaviors must also be considered alongside the "second side of the taxonomic project," imbricated as it was in colonial prospecting: "the search for new specimens."[16] The constant possibility of the new specimen allowed for belief in a form of knowledge that was actually a nonknowledge: eighteenth-century scientists both anticipated and actively sought out "specimens" or forms of life with which they were unfamiliar. This rage for the undiscovered should be understood as an epistemological form that merits a place in our understanding of the history of sexuality in the early national period. More simply put, botanical taxonomy, a science of sexuality that recognizes difference or newness in objects as a moment of the possibility of knowledge to come, allows us to rethink one of the most persistent (and frustrating) problems for historians of sexuality: whether or not Deborah Sampson's contemporaries understood her sexual difference in modern sexual terms. Here I propose that we ask a different question: did Sampson's culture recognize her sexual difference as a possible object of scientific inquiry? In representing Sampson's sapphic behavior, Mann turns to the rhetoric of the miraculous or the exception in nature, citing the unique political and natural environment in which Sampson grew up to explain her aberrant sexual presentation and acts.[17] Mann's reliance on a heavily botanical idiom in his representation of Sampson posits her as something of a "new specimen," in whom the potential for knowledge of her sexual difference is possible, but for whom no nomenclature yet exists. I thus insist that we include the history of opacity as a fungible site of inquiry for the study of early sexuality.[18] Paying critical attention to the way opacity functions as a technique for the representation of socially aberrant sex acts in this late-eighteenth-century text brings into stark relief *The Female Review*'s refusal of characterological or depth-model explanations for sexual behavior; instead, it leans on the language of botany to position Sampson as a miraculous but deeply natural production of her unique historical and physical environment.

A note on terminology: I use the pronouns "she" and "her" when referring to Sampson during periods when she was living as a woman, but I use "Sampson/Shurtliff" (Robert Shurtliff was one of two masculine names that Sampson used while in the military) and both "she" and "he" to discuss the period of Sampson/Shurtliff's life when Sampson lived as Shurtliff. I do this for two reasons. First, I refuse to weigh in on the question of whether Sampson/Shurtliff was a feminist, a queer person, or a trans person. While I absolutely believe that Sampson should be included in feminist, queer, and trans political histories, I am interested neither in using twentieth- and twenty-first-century terms to describe Sampson/Shurtliff nor in reifying the distinctions between *feminist, queer,* and *trans* that the question itself suggests. If Sampson were alive today, he or she might be feminist, queer, *and* trans. Second, I am far more interested in how Sampson/Shurtliff thought about himself or herself, how Sampson/Shurtliff made sense of his/her own behavior, and which of his or her priorities the decision to live as a man reflected. Understanding what motivated Sampson's life decisions seems, to me, a more interesting question. Do not confuse this position with those taken by Sampson's more literal-minded biographers, some of whom have unambiguously asserted that Sampson was not a feminist and was not queer (and thus, by that logic, that she was not transgender, either).[19] Sampson lived as a man both during and *after* the Revolutionary War, and she lived under not only one but two different men's names. Accusations of sapphic or otherwise socially aberrant sexual behavior followed her from some of her earliest positions as an indentured servant, through the end of her life after she married and had children. Even people who later advocated on Sampson's behalf, when she sought an invalid's pension from the government for compensation for her military service, characterized Sampson as strange.[20] It is abundantly clear from the shockingly robust historical record surrounding Sampson's life that her sexual disposition—how she looked (she was allegedly five feet, nine inches tall, very, very tall, by any gendered standard, for her day); her comportment; and how she related to other people, women in particular—was surveilled first in her immediate community and later on a much broader public scale. Regardless of how Sampson might have understood herself or himself in our day, in her own time Sampson was socially suspect.

"A Phenomenon, Which Before Appeared a Natural Object"

Set in Middleborough, Massachusetts, toward the end of the Revolutionary War, *The Female Review* narrates the story of Deborah Sampson, who on May 20, 1782, dressed in men's clothes and enlisted in the Continental Army.[21] She sustained her active enlistment until October 25, 1783, when she was discovered to have breasts and was subsequently discharged. The novel proceeds in the form of a fairly conventional bildungsroman. Born poor and hired out to local families as an indentured servant, Sampson is depicted as a self-educated natural genius coming of age during the rise of anticolonial sentiment in rural Massachusetts, paying particular attention to the steps that she took to secure her education. Some of her early education was fairly traditional in form, even though there were, at best, only uneven educational opportunities for indentured children; she borrowed schoolbooks from the children of the families for whom she worked, for example, and she taught herself. But Mann, himself a self-taught naturalist of sorts, also describes at length how Sampson educates herself through her time outdoors, observing the physical laws of the flora, the fauna, and the cosmos. The remainder of the novel offers an account of Sampson's decision to join the Continental Army, including details of various injuries, battles, and scouting trips she endures before falling ill and being discovered. It concludes with Sampson's discharge from the army and her travel to her uncle's farm, dressed as her brother Ephraim, where she continues to live as a man through the book's conclusion.[22]

The publication and reception of *The Female Review* are framed by repeated acts of recovery. Both Sampson's story and Mann's novel itself sustained multiple revisions as they were published, forgotten, recovered, and republished between 1784, when a short, article-length account of Sampson's military discharge appeared in the *New York Gazette,* and 1916, when William Abbatt published what remains the most recent edition of Mann's *Female Review* in Tarrytown, New York.[23] Herman Mann was responsible for the early editions of the novel that remain available to scholars; he authored the 1797 first edition, wrote an oratorical version of Sampson's story that she delivered repeatedly during a lecture tour in 1802, published another short version of the story in an 1820 edition of the Dedham, Massachusetts, *Village Register,* and authored a 429-page, first-person manuscript in 1829 that never made it into publication. After Mann's death in 1833, *The Female Review* remained out of print until 1866, when John Adams Vinton republished it. Brief biographies of Sampson also appeared in Elizabeth Ellet's 1848 *The Women of the American Revolution* and in an October 1863 issue of the periodical *Beadle's Dime Tales.* Her story was also told as a 1975 Scholastic Biography Series text (aimed at North American second-graders) entitled *The Secret Soldier,* a 2004 biography by Alfred Young, and a 2014 biographical novel by Alex Myers that reads Sampson as transgender.[24] This rather proliferative publication history, while certainly unusual for a novel that boasted of only about two hundred subscribers at the time of its first publication, was a crucial part of the process by which Sampson's sexual behaviors became an object of speculation.

Despite the status that it claims as an autobiography or "memoir" (the subtitle of Mann's text is "Memoirs of an American Young Lady"), *The Female Review* also capitalized on a contemporary trend in popular fiction that Daniel Cohen and Diane Dugaw have called the "cross-dressing narrative" and the "warrior woman" tale, respectively, a genre of popular literature that regularly titillated its readers with allusions to same-sex sexual behavior.[25] From Martinette, of Charles Brockden Brown's novel *Ormond* (1799), to Pulchera of the anonymous, wildly popular *Constantius and Pulchera; or, Virtue Rewarded* (1795), a wealth of American fiction published during the last decade of the eighteenth century features female characters who disguise themselves or even live as men.[26] While Dugaw and Cohen have written book-length studies on the print evolution and transnational cultural origins of these figures, relatively little scholarship focuses on the fascinating genealogy of women in martial attire in popular fiction; recent work by literary scholars of the early national period such as Susan Lanser and Katherine Binhammer have effectively sketched out the relationship between these "warrior women" and British radical feminism's "unsex'd females" so scorned by anti-Jacobins such as Richard Polwhele.[27] By the mid-eighteenth century, however, a handful of popular autobiographical narratives describing the lives of female soldiers had appeared in print in England, the most familiar of which was probably *The Female Soldier or the Surprising Life and Adventures of Hannah Snell* (London, 1750).[28] By the early nineteenth century, many of these ballads and cross-dressing narratives had appeared in print in the United States, and it is likely that Herman Mann, Charles Brockden Brown, Tabitha Tenney, and many more American writers were familiar with this genre of folk imagery.[29]

Like most such narratives, *The Female Review* begins by describing the protagonist's family history, early life, and upbringing. Unlike some of the more popular examples of this genre (such as *The Female Soldier*), however, the narrator's introduction to Sampson's childhood and disposition is rife with biological imagery and metaphor. In describing Sampson's efforts to educate herself, for example, Mann tells us that because the family to whom she was indentured "appeared more eager in the amassing of fortune than of scientific acquisition" and thus did not send her to school, Sampson "was obliged to check the bud" of her intellect.[30] As Mann further warns of the dangers of a lack of education, he muses that "the most fertile genius, like that of soil, which for want of proper cultivation is overrun with noxious weeds, becomes corrupted by neglect and vicious habit: and the inherent beauties that might have eclipsed a more than ordinary show, lie dormant."[31] He refers to intellect as a gift that, unfortunately, nature bestows "unequally . . . on the human *species*."[32] The use of "species," the "bud" of Sampson's intellect, and the "fertile" "soil" of intelligence that can suffer from "noxious weeds" if lacking "proper cultivation," illustrates the naturalist, and often specifically botanical, metaphor through which the narration of Sampson's childhood proceeds. But even more fascinating is that while the narration as a whole is botanically infused, this narratological trend is far more pronounced in descriptions of Sampson's sexual difference, or her romantic exploits with women.

Numerous editors of this text over the past two hundred years have remarked on what one nineteenth-century editor, John Adams Vinton, calls its "prolix and verbose" writing; of primary concern here are the ways in which Mann's narrative style is, quite literally, flowery.[33] Many of the book's passages are extremely confusing or syntactically vexed; the saving grace of the prose is its consistent return to the vocabulary of the natural sciences. Metaphors invoking the natural world are something of a ubiquity in fiction, in the eighteenth century as they are today, to the point where this type of language tends to become unremarkable or invisible, a mere background detail within the narratological environs. But if we shift our overall focus from Mann's explanation of Sampson's reasons for living as a man and engaging in romances with women to the botanical vocabulary that he frequently deploys in depicting these unconventional behaviors, Mann's botanical taxonomy for Sampson's gender difference and sexual variance comes into stark focus.

Botanical taxonomy was still a recent development at the time when Mann penned *The Female Review,* and although Linnaeus's system of binomial nomenclature had achieved a certain primacy by the end of the eighteenth century, more than fifty-two other systems of botanical taxonomy had been invented, debated, and rejected by 1799.[34] Linnaeus's *Systema Naturae* stood out among competing taxonomic systems because it organized the classification of plants along one fundamental axis: Linnaeus anthropomorphically categorizes plants according to their sex and the kinds of "marriages"—singular, polygamous, and incestuous—that they engage in with the flora around them.[35] Botany thus occupied a designation within the popular imagination as a science of sexuality, and it played a crucial role in shaping popular perceptions of sexual diversity in the last half of the eighteenth century in both western Europe and the North American colonies.

Mann consistently uses a botanical idiom as he describes—and often excuses or apologizes for—Sampson's "singular" and explicitly different sexual behaviors. Mann himself was something of an amateur botanist, student of early astronomy, and natural science enthusiast overall, and references to these scientific discourses permeate *The Female Review.* In describing Sampson's efforts at self-education, Mann notes that with "peculiar pleasure I here find occasion to speak of Miss Sampson's taste for the study of Nature, or Natural Philosophy."[36] Here, Mann sketches a causal relationship between Sampson's early botanical interests and the "singular" personality and "passions" that distinguish her adult character:

> [Sampson's] taste for the cultivation of plants and vegetable productions in general, appears to have been somewhat conspicuous in her early years. And she has intimated an idea of this kind, which, from its justness, and the delicate effects it has on many of the softer passions, induces me to notice it. It has been a source of astonishment and mortification to her that so many of her own as well as of the other sex, can dwell with rapture on a romantic scene of love, a piece of painting or sculpture, and, perhaps, upon things of more trivial importance; and yet can walk in the stately and venerable grove, can gaze upon the beautifully variegated landscape, can look with indifference upon the rose and tulip, or can tread on a bank of violets and primroses, without appearing to be affected with any peculiar sensations and emotions. This certainly proceeds from a wrong bias of the mind in its fixing on its first objects of pursuit. And parents cannot be too careful in the prevention of such errors, when they are forming the minds of

their offspring for the courses which are to affect the passions, and give sway to the behavior during life.[37]

The cultivation of plants and other verdure thus directs the cultivation of Sampson's "softer passions," understood in the eighteenth century to be the broad manifestations of love, from compassion to sympathy to tenderness. Importantly, in a decade widely studied for its muscular propagation of the ideology of republican womanhood—a central tenet of which pertains to the importance of proper parenting in the process of raising good male citizens—these "softer passions" are also the feelings associated with republican *motherhood;* contemporaries such as Benjamin Rush advocated publicly at the turn of the century for the proper "cultivation" of new Americans.[38] In this passage, however, Sampson's instincts for cultivation are directed toward unexpected objects: her "peculiar sensations and emotions" come from gazing at *plants,* which the narrator lauds as proper objects to stimulate these feelings.

Here, the narrator draws an analogy between Sampson's experience of nature and cultural understandings of the experience of romance, understanding the development of "passions" or other romantic feelings to be the result of encounters with art or the artistic. This would have been an extremely familiar conversation for contemporary readers of *The Female Review,* especially for female readers who were well versed in both novel reading and the cultural anxieties about what could happen if one read too many novels.[39] Mann's disapproval of women who are moved by "a romantic scene of love" but who "can walk in the stately and venerable grove ... without appearing to be affected with any peculiar sensations or emotions" indicates that what is at issue here is an ethics of the relative appropriateness or inappropriateness of that which induces passionate feeling. Importantly, the narrator implies that what readers might consider the "natural" origins of romantic feelings are not, in fact, *natural,* that deriving "sensations or emotions" from "romantic scene[s] of love" can result in a "wrong bias of the mind."[40] Mann delights in elevating the natural world as the proper object to stimulate "sensations or emotions," and he applauds Sampson's own instinctual pull toward nature. Thus, here and throughout the book, Mann renaturalizes the natural world as central to human romantic and sensual life.

Furthermore, Mann's narrator focuses on the "peculiar sensations or emotions" that might "proceed from a wrong bias of the mind in its fixing on its first objects of pursuit." What this passage insistently communicates is that sensations, emotions, and passions that are derived from looking upon the natural world—the grove, the landscape, or the rose, the tulip, the violet, or the primrose—are also natural or organic feelings and, furthermore, that these feelings should be sanctioned and privileged as such. On the other hand, "peculiar sensations and emotions" that develop from what we understand to be romance ("romantic scene[s] of love, a piece of painting or sculpture," or, in all likelihood, novels) are necessarily questionable because the source from which they are derived is itself suspect. Thus, we should consider how an understanding of the double edge of these "peculiar sensations" (appropriate when derived from nature, and inappropriate when derived from the culture of "romance") might contribute to our understanding of eighteenth-century sexuality.

This passage both identifies the natural world as the only appropriate erotic stimulant and distinguishes Sampson's "peculiar sensations" from those of her peers (based on the objects that induce these sensations). The sensorium, as Jordan Stein has argued, constitutes an important site of inquiry in early American sexuality studies, and in *The Female Review*, Sampson's unique range of sensational responses is proffered as an explanation of her unconventional behavior. At the same time, her unique sensational register is explained in terms of its unmediated relationship to the natural world, and specifically, to the botanical. Given the relationship between the rise of botany as a form of popular knowledge in the second half of the eighteenth century and botany's unique status as a science of sexuality, we must read this description of Sampson's difference—an observation that the narrator announces incessantly throughout the text—as one that even the narrator of *The Female Review* understands to be, in some way, about sex.[41]

This language of sensation also links Sampson's unique gendered and sexual behavior to her experience of the sociopolitical upheaval of her particular historical moment: the Revolutionary War. The narrator details, for example, that early local conflicts, "both before and after the first engagement at Lexington, are well known to have affected the minds, even of both sexes, throughout the Colonies, with sensations and emotions different from whatever they had before experienced"; these "sensations and emotions" provide Sampson the inducement to act on her burgeoning patriotism.[42] Her patriotism, in turn, "filled her mind with sensations, to which she had hitherto been unaccustomed—with a kind of enthusiasm, which strengthened and increased with the progression of the war."[43] In *The Female Review*, what Mann catachrestically glosses as Sampson's patriotism takes the form of sensation, but a set of sensations that cannot be fully described or exhausted by nationalist fervor or mere patriotic sentiment. The midnight scene in which Sampson dons her military uniform and leaves the family to whom she was indentured presents readers once again with an image of Sampson overwhelmed not just by patriotism but, beyond that, by feelings specifically related to the physical process of donning her masculine, martial persona: "But there was none, but the INVISIBLE, who could take cognizance of the effusions of passion on assuming her new garb; but especially, on reflecting upon the *use,* for which it was assigned."[44] Readers understand

Sampson's "effusions of passion" as first a response to her "assuming her new garb" and *also*—but not exclusively—as a response to the patriotic "*use*" of what the narrator very candidly describes as a nice-looking set of men's clothes. Sampson's experience of a seemingly overwhelming effusion of patriotic sensation, then, is inextricably linked to her decision to cross-dress, an act that was widely associated within contemporary popular culture with sexual lasciviousness or promiscuity more generally. Whether or not cross-dressing constituted a sexual act in the eighteenth century is not at issue here. Rather, the fact that Sampson's cross-dressing is explained as her response to sensations induced by her sociopolitical environment, wartime Massachusetts, frames her decision to live as a man as at least partially a result of environmental influence.

Animal Love, on Her Part, Was out of the Question

The internal sexual differences attributed to Sampson throughout *The Female Review* are much subtler, at least to present-day readers, than the far more overt references to Sampson's romantic affairs with women. Indeed, the text tarries on moments where Sampson both courts and is courted by women whom she encounters in her travels with the Continental Army.[45] Like Mann's descriptions of Sampson's unique gender disposition, her liaisons with women—and with one woman in particular, who is referred to as "the Baltimore lady" throughout the novel—are infused with botanical imagery.

Linnaeus's taxonomic system of binomial nomenclature gave botany its reputation as a "sexual science" because of the way he organized his botanical taxonomy according to the relative visibility or invisibility of plants' sexual organs (pistils, stamens, etc.). Yet the 1789 publication of Erasmus Darwin's two-part paean to Linnaean botanical taxonomy, *The Botanic Garden* (1789), served to further establish the popular association of botany with sex, but in a slightly different way.

The Botanic Garden was an immensely popular text, consisting of two epic poems: *The Economy of Vegetation* and *The Loves of the Plants*. While *The Economy of Vegetation*, the first volume of Darwin's oeuvre, is more or less a treatise on popular science that explored new scientific theories (such as the use of electricity or the physics of heat), it was *The Loves of the Plants* that drew readers from all over the Atlantic world to Darwin's text. The book was an immediate success; first published in England in 1789, *The Botanic Garden* was reprinted quickly and often through the first quarter of the nineteenth century. The first American printing was in 1798, but libraries in Boston, New York, and Philadelphia all list copies of Darwin's controversial book in their collections by the mid-1790s.[46]

The Loves of the Plants derived its popularity not only from its blurring of the already tenuous distinction between scientific inquiry and erotic poetry, but also from its exploration of the sociosexual organization of the plant world. Darwin's epic poem mapped the expansive horizon of possibilities for the configuration of sexual relationships among plants. Drawing promiscuously from Linnaeus's taxonomy as well as from his 1787 *Families of Plants,* Darwin figured his plants as husbands and wives, the spectrum of whose sexual relations extended from the extremely conventional to the heterogeneous and socially aberrant.[47] He described plants that had both "public" and "clandestine" marriages, and even "polygamies"; there were "husband" and "wife" flowers "in one bed," "in two beds," or "in separate beds"; marriages in which there were "two husbands in the same marriage"; relationships between "hermaphrodite" plants; and plants that were deemed "feminine males."[48] This riff on Linnaean botanical science was intended as more than just a titillation, however; Darwin "clearly intended that these anecdotes should illuminate human sexual relations," and "these sexual messages came through loud and clear to readers."[49]

Entirely anthropomorphic—the plants feel love, jealousy (or, at times, "unjealousy"), are married or polygamous, male or female, brothers or sisters—*Loves of the Plants* endowed the science of botany with a quietly pornographic aesthetic that participated in wider, transatlantic conversations concerning new possibilities for human sociosexual organization. *Loves of the Plants* was read alongside Wollstonecraft's *Vindication of the Rights of Woman* in political groups and young, mixed-sex intellectual circles in New York in the last years of the eighteenth century; both texts, as Richard Polwhele highlighted in his *Unsex'd Females* a few years later, reflected and propagated radically new notions about both the social and internal significance of gendered behavior and sexual propriety.[50] Thus, while both Darwin and the Linnaean science he drew from explored the possibility of new configurations of womanhood, manhood, and even "hermaphroditic" sexes, Darwin's plants were both unusually sexed and unusually sexual. His text constitutes crucial historical and cultural context through which to approach how eighteenth-century readers would have encountered the moments in *The Female Review* when Mann represents same-sex romance and sex acts between Sampson and her female lover.

As we shall see, in the narrative itself, episodes depicting same-sex sex acts are overblown, dramatic, and often almost unreadable to both eighteenth- and twenty-first-century readers. The overwrought construction of these passages should not be surprising: Herman Mann had to perform some extremely complicated negotiations on behalf of his real-life—and still very much alive—subject, to portray her as not only a legitimate veteran of the Revolutionary War but also a subject available for sympathy rather than derision or even prosecution. Mann did not publish *The Female Review* as solely a fictional enterprise; the text also bears a very explicit objective, as it was published while Sampson was campaigning to receive an invalid pension from the US government after being wounded in the war. The fact of her cross-dressing did not appear unambivalently brave or endearing in her era, as it does to many readers who encounter this text today. Sampson's campaign for a pension had to be levied in terms that legitimated her actions despite being a woman, not because of it. Women who lived in military camps were widely considered socially suspect. Sampson was certainly not the first soldier found to be female during the war; many women followed the army and lived in army camps, and married women not infrequently traveled to war with their husbands.[51] While the presence of women in army camps was fairly common, a woman found living in one was usually considered morally lax or simply assumed to be a sex worker.[52]

Yet, upon being discovered to have breasts while being treated for a gunshot to the thigh, Sampson was not lambasted, tried for fraud, publicly humiliated, or prosecuted. Karen Weyler argues that this was a result of the way Sampson carefully shaped the dissemination of her story, approaching public literary figures such as Herman Mann and prominent newspaper editor Philip Freneau for publicity and support for her case.[53] The records, rumors, and reports upon which *The Female Review* is based, however, still contain multiple accounts of intimate liaisons, intimated to be of an erotic nature, between Robert Shurtliff—Sampson's male persona—and young women. In attempting to cast Sampson as a hero rather than a prostitute, or even, as Philip Freneau putatively referred to her, an "eccentric," Mann had his work cut out for him.

In *The Female Review*, Mann effectively defends Sampson as a soldier worthy not only of his writerly attentions but also of an invalid pension, by returning to the same botanical idiom that characterizes his other descriptions of the cross-dressing soldier. In describing her liaisons with women, Mann again turns to the language of both the "exception" and the "wonder" in nature to craft an apologia for her aberrant behaviors that asserts them as natural, even indigenous to her current political and historical situation. This justification, however, does rely upon the idea that nature sometimes begets life forms that appear supernatural or unnatural—Mann uses the example of a woman giving birth to octuplets—but he takes care to point out that the moniker of supernatural or unnatural is often earned by organisms or behaviors that are illegible to extant taxonomies or classifications. A generous reader might attribute Mann's incomprehensible prose to his effort, in these passages, to describe a wonder or exception in nature for which he had no reliable vocabulary.

In *Revolution and the Word*, Cathy Davidson similarly remarks on the overblown, ridiculous writing that characterizes the narration of the contemporary anonymous female picaresque novel *Constantius and Pulchera* (1795).[54] Even as Davidson asks a very important question about whether eighteenth-century readers could have understood it, she also reminds current-day critics that *Constantius and Pulchera* was enormously popular, overblown prose or not. What I want to caution us against here is reading popularity as an indication that contemporary readers comprehended Mann's writing, especially as the passages in *The Female Review* that represent what seem to be moments of same-sex sex acts or desire are some of the most difficult to understand. We might explain the overwrought narration by simply accepting that Mann was perhaps a bad writer; but it might also be because Mann lacked a vocabulary, or even a coherent epistemological matrix for representing what Martha Vicinus and Judith M. Bennett term "lesbian-like" behavior. Indeed, Mann's narrator often seems unable to explain, apologize for, or even describe the relationship that exists between Sampson and her lover (the text refers to her as "the Baltimore lady"), and at times Mann specifically refuses to try. Sampson and the Baltimore lady's parting scene concludes as follows:

> THUS parted two lovers, more *singular,* if not more *constant,* than perhaps, ever distinguished Columbia's soil.
>
> THIS event, as it is unnatural, may be disputed. It is also rare, that the same passion should ever have brought a woman to bed with seven children at a birth: And I think *eight* would rather be miraculous than natural. But it is said, that though perhaps the colouring is a little exaggerated, that this is a fact that will admit of incontestible evidence. Nor need females think themselves piqued to acknowledge it; as no one denies, she was not an agreeable object when masqueraded; why, by the by, I am sorry to say, is too often mistaken by that sex.
>
> THUS, we have a remarkable instance of the origin of that species of love, which renders the enjoyment of life satisfactory, and consummates the bliss of immortality. The passion entertained by the sexes towards each other is, doubtless, from this source; and will always be laudable, when managed with prudence. But I appeal to the lady's own bosom, if, after discovering her *sister,* her passion had not subsided into a calm, and have drooped, like the rose, or lilly, on its dislocated stalk.[55]

For the first time in the text, Mann's narrator identifies this liaison in socially recognizable terms: he calls the two women "lovers" and continues to describe their relationship using the same botanical idiom that characterizes his representations of Sampson throughout the book. Though "singular" or exceptional, the lovers grow from "Columbia's soil"—"Columbia" being the new nation of the United States—and the "unnatural" qualities that Mann's narrator attributes to their passion seem to describe the social perception, rather than the reality, of the relationship. Analogizing it to a woman who gives birth to septuplets or octuplets, a rare event to be sure, Mann's narrator explains that the singularity of the event (their passion, or the birth of octuplets) would "be rather miraculous than natural." But the effect of the first full paragraph here is that Mann's narrator is defending the unusual, the unlikely, and the seemingly unnatural as, in fact, part of nature. He supports this claim by gesturing to scientific approaches to understanding nature: "This is a fact that will admit of incontestible evidence."

This evidentiary claim for the status of the rare or seemingly unnatural as existing within nature here takes on a specifically sexual charge. Mann's narrator admits, "It is also rare, that the same passion should ever have brought a woman to bed with seven children at a birth." What he is suggesting is that the passion that could beget septuplets is similar to the passion between Sampson/Shurtliff and the Baltimore lady, described in the previous paragraph. The "passion" that begets children is sexual passion, but that "same passion" as Mann's narrator understands the lovers to share is also the passion that begets a rare (and perhaps unnatural) number of offspring. However, the narrator quickly recodes this rare occurrence, not as unnatural, but as "miraculous." Changing the emotional valence of the "singular" nature of the attachment between the young Continental soldier and his/her lover amounts to *renaturalizing* behaviors and sex acts that would ordinarily be rejected from the category of the natural. In this passage, furthermore, exceptions within nature—miracles, octuplets, and "passion" between people of the same sex—are clearly delimited as anomalous, yet Mann's narrator still claims them as instances within nature itself.

The paragraph that follows supports this argument, as Mann's narrator offers a short aside to his female readership, as he is frequently wont to do throughout the text. He once again characterizes Sampson and her Baltimore lover as "a remarkable instance of the origin of that species of love" (anachronistically calling to mind the most famous of Charles Darwin's writings for our time) which even more obviously relies on a scientific or botanical idiom for its explication. Their love is a "species" and, furthermore, a "remarkable" one, which offers the narrator a glimpse into what he considers the "origin" of heterosexual love and, we assume, marriage. The move to argue that "the passion entertained by the sexes toward each other is, doubtless, from this source" is a fascinating one, especially since in so arguing, the narrator effectively posits that this "remarkable" or "miraculous" love can be considered an archetype for marriage or marital happiness. Importantly, however, Mann identifies marital "passion" as "from this source," derived from rather than identical to Sampson and her lover's affection, further differentiating between different classes of love. Mann's narrator seems to sense the almost incredible nature of this claim, too, as he follows it with a short appeal to his female readers, asking them to imagine themselves in the place of the Baltimore lady. In an "appeal to the lady's own bosom," he asks her whether, "after discovering her *sister*"—discovering her presumably male lover to be female—"her passion had not subsided into a calm, and have drooped, like the rose, or lilly, on its dislocated stalk."

This is a difficult passage to parse, but I read this reference to a different form of women "botanizing together" as yet another instance of Mann's disjointed prose, and I interpret it as Mann demanding sympathy of his female readers. This "appeal" asks them to consider whether, upon finding their lover to be a "*sister*," their passion for someone who "no one denies . . . was . . . . an agreeable object when masqueraded" would have actually subsided or not. The narrator preemptively excuses the possibility that ladies might maintain an attraction to a handsome figure whom they find, afterward, to be a woman, "nor need females think themselves piqued to acknowledge it." Later in the text, Mann's narrator actually exhorts his female readership to admit *their own* attraction to his cross-dressing protagonist, admonishing, "REMEMBER, females, I am your advocate; and, like you, would pay my devoirs to the goddess of love. Admit that you conceived an attachment for a *female soldier*. What is the harm? She acted in the department of that sex, whose embraces you naturally seek."[56]

This narratological attempt to naturalize the Baltimore lady's passion for Sampson/Shurtliff helps us read the aforementioned "appeal" to the lady reader, so that she might excuse the Baltimore lady's ostensibly sexual transgressions with another woman because the lady reader can imagine herself falling into the same confusing trap. This reader should understand why the Baltimore lady's passion for Sampson "had not subsided into a calm, and have drooped, like the rose, or lilly, on its dislocated stalk." This obviously phallic image offers a fascinating botanical and metaphorical illustration of Mann's understanding of the possibility of sexual love between a woman and a mannish woman, or a woman living as a man. The "rose" and the "lilly" are both images traditionally gendered as female, although here these flowers are depicted more holistically, with an evocation of their "stalk" as well, expanding the metaphor to suggest that some women bear "stalks" alongside their "blossoms." Appending the "dislocated stalk" to the feminine image of these flowers lends a texture of phallic failure to the relationship between Sampson and her Baltimore lover, which is characteristic of pejorative depictions of tribades and "female husbands" in eighteenth-century literary culture more generally.[57] This passage, which depicts Sampson parting with her lover for the last time, participates in the botanically infused sexual discourse described in this chapter: it deploys the natural science of botany as a means of discussing sexual difference and exposes sexuality as a category of taxonomic knowledge. Nevertheless, in the narration of Sampson's romantic interactions with her paramour in Baltimore, Mann also suggests that no taxonomy yet exists to adequately characterize the "species" of love between them.

Mann's unintentionally difficult prose suggests his struggle to find clear and compelling language through which to describe Sampson/Shurtliff. It was not that there were no eighteenth-century vocabularies for aberrant sexual behaviors; this book explores the surprisingly vast and labile range of words and tactics that writers used to describe sex acts of all kinds. Rather, it is that Mann did not have a coherent vocabulary of associated terms or ideas at his disposal to make sense of Sampson/Shurtliff. His turn to the language of botany, then, makes sense; not only could popular botany accommodate unconventional configurations of sex and sexual behavior, but its explicit interest in taxonomic systems—structures that lent order to the unknown—provided a framework for approaching organisms that were as "singular" or exceptional as Sampson/Shurtliff. But I would also insist that we keep in mind Eve Kosofsky Sedgwick's injunction: a lack of knowledge or vocabulary for sexual behaviors that challenge heterosexist prerogatives can be as indicative of a willful ignorance as of simple silence. Martha Vicinus, following Sedgwick, reminds us of this fact in her polemical work on lesbian history, arguing that "ignorance is not an empty box waiting to be filled by knowledge. Ignorance now and in prior times can be willed."[58]

Ignorance and opacity can be understood as coterminous; for my purposes, however, it is more important to understand the historical conditions under which both ignorance and opacity were able to thrive as categories of a particular kind of knowledge about sexuality. In Mann's inability to make sense of Sampson's participation in her various romantic friendships or sexual relationships, I also think we see some of this willful ignorance. For example, when musing about the details of the nature of Sampson's relationship to other women, Mann refuses to speculate, demurring: "But to mention the intercourse of our Heroine with her sex, would, like others more dangerous, require an apology I know not how to make. It must be supposed, she acted more from necessity, than a voluntary impulse of passion; and no doubt, succeeded beyond her expectations, or desires. Harmless thing! A useful veteran in war!—An inoffensive companion in love!"[59] Describing Sampson's romances as stemming from "necessity" rather than a "voluntary impulse of passion" also tangentially reverberates with the renaturalizing narrative strategies that I have already identified as operating within *The Female Review;* this statement classifies her romances as a biological inevitability, like breathing or regeneration. Furthermore, it renders these romances—alongside Sampson/Shurtliff's unconventional sexed behavior —a natural outgrowth of "Columbia's soil." There is no iteration of Sampson/Shurtliff's sapphism or decision to live as a man that can be understood, in this text, outside of the political and geographical context of the newly formed United States. Indeed, even in the very first scene where the reader encounters Sampson in men's attire, it is not just any suit of clothes that s/he dons; it is a military uniform. The nation thus bears some implicit explanatory power here, and these behaviors, in turn, have something to do with America.

"She Kills Her Indian": Sapphism and the Indigenization of Whiteness

As we have seen, in *The Female Review* Herman Mann both explains and excuses Sampson's unconventional sexed and sexual behavior—both her decision to live as a man and her subsequent romantic dalliances with women—as a specific, natural symptom of her sociopolitical time and place. The combination of revolutionary sentiment and the upheaval of war, the narrator tells us, produced the "sensations and emotions different from what [Sampson] had before experienced" that in turn led to what her contemporaries would have understood as her sapphic behavior.[60] Yet this rhetoric of naturalization is also characteristic of narratives about settlement. By depicting Sampson as a natural wonder, a unique production of the geographical and political specificity of the brand new United States, Mann endows his representation of Sampson with a sense of nativeness, an effect that Scott Morgensen has termed the "indigenisation of white settlers and settler nations."[61] The natural historical framework of botany and natural taxonomy, and its concomitant interest in native flora and fauna, is overlaid here with a series of questions that *The Female Review* poses about the status of Revolutionary War soldiers as "founders" of the United States. What brings these questions together, of course, is the defense of Sampson's unique sexed and sexual behavior.

For a text about a soldier, *The Female Review* contains remarkably few battle scenes. The novel does offer a few brief accounts of skirmishes between the Americans and the French, but the lion's share of the narration of Sampson/Shurtliff's labor in the Continental Army is dedicated to his/her time scouting on the frontiers of the American colonies, from western Virginia up through what is now West Virginia and Ohio. The battle scenes highlighted in *The Female Review* are not conflicts between the Continental Army and the armies of European imperial powers, but rather conflicts between Sampson/Shurtliff's scouting battalion and Native peoples. Indeed, Sampson/Shurtliff's war heroism, in this text, hinges on his/her murder of a Native man while on a scouting mission. It is in this scene that both his/her patriotism and the success of his/her masculinity are cemented.

Importantly, Sampson/Shurtliff is discovered to have breasts *before* s/he is assigned to these scouting missions. After receiving a bullet to the thigh during an engagement with an opposing army, Sampson/Shurtliff digs out the bullet but never fully heals and later falls sick during an epidemic of an unnamed disease that circulates through the army. A doctor by the name of Bana then treats him/her and finds that Sampson/Shurtliff has breasts. Sampson/Shurtliff begs the doctor not to reveal his/her secret and proceeds to his/her next assignment as a scout.

The landscapes that s/he encounters in the Ohio and Virginia frontiers are described in a very different idiom than the one used to portray Sampson/Shurtliff's home of Massachusetts. Mann uses the word "primeval" several times throughout his descriptions of these frontiers and characterizes these places as untouched, areas where "human feet never before trod."[62] This landscape is also filled with both Native peoples and Tories—newfound abjects of the incipient nation—whom Sampson/Shurtliff and his/her scouting group encounter at multiple points throughout chapters 11 and 12 of the book. And it is during these scouting ventures that Sampson/Shurtliff demonstrates his/her military heroism. After being accidentally separated from his/her scouting group, Sampson/Shurtliff finds himself/herself in the company of two Native people, an adult man and a boy. Traveling together, they kill a buffalo and three turkeys. During these scenes, the narrator suggests that the Native man, who is never named, grows more and more irritated with Sampson/Shurtliff, the implication being that he is jealous of Sampson/Shurtliff's hunting prowess; the narrator notes that the "Indian discovered much envy."[63] Sampson/Shurtliff is uneasy about this man and cannot sleep that evening. This, as it turns out, is a good thing, for after dark the Native man makes an attempt on Sampson/Shurtliff's life; the narrator recounts that "without hesitating, she leaped upon her feet, and shot him through the breast, before he had time to beg quarters."[64] That the narrator specifies that Sampson/Shurtliff kills the Native man "without hesitating" is important here; it demonstrates the degree to which Sampson/Shurtliff's self-defense is understood to be instinctual and thus natural. Sampson/Shurtliff has, as the subtitle to the chapter indicates, "killed her Indian"; it suggests that the murder of a Native man is something of a rite of passage for a soldier, and specifically for an American soldier. The possessive "her" (Sampson/Shurtliff has killed *her* Indian, not *an* Indian) implies not only the righteousness of the murder itself but that Sampson/Shurtliff had a rightful claim to the act or experience of killing a Native person. Implied in this possessive language—*her* Indian—is the notion that his/her right or requirement to kill a Native person stems from Sampson/Shurtliff's status as an American. Inherent in his/her status as an American soldier is not only the right but the expectation that s/he kill Native people. At the same time, this murder scene reinforces the reader's sense of the authenticity of Sampson/Shurtliff's performance of masculinity; the success of his/her sexual ruse appears most clearly in this scene of Native death.

The eradication and forced transformation of Native gender roles and sociosexual configurations is a hallmark of settler colonial violence. From Spanish conquistadors who supposedly set their dogs on Native people whom the Spanish perceived to be sodomites, to French colonial administrators who incentivized French settlers to marry Native women to generate a population of stable residents that would maintain possession of colonized territory, the relationship between settler colonial violence and gendered and sexual order is a close and extremely well documented one.[65] Yet often scholarly accounts of the gendered and sexual violence that attended settler colonialism focus exclusively on practices that were eradicated, and not on the largely European-American forms of sociosexual organization that were deployed to forcibly decenter diverse Native sexual and gendered orders. Differently put, scholarship on settler violence has at times failed to consider the specific contours of *settler* sexed and sexual behavior. Scholarship on the intersections between histories of settler violence and histories of sexuality is critically important, but we must be careful, in these efforts, to avoid inadvertently reinscribing the settler sexed and sexual practices as the invisible standard from which diverse Native practices departed. Settler sexual behaviors should themselves constitute discrete objects of inquiry, especially because many of these practices were weaponized—as we see in Sampson/Shurtliff's case—to assist in the project of asserting the indigeneity of European-American settlers and their claim to Native lands.

In particular, sapphism and sapphic behaviors more generally were typically associated with not only European-Americans at the end of the eighteenth century, but with Europeans—full stop. France, and the French aristocracy in particular, was depicted in popular anti-Jacobin North American magazines as teeming with sapphists; Marie Antoinette was perhaps the best-known French figure associated with sapphism.[66] Even the "frenchified" city of Philadelphia, known as such because it was a destination for both French and Haitian migrants, was imagined to have a particular friendliness or predilection toward sapphism. Unlike behaviors such as sodomy, sapphism tended to be associated with the kinds of cultural overcultivation that characterized highly "civilized" parts of the world: the temperate zones and especially cosmopolitan European cities, where the twinned popularity of men's fashion and women's education amounted to a blurring of the lines between masculinity and femininity, men and women. As we have seen, both *The Female Review* and Deborah Sampson herself found in contemporary literary culture, as well as in recent Anglo-Atlantic history, a long series of precedents in female picaresque novels such as *Constantius and Pulchera* and in historical figures such as Hannah Snell. What unites all of these precedents is their situatedness in a particular vision of white, Anglo-American popular culture.

To speak of "white" culture in this era constitutes somewhat of an anachronism, at least relative to our use of "white" today. The 1790s did see a burgeoning identification by white people with other white people that specifically understood this identification to hinge on a chromatist understanding of the term "white," but class, ethnic, and religious differences between various European and European-American peoples were still incredibly important and largely determinative of social order. Furthermore, not everyone whom we think of as "white" today—Italians, for example, and Jews—was considered unambiguously white at the end of the eighteenth century. Yet the 1790s also saw some of the new nation's first restrictions on both immigration and civic enfranchisement through citizenship, and in many of these new restrictions we find an official vocabulary of whiteness emerging that reverberated through later calls for solidarity among white workers organized by the labor movement.[67] Here I simply am offering a reminder of the complexities of imagining a singular "white" culture, in our current moment or in the early national period, and warning against an understanding of that culture that imagines it to be somehow hermetic. As Scott Morgensen reminds us, "Rather than presuming that the [settler culture] is defined by enforcing boundaries to preserve purity, we must consider that the state of exception arises in settler societies as a function of settlers' inherent interdependence with indigeneity."[68] What Morgensen is asking us to consider, then, is that many facets of settler culture were not strictly European or strictly "white," but rather an intentional or inadvertent strategic feature of a process of violent settlement that sought to replace certain (Native) versions of gendered or sexual behavior with other versions.

Reading Sampson's sapphism within a historical frame that interprets the Revolutionary War—and Sampson's own work within it—as a facet of settler colonialism changes the meaning we can attribute to Sampson's unconventional behaviors. How might we imagine Sampson/Shurtliff's sapphism as a form of what might be termed settler gender or settler sexuality? What is the relationship between Sampson's status as a "virago"—a contemporary term for masculine women—and whiteness?[69] Mann's enlistment of the language of botany and the exception in nature to paint Sampson as "the most unparalleled" and "the most singular" kind of being "that ha[s] ever sprung out of Columbia's soil" reveals the sympathies between Mann's apologia for Sampson's sexed and sexual behavior and the refusal of Native claims to sovereignty over the land. If Sampson is a natural product of Columbia's soil, then Sampson's sexed and sexual behavior, too, is a feature of the American claim to Native land. Indeed, it is precisely Sampson's sapphic behavior that provides the foundations for her claim to a "native" status, to be indigenous to America. But it comes, as it must, at great cost to the peoples who possess a rightful claim to the territory that is now, for example, Massachusetts, Virginia, and Ohio, among others. In *The Female Review*, Sampson's peculiar sapphic "sensations and emotions" thus cannot be distinguished from the imperative to "kill her Indian," and these unconventional sexed and sexual tendencies are weaponized in turn to enforce settler claims that assert an original or native right to Columbia's soil.

"We Are in the Dark"

Alfred Young's 2004 biography of Deborah Sampson, *Masquerade*, represents the most recent scholarly effort to shed light on her sexual proclivities. At the book's end, Young unequivocally denies her lesbianism.[70] Regarding several episodes that suggest Sampson's sexual dalliances with women in her town of Middleborough, Massachusetts, however, Young demurs, stating simply that "we are in the dark" as to their significance or veracity.

A central tension of this chapter is whether we are, in fact, in the dark, and additionally, what we might be in the dark *about*. I am interested in this reticence, expressed by Young as well as by many other critics of *The Female Review* and by historians of Sampson's life more generally; these scholars have highlighted, emphatically and repeatedly, all that we do not or cannot know about Deborah Sampson. But that reticence becomes even more interesting when contraposed with historicist readings of her behavior that serve to naturalize or make quotidian precisely those facts of her life that would later ensure its historical staying power. Historians and literary critics alike seem to take great pains to contextualize, to historicize, and to situate Sampson as an unsurprising product of her time—not unlike Mann in his narration of Sampson's life. The attitude with which contemporary critics portray Sampson, however, remains inconsistent with the rather difficult-to-access yet growing archive of materials on Sampson's life and Mann's representation thereof. It is unusual, as many scholars of early American literatures would surely concur, to find such a relatively extensive body of scholarship surrounding the life of a poor, formerly indentured soldier who fought for less than three years during the later portion of the Revolutionary War. The many disavowals of Sampson's singularity are belied by the many attentions that have been aimed in her direction over the past two centuries.

Ultimately, I think this idea that "we are in the dark" about Sampson's life or motivations, expressed almost universally though in variegated terms by everyone from her first biographer, Herman Mann, to her most recent, Alfred Young, is actually a representational trend that has appended itself to historical knowledge about sexuality. For what is not being spoken when we identify ourselves as "in the dark" is precisely what we are in the dark *about*. We know much, much more about Deborah Sampson, for example, than we do about other poor enlisted men who fought during the Revolutionary War. But what we don't know about Deborah Sampson is the thing that fascinates us most: what relationship, if any, can we, as twenty-first-century scholars of the social formation that we call sexuality, draw between Sampson's gender presentation and sexual proclivities—which feel so seductively recognizable to us, even though they occurred so long ago—and those presentations and proclivities that exist today?

Furthermore, in an age of what has increasingly come to be called "white feminism" (and Sampson has become a heroine for proponents of precisely this politics), how do we account for the racialized genealogies of specific gendered and sexual practices? Or for their racialization in the present?[71] Is it possible that we are not just "in the dark" as to what relationship, if any, Sampson's experience of himself or herself, or of his or her behavior, bears to modern-day queerness or transness, but that we are also in the dark about the relationship these histories might bear to settler colonial practices or early nationalisms?

There is another side to what at times feels like a critical commitment to not knowing (the act of knowing, of course, being as historical a construction as anything else) or to simply making the decision to identify Deborah Sampson as a lesbian or a transgendered person *avant la lettre*. The body of critical work that focuses in any way on Mann's *Female Review,* or on Deborah Sampson more generally, contains endless denials of her lesbianism, despite the existence of as many stories, "authenticated" or otherwise, requiring these denials. This critical push and pull—or, perhaps more appropriately, call and response—is important because it emphasizes the ontological chasm between what appear to be identical objects of knowledge. A further epistemological and ethical quandary is posed when we ask whether these objects of knowledge (same-sex sex acts in the eighteenth century and same-sex sex acts today, for example) differ in some fundamental way owing to historically distinct modes of perceiving them, or whether they differ only insofar as separate vocabularies exist to describe them. One of the many fascinating things about Deborah Sampson's story is precisely how much *The Female Review* resembles, at times, the present-day genre of lesbian historical romance, such as books like Isabel Miller's *Patience and Sarah* or the BBC series *Tipping the Velvet* (set in the Victorian era); despite the two centuries separating these texts, they read like literary kin.[72] Similarly, it is difficult to read Mann's novel without feeling drawn to the kind of reading that Jay Prosser offers of the later, but still resonant, protagonist of Radclyffe Hall's *Well of Loneliness:* Prosser identifies Stephen Gordon as experiencing a kind of gender inversion before transsexuality existed as a category of gender experience.[73] At the same time as I recognize, or perhaps misrecognize, reverberations of present-day lesbian cultural representations in *The Female Review,* I am also well aware of the extreme paucity, a practical nonexistence, of records suggesting that anything like modern sexual identity was available in the late eighteenth century. As a scholar of what I increasingly think of as sexual opacity, I would describe my analysis of these exciting but frustrating episodes of historical sex as being precipitous, literally on the edge of two impulses: the first is the wry, Jamesonian exhortation to "always historicize" and reject the idea that modern sexual formations could be represented without the language of modern sexuality; and the second is to submit to the seduction of identification, to see usable history in these moments, and to respond to the call of what many scholars would consider the anachronistic abyss.

Despite the paralysis that the tension between these two impulses creates, I want to return to precipitousness as a way of conceptualizing presexology sexuality that can incorporate both of these relationships to history. The implicit sexual politics of botanical taxonomy and its deployment as a mode of representing unfamiliar sex acts and behaviors at the end of the eighteenth century provides a model that can mobilize both of these disparate historiographic approaches into some kind of joined, albeit messy, history of aberrant presexology sexuality. The cultural conditions that allowed for scientific inquiry into sexuality, initiated by popular botany, simultaneously created a form of knowledge about sexuality marked by illegibility or opacity; scientific inquiry often exposes what we do not know as frequently as it produces new knowledge. Popular botany does offer present-day scholars a new means of conceptualizing how residents of early national America might have come to describe and understand Deborah Sampson; however, popular botany obscured as much as it illuminated, and contributed to the construction of new forms of epistemological opacity as well as to the fetishization of unfamiliarity. Rather than trying to illuminate or eradicate moments when the historical record surrounding sexual and gender aberrance leave us without answers, we must amend our historiographical approach to presexology sexuality and consider opacity not as a veil hiding any "truth" of sexual behavior or experience but as a cultural practice in its own right, bearing its own "characteristic climates and textures."[74]

Epistemological precipitousness offers us a means of thinking about *how to think about* what we may or may not be "in the dark" about, in our research pertaining to Sampson's life and times. Furthermore, this approach to knowledge is one that echoes practices drawn from Sampson's own eighteenth-century moment. As Susan Scott Parrish has demonstrated, epistemological precipitousness (an orientation toward knowledge that makes space for the unfamiliar, the unknown, or the putatively new) was at the heart of Enlightenment-era scientific inquiry, and natural history in particular. Another means of expressing the aforementioned representative trope of opacity, or being on the verge of or to the side of knowledge, is to insist that we carve out a place in the history of sexuality for precipitousness: in individuals, in knowledge, in narrative, and in contemporary cultural vocabularies for assessing difference. Sampson herself retrospectively described her decision to enlist in the Continental Army in terms of precipitousness: "I swerved from the accustomed flowry paths of *female delicacy,* to walk upon the heroic precipice of feminine perdition!"[75] The precipitousness with which Sampson characterizes her experience is defined by absence, loss, or lack—the *perdition* of femininity—rather than by a positive or identifiable change of state; while Sampson describes the "precipice" of "feminine perdition" as "heroic," the precise nature of feminine perdition itself remains opaque. In other words, unfamiliarity and opacity, within both the historical record and the experience of our late-eighteenth-century subjects of inquiry, are both familiar and generative experiences.

Unfamiliarity and opacity, however, may seem so familiar and generative precisely because the sense of absence or lack upon which they turn walks so closely in step with the logic of colonialism and settlement. As we have seen, the narration of *The Female Review* veritably teems with representations of the newness of Columbia (the United States): its soil, its zeitgeist or broader political feeling, and its people. Yet when Mann describes Sampson/Shurtliff, during his/her nearly fatal scouting mission, as walking in a "primeval" wilderness where "human feet never before trod," it is impossible not to see the settler lie at the heart of this articulation: Native feet had been walking that land for hundreds of years, and both the narrator and the reader know this, because Sampson/Shurtliff was accompanied by some of those very Native people in his/her scouting journey.[76] The precipitousness that inheres in the idea of land where "human feet never before trod" is belied by the immediate presence of Sampson/Shurtliff's guide and companions. Thus, we see how epistemological precipitousness also lies at the heart of settler logic, insisting on the unfamiliarity and newness of lands that are neither unfamiliar nor new to those who actually inhabit and control them, because imagining land as untrod imagines it as open to claim and possession. The same knowledge framework that allows Sampson to be understood as a new, natural wonder native to Columbia's soil also asserts the newness of Columbia's soil *tout court* and thus insists on its status as an object that is not only accommodating but inviting of exploration, taxonomization, and possession. The status of the unknown or unfamiliar at the end of the eighteenth-century should not, then, remain understated or overlooked. Understanding precipitousness as a framework that informs the way eighteenth-century sexual behaviors were explained and understood reveals the sympathies between putatively new forms of sexual practice, and both new and emergent forms of state governance and control.

In chapter 5 we travel from Dedham, Massachusetts, to downtown Boston, a city that experienced a massive boom in population in the early national period. That chapter considers a different genre of cross-dressing narratives: those that feature young, unmarried women who dress as men and enlist in the military to escape their former lives as sex workers in Boston's vice districts. These enormously popular narratives—most of them short and cheaply printed in pamphlet form—cite Sampson's cross-dressing military bravery as a forebear but espouse a very different relationship to the sexual politics of the day. Whereas Sampson's unconventional sexed and sexual behavior is explained as the felicitous, if exceptional, result of her exposure to the natural environs of rural Massachusetts, the decline into prostitution experienced by the female protagonists at the heart of *The Female Marine* texts is explained as a result of their prolonged exposure to the degenerating influence of the city. Demonstrating unequivocally that the vicissitudes of environmental influence can produce both positive and negative sexual behaviors, chapter 5 focuses on the municipal leaders, public health officials, and reform organizations who worked to eradicate urban vice districts and who theorized urban space as a defining influence in the rise of the sex trade.

CHAPTER FIVE

"Negro Hill" and the Sexuality of Space

The first that I shall mention, is the melancholy case of a youth bred in the country, and at the age of 16 apprenticed to a respectable merchant in town.... But a few months had he been in town, when a curiosity peculiar to too many in the situation of himself, induced him one fatal night to resort to the seat of perpetual riot and dissipation. As he descended the filthy hill, the discordant sounds of the musical instruments unconsciously drew him to the entrance of one of their dancing cabins; he entered, and mingled with the haggard crew—but, the unsuspecting youth did not long remain an idle spectator—from one of the "marms" most forward pupils (who probably had already marked him as her victim) he received a pressing invitation to join their dance!
— *The Adventures of Lucy Brewer (Alias) Louisa Baker*, 1815

About noon I unconsciously ascended the heights of West-Boston, where I renewed my application for a place, never could the hydra wish for a better chance for his prey, within the walls of a house, the external appearance of which was spacious, I was welcomed and treated with the attention and with that apparent cordiality, which my depressed spirits at that moment really required.
— *An Affecting Narrative of Louisa Baker, a Native of Massachusetts*, 1816

During the first three decades of the nineteenth century, there was no section of Boston more infamous than "the Hill." Located on the north slope of Beacon Hill in Boston's West End, this "seat of perpetual riot and dissipation" was one of Boston's best-known vice districts, and it had been colloquially deemed, from the mid-eighteenth century onward, "the Hill," "Negro Hill," and "Mount Whoredom." Scholars such as Walter Whitehill speculate that the area was initially a favorite of working prostitutes and tavern owners because of its proximity to the Boston ports and the sailors who worked there; "Mount Whoredom" appears explicitly labeled on maps as early as the 1770s, and the epithet was used to designate that area of Boston's West End as early as 1733.[1] By the end of the eighteenth century, "the Hill" was also home to a large and racially mixed community of what we would now term "working-class" black and white residents; "as early as 1810, more than half of Boston's black population resided on or near 'Negro Hill.'"[2] Because it was also one of the first areas to be defined as a "vice district" by the Boston municipal government, it became a persistent target for burgeoning early national reform movements. The Hill thus provides a generative case study of the way "vice"—as a term linked to polity and broader social order—took on increasingly spatial dimensions during this period; policing (with the goal of eliminating) spaces that promoted vice, such as brothels, unlicensed taverns, and bawdy theaters, was understood to be central to the project of eliminating vicious behaviors. And, as we shall see, early national concerns about urban vice enlisted emergent social sciences, such as demography and theories of population, to calculate a prediction for vicious behaviors, and thus strategize for their prevention, at the level of the social rather than of the individual. With the boom in the population of early national cities came a new dominant understanding of vice, a view that understood it as not only a spiritual problem but also a spatial one that could be addressed by strategic urban planning and other forms of city infrastructure. Building on previous understandings of vice as a symptom of the poor spiritual health of a community or congregation, early national understandings of vice reconfigured it as a social affliction that could compromise the economic and even literal health of an urban population. This highly spatialized understanding of vice, steeped as it was in populational science, was critical to the development of sexology as an arm of the human sciences specifically devoted to the study of sex as a discrete item of inquiry.[3]

This chapter examines a broad, historically diverse, and eclectic collection of materials pertaining to Boston's West End, and "the Hill" in particular, to describe how vice was understood to be, at least in part, the result of contact with one's physical and social environments. While many different texts animate the analysis that follows (popular narrative and verse, maps, Bible school association reports, annual statements from prostitution reform organizations, directories of local sex workers, and sermons), I pay special attention to a series of short, cheaply printed pamphlets that were printed and reprinted in the Boston area between 1815 and 1818. *The Adventures of Lucy Brewer* (1815), the first of the series, was followed by a set of variously titled, related texts that were written and printed by the same man and that generally tell the same story (in this chapter, following Daniel Cohen, I refer to them collectively as the *Female Marine* texts).[4] *The Adventures of Lucy Brewer* was reprinted as *The Adventures of Louisa Baker* (August 1815), *An Affecting Narrative of Louisa Baker* (September 1815), *The Adventures of Lucy Brewer* (November 1815), *The Awful Beacon* (May 1816), *The Female Marine* (January 1816), *A Brief Reply to the Late Writings of Louisa Baker* (June 1816), and *The Surprising Adventures of Almira Paul* (September 1816). All of these narratives feature a female protagonist who leaves her home and family in a bucolic town in rural Massachusetts, after being seduced and left pregnant by a local libertine.[5] Running away to Boston, the naive protagonist unwittingly seeks refuge in a brothel in Boston's West End and ends up indentured as a sex worker for three years. After escaping by stealing the clothes of one of her male clients, Lucy/Louisa joins the marines, spends some time at sea, and eventually returns to her overjoyed parents. Capitalizing on the popularity of late-eighteenth-century popular picaresque fiction, and of seduction stories in particular,[6] *The Adventures of Louisa Baker* combines a number of popular generic conventions into a short narrative, as popular as it was inexpensive.[7]

Across numerous reprintings of this one story, an obvious pair of worries announce themselves: an explicit concern with the potentially corrosive effects of urbanization, and broad social fears about the prevalence of sexual immorality (prostitution in particular) in the new nation's cities. Historians of sexuality have fruitfully charted how industrialization and, later, urbanity gave rise to specific sexual practices in the late nineteenth century, the moment of sexuality's putative "modernity." This chapter, however, turns to a moment almost a century earlier, to highlight the way this fear of urban influence, across the nineteenth century, relies on an environmentalist logic that sees the milieu of urban environments—the social, natural, even architectural conditions—as the root of sexual danger.[8] This broad collection of texts voice a fear about the possibly contagious or proliferative qualities of urban vice, a fear that in turn gestures to the way specific forms of sexual behavior (sex work, but also "venery" more generally) were understood to be endemic to particular spatial configurations, such as neighborhoods or architectures. In the *Female Marine* texts, we find several familiar tropes surrounding the corrosiveness of cities: the decreasing space between bodies in urban environments, the higher rate of contact between black people and white people in poorer neighborhoods, and even just the dirtier and more noxious features of urban spaces (a lack of organized systems for sewage treatment, for example). All of these are deployed to explain how urban vice exerts significant influence on the residents of such areas, especially young, unassuming men and women from the country. This chapter takes these concerns seriously and listens closely to the narratives they contain about the sexuality of space, and especially urban space.

The management of urban space was a critical concern in early-nineteenth-century Boston; local politicians, religious organizations, and property owners were all faced with the complicated question of how to manage increasing populations of poor residents. At the same time, the East Coast's most notorious vice districts—Boston's West End and North End, Providence's Hard-Scrabble neighborhood, and New York's Five Points—were historically racially mixed areas; strategies for managing urban poverty were thus frequently shot through with strategies for maintaining white supremacy. Indeed, the conflation of mixed neighborhoods, prostitution, and poverty is a feature of each of the texts at the heart of this chapter, and one that makes them exceptionally difficult to work with, because that conflation represents both a historical reality and a thick, racist cliché.

Because sex provided such a labile tool for early racialization, texts that represented and discussed Boston's "Negro Hill" offer a potent site for thinking about how sexuality was understood to adhere to, and perhaps even reside or originate in, specific and often highly racialized spaces. In these texts, the architectural and social environment of Negro Hill, its sonic and olfactory features, and even the land upon which it sits all combine to exert a heavy influence on the sexual behavior of those who enter its purview. Indeed, in these texts, individual inclination, what is sometimes termed "agency," is importantly vexed; "unsuspecting" youths are frequently drawn to the Hill inadvertently or "unconsciously."[9] This chapter's investigation into the afterlives of eighteenth-century racial science in early-nineteenth-century urban reform considers environmental theories of early sexuality on the eve of the emergence of sexology. Those theories, as I discuss in the epilogue, initiated a new way of thinking about sexual behavior that wedded sex much more tightly to the subjective.

Sex and the City

Like Herman Mann's *Female Review,* the anonymous *Adventures of Louisa Baker* and the subsequent spinoffs of that title belong to a genre of popular print that Daniel Cohen has termed "cross-dressing narratives" and Cathy Davidson has called "the female picaresque," that features stories of young, single women who leave their homes, dress as men, and join the military, often as a result of being seduced.[10] These narratives first became popular in England in the mid-eighteenth century and flourished in late-eighteenth- and early-nineteenth-century North America. The circulation and number of reprintings that the *Louisa Baker* narratives saw in New England, however, was dramatically different from the circulation of Mann's novel. Despite the staying power that Sampson's narrative maintained across different contexts, *The Female Review* could not boast of more than about two hundred subscribers. And although it was reprinted numerous times over the course of the eighteenth, nineteenth, and twentieth centuries, by most standards it was not a particularly widely read or available text. While there were certainly a wide range of printers and a robust market of readers in Massachusetts in the 1790s that assured *The Female Review* of its publication and consumption, when the Louisa Baker narratives appeared just two decades later, the New England print market had changed dramatically. Although the penny press did not come into being until 1830, the means of printing cheaply and quickly had expanded, and short publications like *The Adventures of Louisa Baker* and its related works could boast of more than fourteen editions in a year (1816). The print proliferation of these narratives can be attributed at least in part to two factors: first, the market demand for them, and second, the price. Sold for twelve and a half cents apiece, these short texts were "accessible to all but the most impoverished of readers."[11]

Furthermore, the *Female Marine* texts trafficked in a kind of story for which there had long been a well-established market. As Cohen and DeLombard remind us, the popular demand for cheap literature had in large part grown out of a much earlier print market organized around stories of crime, punishment, and execution.[12] The narrative itself explicitly evokes, and at times simply reproduces, both text and image from other works, among them editions of Susanna Rowson's enormously popular *Charlotte Temple* and Mann's *Female Review*.[13] Borrowing the pedagogical narration frequently characteristic of seduction stories, the *Female Marine* texts announce themselves "worthy of the perusal of young persons of both sexes, and of all classes,"[14] although the primary audience for *The Adventures of Louisa Baker* and the related texts was likely sailors, as well as "'the Ann Street population of that day' (i.e., prostitutes and their compatriots), and 'juveniles.'" These texts became so popular that, as Daniel Cohen speculates, "for a period of a few years . . . [they] must have been among the most widely circulated pamphlets in Boston."[15] It is the wide circulation of these texts that justifies their centrality here; representational tropes and sometimes even passages from *The Adventures of Louisa Baker* and subsequent editions of the *Female Marine* texts were frequently reproduced almost verbatim in publications from missionary societies, sermons, and other forms of more or less "official" communication that were eventually cited by the local government in the 1820s in some of the earliest mobilized efforts to stymie the growth of urban vice districts.

It is no coincidence that the great popularity of these texts, and the myriad uses they were put to, was concurrent with the unprecedented growth of Boston during the first three decades of the nineteenth century and its eventual incorporation as a city. The city's explosive growth looms large over the ideological landscape of this chapter, because the logic undergirding urban development in this moment, and especially that undergirding the development of strategies for citywide, integrated implementation of law and order, was profoundly influenced by early demographic theories such as those of Thomas Malthus, who newly understood the population as the prime target of state governance and management.[16] Early population science was heavily concerned with both sexual behavior and the management of public space, as declining mortality rates, coupled with rapid urbanization, dramatically altered the density and resources of cities. Prostitution, in particular, is consistently represented in both reform literature and jeremiads within an idiom of statistical analysis; for example, after the publication of McDowall's *Magdalen Report* in 1831, Asa Hopkins wrote in 1834, "It is morally certain that we have, in the *centres of influence* in our land, fifty thousand female devotees to this vice [prostitution], sustained at an expense of thirty millions of dollars per annum."[17] Furthermore, the population boom in Boston by the time of the 1822 incorporation of the city came on the heels of an international interest in theorizing the population as the constitutive object of governance more generally. By the early nineteenth century, Boston had grown from a town of 16,382 in 1743, to 18,038 in 1790 (an increase of fewer than 2,000 people in forty-seven years), but then it swelled to 43,000 by 1820 and to 57,000 by 1825.[18]

This rapid growth was especially evident in the North End and West End districts, which quickly became the mostly densely populated of the various neighborhoods of the city.[19] The West End, already quite visible as the longtime home of Boston's largest African American community, had long been associated with sex work and drunkenness, owing to its location (near the North End, later rebranded "Mount Vernon") and to its having "the most transients and the highest concentrations of taverns and brothels."[20] As early as the 1770s, Boston had accrued a reputation for housing a thriving sex economy; one "Lieutenant Williams of His Majesty's Twenty-Third Regiment, who . . . arriv[ed] in Boston in June 1775, noted in his journal: 'No such thing as a play house, they were too puritanical a set to admit of such lewd Diversions, tho'. . . ther's perhaps no town of its size cou'd turn out more whores than this cou'd. They have left us an ample sample of them.'"[21] The status of the West End as a vice district was further solidified when the nearby area, Mount Vernon, was increasingly developed by wealthy proprietors during the first quarter of the nineteenth century.[22] As Walter Whitehill writes of Negro Hill, "this mangy area on the northern slope of the hill—well isolated from the handsome development of the Mount Vernon proprietors—continued as a source of disorder and subject of complaint."[23] The North End, on the other hand, "the oldest part of the city, had once been the home of the illustrious Mathers, Paul Revere, and some of the richest colonial Bostonians, but by the late eighteenth century it was already in a state of decline."[24] In a story that will ring familiar to many present-day readers, Boston's North End and West End did not become a target for municipal intervention until property values rose; "For years citizens' complaints of disturbances on the Hill had been virtually ignored by the selectmen. Brothels, dance halls, and gambling houses operated seven days a week without interference."[25] These areas that had historically housed a thriving sex work industry with relatively little institutional intervention (a fact that also reflects the status of Boston as an unincorporated city before 1822) became targets for reform as Boston slowly filled in the Back Bay and wealthier residents returned to Mount Vernon; "for West End residents, disorderly houses represented more than a nuisance; they were a threat to property values, a point made in a petition signed by over a hundred persons in 1820."[26] The values at stake, for both local residents and early reform organizations, concerned far more than just property, though. By 1815 or so, groups of artisans and guildsmen were pressuring local officials to address the perceived problem of the "brothels, dance halls, and gambling houses," sometimes dressing in blackface and assembling in mobs to riot and destroy both brothels and houses of African American residents of the neighborhood.[27] Evangelical reformers had organized a system of Bible schools in an effort to save the souls of the youth of these neighborhoods, and several charitable organizations were founded to save women from prostitution.[28] These developments reflect a trend seen throughout urban areas up and down the East Coast that understood even the most quotidian elements of urban life to be a threat to both individual and social morality. In New York, Providence, Philadelphia, and other cities, there were similar attempts to eradicate urban vice, and ministers in rural areas all over the Northeast warned parishioners of the dangers of city living. Thus the city is portrayed quite similarly, as a potential spiritual and physical pathogen, in both religious and public health texts and in licentious popular literature.[29] During the first three decades of the nineteenth century, then, vice districts became both objects of popular interest and representation and targets for intervention in a way they had not been before.

But to describe the Hill and Boston's North End as "vice districts" in the early nineteenth century is perhaps to get ahead of ourselves, since it is precisely during this period when vice districts came into being. This is not to say that specific neighborhoods or even entire cities were not, at earlier times, associated with or characterized by the proliferation of sex work, drunkenness, gambling, or other illegal or socially problematic behaviors practiced therein. In the early national period, however, vice, alongside the less problematic formation with which it was frequently coupled, virtue, described not only a set of behaviors but also a legal and extralegal political sphere that was framed in not only characterological, but also economic, terms.[30] Scholars such as Carroll Smith-Rosenberg and Wai Chee Dimock have carefully traced the transformation of virtue from a masculine politics of character in the late eighteenth century into a feminized politics of sexuality in the nineteenth, a transition that Dimock has termed "the feminization of virtue."[31] Vice, as virtue's shadowy twin, was similarly feminized by the early nineteenth century and also became both a characterological and a spiritual problem. For example, let us return to Asa Hopkins's panicked claim mentioned earlier: "It is morally certain that we have, in the *centres of influence* in our land, fifty thousand female devotees to this vice [prostitution], sustained at an expense of thirty millions of dollars per annum."[32] He frames his jeremiad against prostitution in economic as well as moral terms, a rhetorical tactic common in discussions of sex work from this era. Prostitution, for Hopkins, represents not only a spiritual danger but also broader economic loss; in his vision, sex work is not only a crime but also a waste of public moneys. With the rise of economic liberalism in the young United States, vice of all kinds—sex work, drunkenness, gambling, and certain forms of poverty (that of the so-called "vicious" poor)—became a social problem, and one necessitating state intervention.

In the early-nineteenth-century United States, "vice" increasingly came to describe specific modes of life, habits, or behaviors, many of which were subsequently criminalized, rather than simply a broad, conceptual sinfulness.[33] The *Oxford English Dictionary* defines "vice" as "depravity or corruption of morals; evil, immoral, or wicked habits or conduct; indulgence in degrading pleasures or practices," and its nineteenth-century citations link vice specifically to poverty and to "beggary and plunder." Subsequent definitions refer to vice as "a habit or practice of an immoral, degrading, or wicked nature."[34] I emphasize these nineteenth-century uses of the term to highlight that "vice" was linked to specific behaviors, habits, forms of labor or of self-sustenance, and modes of living and was broadly associated, in very explicit terms, with poverty or the lives of the poor. Before becoming the first mayor of Boston, Josiah Quincy, who as mayor would be remembered for his targeted interventions into Boston's sex and alcohol economies, served as a municipal court judge and contributed to "two studies of pauperism in Massachusetts, one sponsored by the state legislature and the other by the Boston City Council." These reports contended that "vice, crime, and poverty were so often found together that they were in some sense inseparable; individuals who exhibited a propensity for one vicious trait were most likely to have a propensity for other vices"; such reports advanced "a new terminology that classified the poor into two groups. The sick, infirm, or widowed, who could not be blamed for their poverty, were the worthy, deserving, or virtuous poor. Those whose poverty was the result of indolence, sensuality, and vice belonged to the unworthy, undeserving, or vicious class of poor."[35] By the time of the incorporation of the city in 1822 and the beginning of Josiah Quincy's mayoral term in 1823, the challenge of supporting the city's impoverished residents and eradicating urban vice had become mutually imbricated efforts, frequently targeting the same neighborhoods and the same people.

This new definition of vice that gained traction as an item of concern within burgeoning, Boston-based municipal politics is important for two reasons. First, it linked vice to modes of living, including both good and bad habits (drinking, gambling, churchgoing), labor (whether one was working or "indolent"), physical environment (clean air and water, urban miasmas), and social milieu (beneficial or harmful relationships or influences), which meant that vice was conceptualized in collective and environmental terms, outside of or to the side of extant modes of theorizing vice as a spiritual problem at the level of the individual or the community.[36] The turn to demographic analysis in the wake of late-eighteenth-century theories of population further allowed for specific forms of vice, namely prostitution, to be quantified in large-scale, collective terms, rather than on an individual or even a local basis; as Asa Hopkins put it baldly in 1834, lewdness "depopulates heaven. It peoples hell."[37] As early as 1817, religious reformers conducted surveys to assess the extent of the "problem" of prostitution, and by the time John R. MacDowell's *Magdalen Report* was published in 1831, sex work was routinely represented in abstract statistical terms, and the "cost" of the oldest profession was being rendered in financial rather than spiritual terms.[38]

Second, one of the animating impulses behind the transformation of vice into a legal structure and a category of poor people ("the vicious poor") in Boston and other cities in the early-nineteenth-century United States was the intention on the part of religious reformers, concerned business owners, and city leadership to curtail the existence or spread of what they deemed "vicious" practices. Interested parties developed a range of tactics to intervene into what they perceived to be the spreading contagion of vice, including the targeting of specific geographical areas for reform through policing and terrorizing both residents and frequenters of vice districts through mob violence (an approach often deployed by groups of white guildsmen). Vice, then, at least when it came to municipal politics, was most strongly associated with *spaces* and *places*, rather than specific peoples. Although it is certainly no coincidence that poor, mixed-race neighborhoods became the first target for antivice reform efforts, the municipal interventions themselves were overwhelmingly infrastructural. Instead of targeting individuals, organizations, or classes of people, the new city of Boston targeted *neighborhoods*, and in so doing it followed a path that had been trod by religious reform organizations and angry mobs of artisans and guildsmen for more than two decades. Baptist reform organizations, for example, had organized a system of Sunday schools (specifically targeting the children of the African American community on the Hill in November 1816) aimed at curbing the proliferation of vice among youth.[39] In 1817 the Boston Female Society for Missionary Purposes commissioned two preachers to minister specifically to Boston's West End and North End neighborhoods. And Dan Cohen writes that in "September 1815, just a month after the publication of *The Adventures of Louisa Baker,* a mob of over twenty local laborers and others pulled down a 'disorderly house' in Boston, the first of several documented anti-brothel riots to take place in that city between the mid-1810s and mid-1820s."[40] These earlier efforts by both reformers and mobs of local residents laid the groundwork for later municipal interventions—namely, policing—that began with the incorporation of the city and intensified throughout Quincy's mayoral term. Quincy organized and carried out numerous raids on the Hill's brothels, which were frequently followed by mob violence against those same disorderly houses, such as the Bee-Hive Riots of 1825. In addition, "the North End and West End were each assigned a permanent constable to patrol their streets daily," and by the middle of the 1820s, "the number of prostitution cases reached three hundred a year. . . . The prosecution rate for Boston's population would never be higher."[41]

It is important to note that these raids did not target brothels alone, but also "disorderly houses" of all kinds, including unlicensed taverns, halls for gambling and dancing, and at times simply residential areas inhabited by poor people. While "fallen women" or prostitutes were certainly identified, in the popular literature and the reform efforts of this period, as some of the most tragic victims of vicious practices, sex workers were not yet isolated as a distinct class of criminals but were subsumed into the larger category of the vicious poor: the idle, dependent, and dissolute. Nineteenth-century understandings of vice, furthermore, saw vice less as inherent in specific behaviors than in ways of living: "The criminal law applied to prostitutes in the first wave of repression defined prostitution as a status or style of life rather than as the act of selling one's sex. Workhouses and other penal institutions that came of age in this era were aimed at the potential as well as the practicing prostitute."[42] Clearly, individuals were themselves just one facet of the broader problem of vice.

The advent of municipal disciplining of sex work is thus a hinge in our understanding of vice and of sexuality more generally. Many scholars consider these early efforts to understand and monitor sex work, by reform organizations, early public health institutions, and the police, to be the earliest instantiation of a formalized or institutionalized sexology, which in turn became one of the foundations of depth-model understandings of the subject advanced and popularized by psychoanalysis. Sexology, according to these scholarly narratives, gave birth to sexuality as a modern formation bound to the subject and to subjectivity more generally,[43] and our exploration of the Hill allows us to glimpse the moment just before this birth. Both the *Female Marine* texts and the surrounding reform literatures were printed, circulated, and discussed in the two decades directly before the birth of sexology. By directing our attention to these specific early-nineteenth-century histories of the ways prostitution, specifically, and vice more generally were conceptualized by religious reformers, municipal government, local business owners, and popular writers in specifically spatial or environmental rather than individual or internally spiritual terms, we can chart this early American vision of sex just before its imbrication in the subject.

The Sexuality of Space

Conceptualizing nineteenth-century cities as sites of danger or moral contagion is well-trodden scholarly ground.[44] The numerous *Female Marine* texts provide a remarkably consistent mirror of growing cultural concerns about cities. Throughout, these texts capitalize on the rhetoric of urban danger, as each narrative follows the same arc: the protagonist, usually named Lucy or Louisa, absconds from her parents' home in rural Massachusetts after being seduced by her lover and left pregnant and ashamed. She walks miles to Boston in a snowstorm and unwittingly finds respite at a house of ill repute. Evoking the narrative conventions of popular seduction fiction, the *Female Marine* texts and others like them relocate the site of seduction from the country to the city. While seduction fiction warned its ostensibly young, female audience of the dangers of falling prey to the charms of local libertines, the *Female Marine* texts and their contemporaries target both young men and young women—anyone at a distance from the rural safety of the parental home—and identify the threat of seduction as the city itself, especially the specifically urban figure of the marm or the prostitute. Rapid growth in urban areas, with the commensurate increase in proximity between bodies that attended this growth, exacerbated fears about cities as areas of both physiological and moral or spiritual contagion. Asa Hopkins refers to the vice-ridden as "the infected";[45] to "lewdness" as a "contagion [that] is spreading wider and still wider," a "canker" that "is eating deeper and still deeper"[46]; and to New York as a city with no fewer than "fifteen thousand dissolute women."[47]

This understanding of urban areas as categorically fostering individual inclinations toward vice also contributed to earlier efforts by religious communities—especially Baptists in Boston—to develop a system of Sunday schools for children, which they imagined as an attempt to counteract the lessons in viciousness and licentiousness that urban youth absorbed from their families and neighborhoods. The 1817 *Report of the Union Committee of the Sunday Schools of the Three Baptist Societies in Boston* suggests that urban youth, including but not limited to the youth inhabiting the communities in the West End and the North End of Boston, "have been accustomed to neglect the means of religious instructions" and "have contracted habits of vice which have led them to constantly violate the rest of the Sabbath, by engaging in idle and profane amusements."[48] The Sunday school movement, supported primarily by Baptist churches in Boston, tried to ensure that "instead of being in the streets the whole of the Sabbath, blaspheming in the name of God, fighting, pitching cents, &c. the children are in school at the ringing of the first bell."[49] The Sunday school movement had specifically targeted both the North End and the West End of Boston by 1817 for Sunday schools,[50] and by 1820 "missionaries [had] descended upon the city's poor neighborhoods like a swarm of locusts."[51]

Suspicions about the corrupting influence of urban areas were only heightened when it came to public perception of areas that supported large numbers of disorderly houses and robust sex economies. In both the *Female Marine* texts and missionary society reports, a specific set of tropes describe the highly specific environment of the Hill. Across these literatures, the Hill is represented as bearing a particular set of scenic, sonic, olfactory, geographical, architectural, and social qualities. By turning our attention to these tropes, we are better able to identify how urban space itself, and the interracial space of Negro Hill in particular, came to embody a form of sexual experience or behavior or, at the very least, became a container of sorts for one.

Sound is perhaps the most powerful sensory quality used to characterize the Hill across both the *Female Marine* texts and reform organizations' descriptions of the area. In the *Female Marine* texts, readers are not offered a representation of the soundscape of the Hill until the end of the text, when the narrator, now disguised as a man, returns to the Hill, to the brothel where she worked for three years, to see whether she will be recognized. In *The Adventures of Lucy Brewer* (the third, 1815 edition, which was printed under a slightly different title than the first edition, *The Adventures of Louisa Baker*), after a return to Boston in the suit of clothes that she donned during her time in the military, passing as a male marine, the narrator offers a powerful description of the sonic environment of the Hill:

> In the evening I resorted to the *Hill*, to take a final leave of this infamous seat of riot and dissipation, which I never calculate to visit more, whatever may be the character I may hereafter assume—the terrific yells of the blacks—the vile imprecations of the sailors, and their intoxicated strumpets—the discordant sounds of violins, claranets and tambarines, issuing from their stenchified "dancing halls"—and the perpetual howlings of their affrighted dogs! as I ascended Garden and passed Buttolf Streets, could not fail to remind me of days when I took myself an active part in these nocturnal revels![52]

The scene is, to say the least, cacophonous. The noises of people, musical instruments, and animals permeate the soundscape, contributing to the area's affective texture as well as hinting at the confusion, the disgust ("vile imprecations"), and the fear that, according to the narrator, fill the air in this neighborhood. More importantly, in this passage, fornication, lewdness, and perhaps even sex work are translated into sensory experiences: "the vile imprecations of the sailors" with their "intoxicated strumpets" are folded into a list of sonic qualities of the neighborhood. In other words, sex becomes dissipated as it is spatialized as *sound*. Blackness, too, is rendered in sonic terms: the passage identifies the black residents of Negro Hill by their "terrific"—terrifying—yells before it identifies them by their skin color or race. Another edition of this story, printed a little less than a year later, reproduces this scene but renders both blackness and lewdness more implicitly as elements of the sonic landscape: "The terrific and perpetual yells of these nocturnal disturbers of the public repose ever attended by the howlings of their affrighted dogs, is not perhaps equaled by that of the numerous hordes of the wild inhabitants of the Ganges: this continues with little intermission from the setting until the rising of the sun. The unsuspecting stranger who accidentally becomes a spectator of these midnight scenes is not only filled with disgust but trembles for his personal safety."[53] Although the text repeats specific words and images ("terrific . . . yells," "howlings" of dogs), this 1816 version has allowed the more explicit sexual implications of the 1815 version's "vile imprecations of the sailors" and "their intoxicated strumpets" to become subsumed as an implicit element of the soundscape. Sound carries the "race" of the scene here too, as the noises that characterize this neighborhood are "not perhaps equaled" by those of "the numerous hordes of the wild inhabitants of the Ganges." While the text is not immediately clear about who "these nocturnal disturbers of the public repose" are, the sounds that they make exert influence on any "accidental" spectator or passerby. The implication here is that this is a "dangerous" part of town and thus one should "tremble" for one's safety. But the passage can be read in another way as well: as evidence of the animate potential of the *space* of the neighborhood itself, of its ability to exert a corrosive influence on those who pass through it, or as simply a moment in which the text registers the vulnerability of the human body to its environment.

The olfactory dimensions of this scene exaggerate this effect, as smells are able to alter the sensate landscape of the Hill subtly and also at a distance from where they originate. The smells of the area emanate at times from very local sources, such as the mouths of local sex workers (with their "rotten teeth and stenchified breath"),[54] and at others from much larger spaces (the "stenchified 'dancing halls'"),[55] but they combine with the dissipated, ubiquitous qualities of the sonic landscape to create an all-encompassing sensate experience that is independent from and not reducible to any one source, body, or group of bodies. As we have seen, the endangered and "unsuspecting" spectator who accidentally witnesses these scenes notably does not affect the environment into which he or she wanders but rather is changed by it;[56] Negro Hill is dangerous precisely because it contains greater, or simply more animate, force than the bodies that circulate therein. While the noise and smells of the area are attributed to the bodies within it, the aggregate effect of these noises and smells, combined with the physical, architectural, and social qualities of the area, is greater than the sum of its parts. In a way, it assumes a life of its own, promiscuously touching and changing the human forms that move through and within it.

The same 1816 edition of *An Affecting Narrative of Louisa Baker* further ties the *architecture* of this neighborhood to the sex economy that proliferates there. The narrator describes Negro Hill as a "corrupted spot, which fortunately for the town comprises but a few acres, [that] appears to have been a department allotted to the people of colour, as most of the inhabitants are of this description [and] the tenements erected thereon, like its wretched inhabitants, are of every description, from the comfortable dwelling to the most miserable cell, apparently erected and proped up for temporary purpose only; in those of the first class the ladies, strumpets, of distinction claim a residence, while the latter are inhabited by a lower order, apparently composed of the worst of creation."[57] In this passage, the temporary nature of the neighborhood's architecture metaphorically describes the inhabitants of this area—both the racial makeup of Negro Hill and its status as a vice district. This scene tautologically describes the neighborhood architecture in terms of its inhabitants, and these inhabitants in terms of the architecture. The logic of this description proceeds by employing four different descriptions of the neighborhood to explain each other, even if this explanation is at best catachrestic: Negro Hill, a "corrupted spot," is an area "allotted to the people of colour," and the "tenements" there, like the people of color that inhabit it, "are of every description," but the passage concludes by effectively glossing all of the "tenements" in Negro Hill as imbricated in the sex industry ("in those of the first class ['the comfortable dwelling'] the ladies, strumpets, of distinction claim a residence, while the latter ['the most miserable cell'] are inhabited by a lower order, apparently composed of the worst of creation"). In other words, the entire neighborhood is linked as much to blackness as to the sex industry, which, as historians of Boston all concur, was not actually the case.[58] Even the narrator of *The Affecting Narrative of Louisa Baker* later admits that "the respectable part of the people of color, however, of whom there is a considerable portion who reside in the vicinity, as much dispise and avoid the company of these filthy strumpets, as the gentleman does that of a convicted jail bird!"[59]

It is not exclusively residential architecture that exerts its negative influence on the neighborhood, either; in the 1836 *Case of Young Robinson,* a cheap pamphlet telling the story of the eponymous Young Robinson's descent into dissipation, the narrator is careful to warn young men, in particular, about the dangers of the theater as effectively a gateway to the brothel. "To the patrons and upholders of theatres, especially, does this case come home as one of solemn warning," the narrator intones. "To the theatres of this city, above all other places, is the iniquity that abounds to be traced. We stop not to inquire what harmless and even valuable places of recreation, or of intellectual enjoyment, they might be made. Our design is to speak of them as they are, in their existing condition of debasement. They are sinks of vice and pollution—houses of assignation and incipient prostitution—in four words, 'THE VESTIBULES OF HELL!'"[60] While the narrator is primarily concerned with the activities that transpire *within* theaters, she describes them in a fully architectural idiom: "sinks of vice and pollution," "houses of assignation," and "VESTIBULES OF HELL." The vicious character of theaters cannot be separated from their spatial characteristics.

The combination of the neighborhood's definitive blackness and its association with the sex economy absorbs the lion's share of narrative attention in most descriptions of the Hill that appear in the *Female Marine* texts; the promiscuously interracial social milieu of the Hill bears as much potential to negatively influence those that circulate within it as its smells, sounds, and spaces. These texts deploy scenes of interracial sexual interactions to stage the fears of sexual, racial, and cross-class contagion in which these narratives are steeped. Sex workers in these stories are frequently referred to as the "false fair," invoking contemporary critiques of femininity that linked its techniques (makeup, dress, etc.) to artifice. But "false fair" also bears much older racial implications, especially in the early nineteenth century when natural history and other forms of racial science were beginning to codify race more explicitly in terms of anatomy and skin color. Especially in the context of the mixed-race neighborhood of Negro Hill, "false fair" gestures toward both the deceptive nature of the sex workers' beauty and the possible ambiguity of their racial status. *An Affecting Narrative of Louisa Baker* describes the lifestyle of the neighborhood's "false fair" as follows:

> The girls seldom quit their beds till noon, when they hobble down stairs and refresh themselves with bitters, toast and coffee; at three they commence the arduous task of dressing themselves for their evening revels; their object is to disguise themselves as much as possible from what they really are; which they so effectually do, with the aid of paint, patches, false teeth and hair, that a stranger to view them by candle light, would suppose that some modern Solomon had been collecting beauties from the four corners of the world; but could he but have a peep at these bewitching girls in their dishabille, their awkward gestures, their blotched faces, their crimsoned eyes, their rotten teeth and stenchified breath, would, I think, effectually wean him from every thing like an amorous assault.[61]

The attention to artifice and a wariness of the guiles of femininity appear persistently throughout the *Female Marine* texts, indexing a broader concern about the dangers of external appearance and the latent power of *surface* to exert influence—negative influence, in this case—on those who come into contact with it. Like the circular logic in these texts that considers the neighborhood architecture to be a reflection of its inhabitants and its inhabitants to be a reflection of the architecture, the *Female Marine* texts express ambivalence about economies of influence. Although the brothel inhabitants' attempts to deceive their patrons are effective only by candlelight, the narrators of these texts imagine these "girls" to bear enormous potential to corrupt those who come into contact with them. While the text pokes fun at their "paint, patches, false teeth and hair," as well as their "awkward gestures, their blotched faces, their crimsoned eyes, [and] their rotten teeth and stenchified breath," it nonetheless paints these women as some of the most dangerous personae in the city of Boston.

The 1815 edition of *The Adventures of Lucy Brewer* expresses similar concerns about the external, in a narratological interjection aimed explicitly at the young, and ostensibly young men: "O! My dear Youths, may the fate of this unfortunate young man, who so early fell a victim to his own imprudent indulgences, serve as a beacon to warn you of the danger of resorting to those vile haunts of disease, robbery and murder! Beware that you are not allured by external embellishments—permit one, who has been an eye witness of the fact, to assure you, that those *artificial decorations,* cover a form, which could you but see in its natural state, could not fail to fill you with the utmost disgust!"[62] While the narrator is careful to locate the dangers of "external embellishments" in the disparity between external form and inner content, it seems worth noting that the external or the artificial is, in these texts, precisely what makes these sex workers alluring to uninitiated young people of all stripes. The detailing of the threat of the "false fair" thus registers broader textual fears about the power of the external to exert corrosive influence.

Of course, not all of the sex workers who ply their trade on Negro Hill are, in fact, members of the "false fair." As the narrator explains:

> There is still another [class of sex worker], who, if possible, are not only far more disgusting to their sight, but whose actions may well be said to vie in wickedness with the evil one; an impudent air being the only charms of their countenance, and a lewd carriage the studied grace of their deportment. . . . They are of every age and complexion; but black is far the most respectable color among them! It is not unfrequently that you here see white girls of not more than 17 years of age throwing their scarified arms around the neck of an ugly looking black, and bringing their ulcerated faces in contact with his! . . . Their companions are principally composed of sailors of the lowest grade and stragling mulattoes and blacks, and who generally are never without those unpleasant proofs of their imprudence and folly in forming so dangerous a connexion. There are but few of these girls, however young, but can boast of having been a half a dozen times

under the hands of a doctor, as many times in prison, and probably three or four times an inhabitant of the alms house.[63]

In this scene, what becomes dangerous about the sex economy on Negro Hill is less the risk that an "accidental" passerby will be lured in by the artifice of feminine beauty than the general threat of contagion that characterizes this scene of interracial sexual contact. This moment offers an ethnographic and prurient look at the ostensible predations of the sex industry, while creating an association between interracial contact and the inevitability of contagion. In this passage, the blackness of the skin of some individuals who move through the neighborhood ("sailors of the lowest grade and stragling mulattoes and blacks") becomes indistinguishable from the darkened skin symptomatic of illness or disease ("scarified arms," "ulcerated faces," "unpleasant proofs of their imprudence and folly in forming so dangerous a connexion"). Dark skin, in these passages, tends in the direction of becoming a metaphor for disease or contagion, and specifically for venereal disease. The blackness of the neighborhood residents is represented as indistinguishable from the darkness of disease and the dye of sexual vice, and together these elements become the definitive qualities of the contagious or infectious environment of the Hill itself. Through the rhetoric of contagion, blackness takes on an active, spatial dimension; deployed as a scenic detail, blackness aestheticizes vice. In so doing, blackness thus takes on an almost animate quality as it is diffused throughout the space of the neighborhood. The scene in which young white girls "throw[] their scarified arms around the neck of an ugly looking black, and bring[] their ulcerated faces in contact with his" is thus positioned as a scene of transfer, the blackness of the white woman's lover narratively linked to her "scarified arms" and "ulcerated face." Blackness is not only rendered contagious here; it operates in these texts as something of an animate entity.

Blackness is presented as neither an anatomical nor a characterological detail, but rather a social one, with all of the potential for promiscuous movement, transfer, and even growth, proliferation, or reproduction that that might entail. Especially insofar as blackness is equated to contagion or disease, it becomes an organic, bacterial, or vital entity in and of itself: an active element in the narrative, coloring the architecture, the social landscape, and even the people.

Yet it is not only the manmade or human qualities of the unique environment of Negro Hill that endow it with its ability to leave its mark on those who pass through it. The Hill not only sports specific social and architectural qualities; it is even made up of a very specific kind of *dirt*. Visitors to the Hill—patrons of the sex workers there, as well as nighttime revelers of all kinds—leave the neighborhood changed or touched by their time there in ways that prove visible to those outside of the Hill. In the 1815 *Adventures of Lucy Brewer*, after the narrator returns to the Hill dressed as a man to reexperience her old haunts from a safe vantage, she spends the night in a nearby inn on Marlborough Street. The next morning, she notes:

> Observing that my boots had received a plentiful coat of the white coloured clay, peculiar to the Hill, I took them to a boot black in the neighborhood—Pompey (whose occupation it appeared had taught him to determine with some accuracy, the places of resort of his customers, by the state of their boots) in a jocular manner, observed to me, that "if master had not the appearance of a gentleman of too much delicacy, he should judge by the appearance of my boots that I had recently visited a part of the town not the most respectable!"—Here I could not but remark, that while this *man of colour* spoke most contemptuously of this corrupted place, yet, it was actually the favourite place of resort of many youths who claim the appellation of *"young gentlemen!"*[64]

In this remarkable passage, the Hill's "corrupted" influence materializes on the narrator's boots; the neighborhood literally rubs off on her. And neither is it only the narrator who is able to observe and identify this influence; upon seeing the clay on her boots, Pompey immediately knows where she has been.[65] A curious reformulation of popular minstrelsy haunts this scene: Pompey (also the name of a stock character in minstrel shows), the bootblack, polishes the white clay off of Lucy's shoes, returning them to their original dark color. Evoking the cork that white minstrel performers used in blackface makeup, the clay leaves a patina of whiteness—another iteration of "false fairness"—on the narrator's shoes that is rubbed away by this respectable *"man of colour."* Here, the whiteness of the narrator's shoes is revealed to be a deceptive finish accrued from her evening of rubbing elbows with unsavory company, and one that can only temporarily hide the blackness underneath. Beyond merely a reversal of minstrelsy's performative layering of blackness and whiteness, the sticky yet removable white clay endemic to the Hill indexes the way whiteness acquires a kind of contingency when exposed to the degenerating influence of that neighborhood. If the environment of the Hill exerts its corrupting influence on those who move within it, exacerbating and giving license to their vicious tendencies, it also threatens to racially degenerate its residents. In this sense, the *Female Marine* texts reproduce concerns that natural historians such as the Comte de Buffon express about the racially degenerating effects of specific environments (see chapter 1).

Furthermore, it is not just any kind of white dirt that adheres to the narrator's feet, but clay, an overwrought material, often metaphorically endowed with forms of animacy that similar materials—silt, slate, and so forth—are not. This metaphor earns its staying power at least in part because of its biblical derivation: Isaiah 64:8, in the King James Version of the Bible, reads, "But now, O

Lord, thou art our father; we are the clay, and thou our potter; and we all are the work of thy hand." In the eighteenth century, the malleability and durability of clay is a metaphor for human development, but clay metaphors proliferate far beyond conversations about human progress. Calvinist James Hervey's 1745–46 devotional treatise *Meditations and Contemplations,* for example, opens with the narrator's meditation on a pavement floor inscribed with the names of the dead who lie beneath it. The "lettered floor" itself is anthropomorphized such that the tiles "seemed to court my Observation, silently inviting me to read them. And what would these *dumb* Monitors inform me of? 'That, beneath their little Circumferences, were deposited such and such *Pieces of Clay,* that once *lived,* and *moved,* and *talked:* That they had received a Charge to preserve their Names, and were the remaining *Trustees* of their *Memory.*'"[66] In Hervey's important work (it saw fourteen editions in fourteen years and was eventually reprinted in Boston in 1750), clay is not a metaphor but simply what human bodies become after death; humans become "such and such *Pieces of Clay.*" Thomas Malthus also used clay as a metaphor for human impressibility, understanding some humans as more impressible, sensible, and by implication more civilized, than others. In terms of the white clay specific to Negro Hill, this imagery provides us with an appreciation of clay's special animacy, and by extension the animacy of the natural environment of the Hill itself.

The vigilance and virulence of these inanimate elements—clay, architecture, sound, smells—are reminders of the power of the Hill. Once again, we must take seriously the attribution of unconscious force, pull, or draw to the neighborhood. When narrators or hapless youth describe being "unconsciously dr[awn] ... to the entrance of one of [the Hill's] dancing cabins"[67] or "unconsciously ascending the heights of West-Boston,"[68] they are alluding to a neighborhood that was understood to bear not merely metaphorical force, as magnetic as it was corrupting. What might be the stakes of theorizing the neighborhood of the Hill, both the matter that constitutes it and the area in its aggregate form, in the same terms as the nineteenth-century readers and writers of these texts did, as a form of, as Jane Bennett might have it, "vibrant matter"?[69] What if we were to theorize the etiology of the vice that defines this vice district as deriving from the architecture, the social landscape, the types of institutions found there, and even the white clay that makes up the physical environment of the Hill, rather than deriving from the individuals who inhabit this area? How might we characterize the vice (in this case the sex work and the "lewdness," as Asa Hopkins put it, that drives the sex economy) outside of, or at a distance from, the subject? From the human?

Reform efforts on the Hill and the archive of pamphlets, reports, and plaintive fund-raising campaigns that they produced belie an awareness that the putative problem of the sex work industry was much bigger and more complicated than merely human folly or innate depravity. Organizations like the Boston Female Society for Missionary Purposes, founded in 1800 by women of both Baptist and Congregational affiliations, dedicated themselves to saving women working in the sex industry on the Hill as early as 1817, when they first dispatched a minister to work there.[70] They, too, understood the problem of sex work in environmental terms. In an 1818 report, the Boston Female Society described the Hill in terms eerily similar to those that appear in the *Female Marine* texts. The report, titled *A Brief Account of the Origin and Progress of the Boston Female Society for Missionary Purposes,* includes this description of the Hill:

> With propriety it may be said, *there* is the place where *Satan's seat is. There* awful impieties prevail; and all conceivable abominations are practiced; *there* the depravity of the human heart is acted on; and from this sink of sin, the seeds of corruption are conveyed into every part of the town. Five and twenty or thirty shops are opened on the Lord's days from morning to evening, and ardent spirits are retailed without restraint, while hundreds are intoxicated, and spend the holy Sabbath in frolicking and gambling, in fighting and blaspheming; and many in scenes of iniquity and debauchery too dreadful to be named. The street is filled during the day with old and young of all complexions, numbers drunken and sleeping by its sides and corners; and awful noises and confusions are witnessed. Lord's Day evening is the period when greater numbers collect than at any other season of the week; hundreds of boys from all parts of the town, on this evening repair thither, where their ears are assailed with the dialect of the dark world; while all the oaths are uttered, which the powers of the mind, long exercised in the service of the adversary, and excited to action by a totally depraved heart, could possibly invent.[71]

The report reads as a feverish distillation of the *Female Marine* texts, a grab bag of sensations, each more awful than the last. Once again, the Hill is a total sensory experience and, importantly, an experience that is endemic to the specific location of the Hill—"where *Satan's seat is.*" While they are, of course, concerned with the "totally depraved heart" of the people who frequent the area, it is *the area itself,* this "sink of sin," that provides the occasion for "act[ing] on" the "impieties," and generates the "seeds of sin" that are subsequently "conveyed into every part of town." Here too we glimpse strains of natural historical language, from said "seeds of sin" to the description of the many different "complexions" of the people milling in the streets.

Sex workers, in this period, were not understood as a class of people distinct from the "vicious poor." The commonsense of the day understood sex work as a practice of living or behavior. Efforts to reform working prostitutes thus focused extensively on removing women from their immediate environments, and many missionary organizations established residences, effectively workhouses, for the reform of "penitent" sex workers, as long as they were neither pregnant nor "diseased."[72] Workhouses were established both in Boston proper and in relatively remote locations, such as Worcester, Massachusetts, to isolate women from their putatively corrosive environments, sometimes for long periods of time, to allow them both reprieve from the influence of their environments and time to develop virtuous habits through work and prayer.[73] The 1832 report of the Penitent Female's Refuge of Boston, for example, revised its charitable model after "a few years" of operation, when "it became apparent that a longer residence of the inmates in the Refuge was necessary, either to establish such habits of right feeling and conduct as should secure their permanent reformation, or to gain the confidence of the community, in the reality of such a reformation, to such a degree, as to enable the Directors to provide suitable places for them." Indeed, part of performing the appropriate "penitence" that was necessary to gain access to the support that the Penitent Female's Refuge offered was being willing to live in relative isolation for long periods of time; the same 1832 report states that one of the "condition[s] of admission" was that "the candidate should possess so fixed a desire for a thorough reformation, as to be willing to give herself up to a long seclusion for the sake of obtaining it."[74]

These reform efforts predated, and later worked in consult with, municipal efforts to crack down on the sex industry in Boston. By the time Boston was incorporated and Josiah Quincy was elected mayor, the city government was under immense public pressure to address Boston's sex industry. During his term as mayor, Quincy instated "a policy of mass arrests of prostitutes and brothelkeepers [that] represented a radical break with the past. Throughout his years as mayor, court dockets were filled with consecutively numbered prosecutions for prostitution and brothelkeeping, many of them with the same arresting officer, evidence which suggests that there were periodic sweeps of prostitutes in vice districts."[75] Part of this new, intensified policing campaign was the approval of funds in 1822 for the House of Industry, an institution in which many women, sex workers and unwed mothers in particular, were confined upon arrest or application for public charity. The workhouse model for supporting all "classes" of poor people (virtuous or vicious) was in step with institutions that had already been founded by private organizations for the reform of fallen women. In the workhouse model, we can retroactively see the policing of vice (especially the sex worker), the criminalization of poverty, and the intensification of state surveillance of people of color come together, as areas like Negro Hill and the North End were frequently targeted for "sweeps." On the heels of the founding of the House of Industry, a prison—the House of Correction—

was founded in 1823, "and the city council set aside funds for building a larger facility."[76]

It should come as no surprise that these workhouses supposedly dedicated to restoring fallen women relied on a model that understood a spiritually healthy environment as central to the success of reform efforts. This model theorized spiritual betterment as a process that worked from the outside in, and it began with removing women from their immediate environments and exposing them to new (ostensibly healthier) food and social landscapes, altering their habits (teaching women to read, giving them Bibles), and insisting on particular forms and environments (single-sex spaces) of labor or industry. In this insistence on relocating women from putatively unhealthy environments to healthy ones, we can identify the power that early-nineteenth-century reformers attributed to environmental influence and glimpse the relationship that these same societies drew between one's environment and the sexual behavior, whether aberrant or appropriate, that would ensue.

Conclusion

The 1819 *Constitution of the Society and Directors of the Penitent Female's Refuge* represents the society's missionary work in Boston as precisely that: Christian intervention into unchristian lands and communities.[77] After offering a brief sketch of the particular dangers of Negro Hill, the *Constitution* goes on to compare the neighborhood and its inhabitants to "heathen" peoples of far-away lands:

> It may also be urged as a christian duty upon the principles and propriety of missionary exertions, whether foreign or domestic. In what respects are these characters different from the heathen? Are the heathen very ignorant? so are many of these. Are the heathen vicious? so are these. Are the heathen cruel? So are these in many respects; "fightings" and even murders are not uncommon in these haunts of sin and misery, as the effects of their vice and ignorance. If it be admitted a christian duty, to go and preach the gospel to the heathen, and thereby to make them better and happier, ought we not to use similar exertions for those at our own doors, whose characters are nearly the same?[78]

The comments from the 1819 report thematically echo the representation of the inhabitants of Negro Hill that appears in the edition of *The Affecting Narrative of Louisa Baker* of three years earlier, which refers to noise of the Hill as "not perhaps equaled by that of the numerous hordes of the wild inhabitants of the Ganges."[79] The year before, the Boston Female Society for Missionary Purposes had released a report that combined "extracts" of its annual reports from 1817 and 1818—mostly sentimental stories of white women fatally seduced into the Boston sex industry—in an effort to raise funds for its work of spreading the gospel. It, too, saw its work in Negro Hill as akin to a foreign mission, reminding readers:

> Without boasting, it may be asserted, that many thousands of dollars have been raised in this town to spread the gospel in regions not favored with its heavenly influence; we may therefore conclude that the friends of religion and virtue in other places would cheerfully contribute to an undertaking like this; particularly when it is considered, that but a small proportion of these unhappy creatures are natives of this place. They are collectioned from almost all parts of the country, and some even from foreign climes. The calamity is a *public* calamity; the cause of virtue is a *public cause,* and if good is done it will be felt by the community.[80]

All of these texts deploy the eponymous blackness of the neighborhood and its residents to code it as foreign, a far-away land in the backyard of many of the wealthy Boston reformers whose names appear on the subscription lists of these charitable societies. The residents of Negro Hill are akin to the "wild inhabitants of the Ganges," only "a small proportion . . . are natives of this place," and they are "from almost all parts of the country, and some even from foreign climes."[81] In the 1819 *Brief Account of the Origin and Progress of the Boston Female Society for Missionary Purposes,* the international slave trade is cited as the way many of the residents of Negro Hill even came to be in the United States, and slavery becomes the explanation of this community's internal foreignness and a compelling reason for its need for support; the report admonishes readers, "It should be remembered that this unfortunate people were introduced into our country by the cruel hand of avarice and barbarity; and if there be a race of human beings, which have a right to *demand* our charity, next to the descendants of Abraham, these must have a claim."[82] Combined with the missionary tactics of these organizations—establishing Bible schools, distributing Bibles and pamphlets, organizing services for worship—the logic behind the work of awakening and conversion to which these societies dedicated themselves was very much in step with that deployed in both established and concurrent colonial and missionary ventures in locations from India to Hawai'i. However, in the case of Negro Hill, such efforts were combined with and at times animated by new theories and tactics of public health administration that had begun to arrive with developing forms of both state-based and national infrastructure.

In this sense, blackness in these texts indexes two idioms of foreignness that derive from rather different sources: first, missionary and travel writing (which both were informed by and contributed to natural history), wherein the difference of blackness was still being understood in some of the earliest of natural historical terms, discussed in the introduction and chapter 1. The most obvious of those terms are geographical ("this unfortunate people were introduced into our country by the cruel hand of avarice and barbarity" and "but a small proportion . . . are natives of this place"), climatic ("some even from foreign climes"), and religious ("the descendants of Abraham," "heathen"). Second, there is a distinct concern for public health in these conversations: improving the contagious and morally toxic environment of the Hill (a "public calamity") will in turn be a public *utility,* a notion of the public good that moves in lockstep with at least two decades' worth of community efforts to curb the vicious practices that both lowered property values and conveyed "the seeds of corruption . . . into every part of the town."[83] These two understandings of the foreignness of blackness are not in competition; both contribute to the complex meaning of urban racialization in the early nineteenth century, a moment when "the connection between black women and sexual disorder was pervasive throughout the northern states as free black communities grew and urban populations exploded," and when we begin to see a shift in racist sensibility "away from the cautious natural history of human variation of the eighteenth-century Atlantic world and toward the confident scientific racism of antebellum America."[84] With this shift away from environmental frameworks and toward anatomical understandings of human physical variety came a concomitant shift in the way sexual behavior was understood. If human variety derived predominantly from internal, anatomical, or otherwise fixed attributes, then sexual behavior also became a more individuated and embodied phenomenon. And as public health campaigns in both Europe and the United States began to study sex workers as a discrete "class" or population, they began to imagine the inclination toward specific sexual behaviors not as an epiphenomenon of racial variety but as a category of human experience in and unto itself.

Reform organizations dedicated to the rehabilitation of sex workers were simultaneously also dedicated to managing and preventing the proliferation of vicious spaces and practices. In this sense, antivice reform organizations were concerned with the reproduction (although that term was not used in its modern sense until the later nineteenth century) and surveillance of racialized milieus in the early United States. As reformers worked alongside the municipal government, local business (in the form of guilds of white laborers), other reform movements, and the new city police, their approaches to the problem of urban vice anticipated—or merely reflected—an increasingly biopolitical framework for understanding the management of people and populations. The rise of biopolitical tactics of governance, as we shall see in the epilogue, simultaneously drew from extant environmental understandings of the body and behavior and was aimed not at individuals but at populations. While the transformation of the human body from a humoral or primarily environmental body to an anatomical one might appear to be something of a sea change in scientific understanding, attention to the emergent logic of biopower reveals this transformation to be a much less radical shift. We shall see how, by subjecting the newly biological body to carefully selected environmental conditions that promote its life or death, emergent biopolitical strategies of control offer a more nuanced account of how both sexual behavior and racial difference hardened into primarily internal categories of human variety.

EPILOGUE

Thinking Sex—without the Subject

> The ambition of performing sex as only power is a salvational project, one designed to preserve us from a nightmare of ontological obscenity, from the prospect of a breakdown of the human itself in sexual intensities, from a kind of selfless communication with "lower" orders of being.
> —Leo Bersani, "Is the Rectum a Grave?," 1987

> If we ask decorous questions of history, we will get a genteel history. If we assume that because sex was a secret it did not exist, we will get a sexless history.... We create history as much as we discover it. What we call history becomes history and since this is a naming time, we must be on guard against our own class prejudices and discomforts.   —Joan Nestle, "Thinking about History," 1981

The emergence of biopolitical governance, as the story goes, begins with a shift not in the tactics of state management but in the objects of state management. In this transition that, by most accounts, occurred at some point between the early modern period and the end of the eighteenth century—for historical specificity tends not to be one of the foremost concerns of most theorists of the biopolitical—statecraft newly set its sights on the governance of populations instead of individuals and on the management of life instead of the meting out of death.[1]

Biopower "recast the role of the state, changing its raison d'etre from that of a saver of souls to a governor of bodies."[2] This shift in tactics of governance both heralded large-scale shifts in the understanding of sexual behavior as a potential site of state management—we see this in urban antivice campaigns aimed at eradicating sex economies, for example, as in chapter 5—and, importantly, it also put sexual behavior to work as a tactic of governance itself. This is, of course, Foucault's field-defining hypothesis about what he terms *sexuality;*

briefly, he argues that sexuality materializes a newly dominant vision of disciplinary power that exceeds the state and is both productive *and* prohibitive, instead of exclusively prohibitive. Sexuality, in this account, aided in the process of changing the layout or organization of state management, such that it shifted the task of disciplining both human behavior and human desires off of the state and onto the individual person. Knowing one's sexual desire, now arranged in a predictive relationship to one's behavior, became a means of knowing oneself. This new form of truth of self more or less provided the basic architecture of the subject, with sexuality lending the subject its defining form and the body being the privileged and highly delimited location of the subject. Importantly, this still-definitive account of the emergence of modern sexuality, itself a fundamentally biopolitical framework, is also a story about a new, modern ontology of the human body and its capacity to house sexuality. This triangular equation, in which modern sexuality, modern subjectivity, and modern concepts of the body are all more or less defined by each other, constitutes the vocabulary through which sex, for most present-day readers, makes sense.

To imagine sex otherwise is difficult. Despite the ways Foucault's account of modern sexuality has been importantly challenged, problematized, and historicized by scholars coming from a range of fields, the enforced situatedness of sex in the subject, and of the subject in the body more generally, has been hard to shake. Even among studies of sexuality in the period before 1800, the conceit that sexual behavior is located in the body and is animated by—or is an expression of—desire has nonetheless informed or shaped many of our questions. Yet there are great intellectual losses that attend this persistent imbrication of sexuality in subjectivity, and vice versa, particularly for scholars of the early modern and eighteenth-century worlds. And perhaps one of the greatest scholarly foreclosures effected by this framing is the unnatural cleaving of early racial science and its attendant environmental understanding of the body from the genealogies we have crafted of modern sexuality. As modern notions of the hermetic, biopoliticized body, fighting off its environment rather than coexisting ecologically with it, rose to dominance, eighteenth-century genealogies of the environmental body became harder to see.[3] Commensurately, early modern and eighteenth-century theories of the power of environmental or other external circumstances to shape racial morphologies or sexual behavior were rendered atavistic and lost their formerly quite considerable explanatory purchase.

If these genealogies of the environmental body are indeed lost—and I am not so sure that they are—then what was lost, with them? I propose that the conversion of the human body from an environmental entity into a biopolitical one has actually forced an artificial chasm between how we understand "race" and how we understand "sexuality." To be sure, these terms are neither identical nor self-identical, but in the scholarly insistence on refusing the kinds of flat, "like-race" arguments about sexuality that I address in the introduction, we may have inadvertently threatened to throw the baby out with the bathwater. While sexuality is not "like race"—an analogy generally strategically invoked in the legal arena to make LGBT civil rights claims—current understandings of both sexuality and racial difference might actually share more than we think. Studies of racialization have been increasingly attentive to the way race shapes social understandings of both gender and sexuality; and, increasingly, work in gender and sexuality studies has demanded that we center the particularities of racial and ethnic experience in scholarship on men, women, trans people, and the weaponized wielding of gender by the state. In the wake of about two decades of often unheeded critiques of queer studies for its willfully self-centered whiteness, as well as similar critiques of studies of racialization for their lack of interest in gender and sexuality, it is becoming a more and more serious expectation that each of these fields think with the other.[4]

But I wonder if it is not *only* racism or not *only* sexism or homophobia that has informed these oft-criticized disciplinary lacunae. Indeed, I wonder whether the modern focus of each of these fields, with their concomitant interest in and focus on the modern body, has also, in fact, aided in the process wherein scholars have imagined themselves able to theorize race at a distance from gender and sexuality, and gender and sexuality at a distance from race. I have demonstrated the shared origin of the emergent sciences of racial difference and sexual diversity in the environmental body, and my efforts have been animated, at least in part, by an eighteenth-century Americanist's perspective on the absolute inextricability of these ideas and the importance of having these earlier understandings of racial difference and sexual diversity available to scholars working on race, gender, and sexuality today. And this is not just a restatement of black feminist intersectionality theory or standpoint epistemology, although this project is certainly indebted to those ideas and their thinkers. The stakes of bringing eighteenth-century sexual politics of racialization and environmental theories of sexuality to twenty-first-century scholarly conversations about sexuality, in particular, are twofold: first, these histories reveal different genealogies of both ideas and theoretical frameworks that are currently seeing a lot of traction in the humanities, such as biopower, or the kinds of skepticism about the utility of the subject that we see in both queer theory's antisocial turn and in Afro-pessimism. Second, these distinct eighteenth-century environmental understandings of human variety and sexual diversity allow us to imagine sexuality differently, to ask different questions about what it is, where it is, what it might mean, and how to better imagine its politics.

If the preceding five chapters point to a range of possible iterations of modern sexuality that are not coterminous with subjectivity, cannot be exhaustively housed in the body, and do not insist on being categorically divorced from even the most ontological questions about race, in what follows, instead of proceeding from the assumption that these older, eighteenth-century modes of theorizing human variety and sexual diversity have been lost, I will chart trajectories of both some losses and persistences of these ideas into twentieth- and twenty-first-century sexuality studies. Numerous historians of sexuality have attempted to catalogue modes of theorizing sexuality that did not survive what Eve Kosofsky Sedgwick wryly termed "the great paradigm shift" (a reference to Foucault's assertion that homosexuality arrived sometime around 1869) that gave us modern sexuality around the turn of the twentieth century.[5] Yet few have taken seriously the question of what frameworks for making sense of sexuality persisted. Here I am especially interested in modes of theorizing sexuality that *did* survive that shift but were not able to be seamlessly absorbed by this new dominant framework: ways of understanding sexuality that are less tightly bound to the subject, or bound to the subject in more vexed ways (e.g., popular interest in the "down low," or in men—usually men of color—who have sex with men who do not identify as queer or LGBT);[6] environmental accounts of sexuality or desire (e.g., tourism advertising campaigns like "what happens in Vegas stays in Vegas," or imagining the spatial distance of "vacation" itself as a delimited site of potential hedonism); or even the persistence of environmental frameworks that do not at first glance appear to speak directly to or about sexuality (e.g., the heavily racialized mythos of the "bad neighborhood"). At the same time, the loss or occlusion of environmental frameworks for making sense of sexual diversity that attended the rise of biopower, with its modern understanding of the body as the delimited site of both sexuality and subjectivity, cannot and should not be ignored, for by calculating these losses we can better see both what might have been and, perhaps, what might still, quietly, be.

Persistences: Natural History, Population Science, Sexology, Biopower

In the conclusion to chapter 5, I pointed to the way urban reform organizations increasingly relied on population-based analyses of sex workers to develop tactics for intervening in and eventually eradicating the sex economy. Alongside reform efforts, broader concerns among ministers, public health and municipal officials, and popular writers about the proliferation of vice itself also became increasingly codified in the idiom of statistical or demographic analysis. Indeed, population science and demography loom large in accounts of the emergence of biopower, as these sciences, aimed at anatomizing life in aggregate or collective terms, provided the basis for developing a form of governance organized around the commodification and maximization of life itself as an extractable resource.[7] Yet, despite the rather obvious sympathies between this structure of governance and, say, institutions dedicated to the maximization of life and extraction of labor such as slavery, the earliest and arguably most influential accounts of biopower have been only sloppily attentive to questions of race, colonialism, or early racial sciences more generally.[8] Yet natural history, and the hypotheses about race and racial difference that it offered, actually significantly shaped early population science.[9]

This section imagines one particularly salient way that environmental accounts of racial difference persisted through and even after the emergence of biopower and its reorganization of the body into a hermetic rather than environmental entity. Here I turn to population theory, the base unit of biopolitical strategies of governance, to offer an amended account of the role of racialization more generally, and of natural history in particular, in the emergence of this new approach to statecraft. Population theory and demography more broadly, as demonstrated in chapter 5, significantly shaped early urban public health campaigns, especially those organized around the eradication of sex work. Sex workers were among the first groups of workers to be studied as a population, and the science that these public health studies yielded provided the basis for the emergence of sexology in the 1830s. Sexology, then, was fundamentally a biopolitical science, dedicated to the protective management of certain populations (white, reproductive, and often married people, who were able to feed and sustain themselves and their offspring) and to the limitation of others (sex workers, poor people, and people of color). And, just as importantly, it was also a field of study organized around the human body, deeply invested in anatomy and, later, in structures of identification and desire as well. Yet sexology was always also a racial science, inheriting from its reliance on population theory many of natural history's most cherished forms of imagining human difference.[10] Indeed, I argue that eighteenth-century natural history constitutes a critical intellectual and cultural forebear to the development of sexology in the beginning of the nineteenth century, and thus I outline a corrective genealogy of modern sexuality that accounts for the centrality of eighteenth-century racial science to its emergence.

Most historians of science attribute the beginnings of population theory to Thomas Malthus, whose *Essay on a Principle of Population* was actually at least in part a response to earlier writings by William Godwin and the Marquis de Condorcet, and which was also inspired by some of Benjamin Franklin's work on demography.[11] For Malthus, the term "population" effectively describes one of the central challenges of what we now call political economy (indeed, Malthus became the first professor of "political economy" in 1805), meaning a way of calculating, explaining, predicting, and engineering human production and consumption. Human population, at the time of the first printing of Malthus's *Essay on a Principle of Population* (1798), was also a pressing contemporary political problem specifically in relation to urban poverty; spurred by the existence of ostensibly unsupportable numbers of poor people in England, Malthus in the *Essay* ponders the seeming inevitability of poverty and speculates that as the population continues to grow exponentially, the production of necessary resources such as food will not increase at the same rate. Population did not yet refer to specific groups of people drawn along the lines of race, ethnicity, class, and so forth, that are hierarchized and allowed to "make live or let die," a development that Foucault narrates in his lectures at the Collège de France and in the second half of the first volume of *The History of Sexuality*.[12] Instead, as Maureen McLane argues, theories of population (Malthusian and otherwise) are also theories of history, theories of human development, and, more broadly, theories of time.[13] Highlighting this element of populational thinking brings into sharp relief Malthus's reliance on progressive theories of human development central to eighteenth-century naturalist conversations about racial difference. Indeed, the *Essay* both explicitly and implicitly invokes writings by Scottish Enlightenment philosophers whom Malthus knew personally, that were produced and circulated in the decade prior to its publication, revealing the intellectual linkages between natural history and demography. Ultimately, I argue that natural historical thinking—stadial theory in particular—provided some of the critical intellectual foundations and logical structures for the development of Malthus's notion of the population.

Most historians of demography acknowledge the sympathies and even direct influences that exist between Malthus and natural historical thought. However, when discussing the relationship between the development of early demography and natural historical thought, scholars tend to tell this story in a way that *begins* with Malthus and early demography (or that begins with early modern mathematics and then moves through late-eighteenth-century demography) and ends with nineteenth-century theories of evolution (Darwin et al.).[14] Yet, at moments in *An Essay on the Principle of Population,* Malthus borrows directly from oft-repeated or reprinted ideas and passages in natural historical texts that directly invoke stadial theory, especially in his development of the theory of "positive" and "negative" checks to population growth. In chapter 3 of the first edition, Malthus speculates about historical processes by which populations have been checked, specifically within the context of the "savage" and the "hunter" state and then in the context of the "shepherd" state. While he assumes that contemporary Native people (he uses the generic example of indigenous peoples of North America) are still living in the "savage" or "hunter" state, he assumes that European peoples moved through this first stage of human development at a much earlier period. For example, his discussion of the second phase of human civilizational progress, the shepherd state, dates this stage to "the tribes of barbarians that overran the Roman Empire," a period almost a millennium before his own time.[15] Malthus's chapter 4, on what he terms "the state of civilized nations," specifically addresses "tillage," or agriculture, and commerce, the third and fourth stages of stadial theories of human civilization.[16] In so doing, he explicitly invokes "four-stages" theories of human development that were first advanced by Samuel von Pufendorf, Adam Smith, the Comte de Buffon, and other philosophers whose writing was central to naturalist discussions about the origins of the world and of human racial difference.[17]

Perhaps more interesting than Malthus's explicit reliance on stadial theory, however, are what appear to be his explicit though uncited references to either the Comte de Buffon's *Histoire Naturelle* (1750) or Thomas Jefferson's *Notes on the State of Virginia* (1781–85), and specifically to the perplexing passage discussed at length in chapter 1 of this book. Chapter 3 of *An Essay on the Principle of Population* opens as follows: "IN THE RUDEST STATE OF MANKIND, in which hunting is the principal occupation, and the only mode of acquiring food, the means of subsistence being scattered over a large extent of territory, the comparative population must necessarily be thin. *It is said that the passion between the sexes is less ardent among the North American Indians than among any other race of men.* Yet, notwithstanding this apathy, the effort towards population, even in this people, seems to be always greater than the means to support it."[18] Here, we see Malthus parroting a relatively widespread idea among naturalists: that "the passion between the sexes is less ardent among the North American Indians" than among other peoples. He offers this "fact" as some of the supporting evidence for his theory that the rate of population growth will exceed the rate of food production or availability in most communities. In this moment, he invokes a passage from Buffon's *Histoire Naturelle* that Jefferson cites (in French) thirty years later in *Notes*:

> Although the savage of the new world might be of about the same size of that of man of our world, that is not enough to make an exception to the general fact of shortening of life throughout that continent; the savage is weak and small in the organs of generation; he has neither body hair nor beard, & no ardour for his female.... Do not look any further for the cause behind the itinerant life of savages or their removal from society: the most precious spark of the fire of nature was denied them; they lack passion for their female, and therefore love for their fellow men.[19]

Although the idea that North American indigenous peoples experienced less sexual "ardour" than European peoples was fairly commonplace within naturalist writings of the eighteenth century, the next few sentences from chapter 3 of Malthus's *Essay on the Principle of Population* further suggest that he is citing Buffon or Jefferson, or both. Malthus writes: "The North American Indians, considered as a people, cannot justly be called free and equal. In all the accounts we have of them, and, indeed, of most other savage nations, the women are represented as much more completely in a state of slavery to the men than the poor are to the rich in civilized countries."[20] As I discuss in chapter 1, civilizationalist narratives about the relative freedom of women of various societies—and the way the relative empowerment of women tended to index a cultural capacity for the ostensibly more evolved project of democratic governance—constitute one of the major narratives of naturalist engagement with stadial theory, and it is not unlikely that some of the "accounts" with which Malthus admits himself familiar would include the enormously popular *Histoire Naturelle* or some of Jefferson's work. Furthermore, in the passage that I believe Malthus to be citing from Jefferson's *Notes,* Jefferson's citation of Buffon continues, describing the putatively degraded status of women in North American Native communities: "They regard their women only as domestic servants or beasts of burden whom they task, without management, with the burden of their hunt, and whom they force without pity, without gratitude, to do work that is often greater than their strength: they only have a few children; they lack care; everything stems from this first flaw; they are indifferent because they are not strong, and this indifference for sex is the original sin that withers their nature, that prevents them from flourishing, and that destroys the seeds of life, cutting off the root of society at the same time."[21] Jefferson's interest in demography is quite well documented, and we know that he read Malthus's *Essay on the Principle of Population.*[22] Malthus, in turn, read the population theories of Benjamin Franklin, but there is no work to my knowledge that considers the influence of any of the major works in natural historical thinking, including Buffon and Jefferson, on Malthusian demography. What, then, might be gained from sustained attention to the influence of naturalist philosophy—so apparent in Malthus's thinking—on emergent theories of populational science and demography? What might the intellectual and cultural sympathies between naturalist narratives about the emergence of racial difference and demographic modes for collecting and representing information tell us about how readers, writers, naturalists, and reformers relied on extant logics of human racial differentiation to think about sexual behavior?

First, and most simply, emphasizing the relationship between that strain of natural historical thought, so concerned with the origins of human racial difference, and early demography and theories of population, so preoccupied with the motivating factors behind reproduction or the lack thereof, exemplifies the way theories of human racial difference lent the emergent sciences of sexual diversity some of their principal governing logics (for example, explanations for why population growth booms in some cultures and wanes in others). Furthermore, putting these two bodies of information alongside one another allows us to note that both scientific and popular interest in sexual diversity, or specific, demonstrable patterns in human sexual behavior, was beginning to "catch up" to popular and scientific interest in racial difference by the late eighteenth and early nineteenth centuries. In chapter 1, I consider the challenges of how to draw out the quiet narrative about human sexual diversity embedded in natural history's profoundly *loud* interest in racial difference; here I emphasize how theories of human sexual and reproductive diversity began to emerge as objects of interest in their own right, apart from but still indebted to earlier racial science. Situating natural history as a critical contributing intellectual framework for the development of demography and theories of population thus suggests that we cannot and should not theorize the emergence of sexology outside of the long shadow of the epistemological power of theories of human racial difference. While this might be intellectual commonsense to historians of science and sexuality who focus on the nineteenth century, and on nineteenth-century sexology in particular, insisting on formalized sexology as an intellectual assemblage ineluctably indebted to both natural history and demography draws the historical epistemological overlay between racialization and sex backward in historical time, rendering this genealogy much older than even the best scholarship on the origins of nineteenth-century sexology has suggested.[23]

Second, putting natural history and demography into conversation actually demands a historically earlier narrative of the emergence of demography, population science, and biopolitical strategies of governance in the British colonies more generally. More than forty years before Malthus's *Essay on the Principle of Population*, Benjamin Franklin wrote (and four years later published) his *Observations concerning the increase of mankind, peopling of countries, &c.* (1755) in Boston. His colonizationist tract offers suggestions for both how to produce and how to maintain strong, healthy populations. Franklin's essay is plagued by the specter of interracial coupling and its attending offspring and the subsequent likelihood of the racial weakening of white populations by the British colonies' reliance on slavery, reminding us that theories of population have always worked at least partially in step with what we now term eugenics, the strategic management and engineering of specific populations.[24] Indeed, as scholars such as William Max Nelson have demonstrated, theories of population girded early Enlightenment experiments with strategic "breeding" programs, for both enslaved peoples and nonslave white populations in the French colonies.[25]

Third, thinking about the constitutive early modern environmentalism of natural historical writing alongside the critical role that space plays in population theory and demography more generally brings into sharp relief the different impact that the external or the environmental had upon these very different ways of theorizing human diversity. Whereas for many natural philosophers, the physical environment—exposure to the sun, availability of certain types of food, one's location vis-à-vis the torrid, frigid, or temperate zones—exerted the most powerful influence on humans, population theorists saw the physical environment as capable of affecting human societies but *also* as an object for strategic human intervention and engineering. Population theorists such as Franklin and Malthus both express their faith in the determining impact of the space one inhabits, whether the city, the nation, the culture, or the natural environment, but they also express great faith and interest in human ability to arrange the environment in a way that allows it to bear the best, most felicitous impact on those living within it. For natural historians, the torrid zones might be those most likely to produce cultures with sodomitical tendencies; but for population theorists, it is the particular milieu, as Foucault would put it, of manmade spaces that is increasingly likely to both positively and negatively determine human behavior. Despite the major differences between, for example, William Falconer's *Remarks on the Influence of Climate* (1781) and John McDowell's *Magdalen Report* (1830), by putting these different epistemological approaches to theorizing environmental influence into conversation, we are able to map the shared geography between the torrid zone and the vice district and chart the imbrication of each of these locations in long-eighteenth-century discussions about racial difference and sexual diversity. Natural history and population theory thus offer two visions of the problem of the external, two vocabularies for cataloguing long-eighteenth-century ways of understanding sex, without or at a distance from the subject.

Noting their shared intellectual genealogy brings into sharp relief the increasing attention to the specificity of sex as an element of human experience and as a social concern as this attention waxes and wanes over the course of the century. By the time Malthus writes *An Essay on the Theory of Population* in 1798, "sex" emerges in his work as one of the two most basic and undeniable priorities of human lived experience, the other being food.[26] The inarguable centrality of sex to theories of population, then, and the influence of theories of population on demographic studies that animated early public health efforts (like those that gave rise to the policing of Negro Hill) rendered population management and governance more generally a project fundamentally linked to sex and racialization.[27] Thus it is no wonder that sex workers operating within mixed-race districts became the first occasion for large-scale studies (and subsequent interventions), now cited as the earliest murmurings of sexology as a discrete arm of the human sciences. Environmentalist explanations of sexual behavior, however, gave way to new modes of explanation that relied heavily on both anatomical science and burgeoning theories of character. Population science became one of the epistemological conditions for the emergence of sexology, a biopolitical knowledge framework through which sex became embodied, human, modern, and expressive.

Highlighting the staying power of environmental theories of racial difference as they quietly persisted in the development of population sciences reveals unexpected routes through which biopower absorbed some of the defining frameworks of eighteenth-century racial sciences. Nowhere is this more evident than in the way biopower targets the environment itself as a site of intervention or, perhaps more insidiously, nonintervention in the management of people and populations; one need only think of urban "food deserts," the differential exposure of poor people and people of color to environmental toxicity, or the tendency to expose poor people and people of color to greater degrees of infrastructural risk (e.g., the routing of the Dakota Access Pipeline through the lands of the Standing Rock Sioux instead of through the wealthier and whiter communities in Bismarck, North Dakota, where it had been blocked). The biopolitical environment, in our time, is one of several particularly prominent loci of undermanagement through which deprioritized populations are "let die."

Losses: Sex and the Subject

Despite the importance of population science to the emergence of biopower, however, biopolitical governance does not ignore the individual human body. If the population—what Foucault termed the "species body, the body imbued with the mechanics of life and serving as the basis of the biological processes: propagation, births and mortality, the level of health, life expectancy and longevity"—constituted a critical site of intervention for biopolitical governance, so too did the individual human body provide an important location for the consolidation of disciplinary power.[28] In volume 1 of *The History of Sexuality,* Foucault "map[ped] this new domain," biopower, "across *two* axes: the 'disciplines: an anatomo-politics of the human body' and 'regulatory controls: the bio-politics of the population.'"[29] This "anatomo-politics of the body" describes how the body became the target for a number of different disciplinary regimes, from medicine to law to education, that worked together to observe, document, and separate a people into individuals, a subtle set of controls aimed at maximizing the utility and self-discipline of the singular person. I will refer to this body as the biopoliticized body.

For Foucault, the importance of the consolidation of disciplinary regimes of power such as medicine, education, and law into the delimited form of the human body cannot be overstated. Whereas in early modern and eighteenth-century medicine, for example, the body constituted a "porous envelope," eminently vulnerable to the power of external influences such as climate, diet, society, and form of governance, the biopoliticized body is an entity with a distinct inside and outside, a living form that provides the privileged location of individual life, a vision of personhood situated in the singular body rather than in a community or an ecology. The modern body, as Ed Cohen succinctly puts it, "is a hybrid biopolitical formation which we must *have* in order to be a person"; the hegemony of this modern understanding of the body has the effect of "rooting the human being—and human beings—both spatially and temporally in the localized domain that we call the body."[30] This shift "inaugurate[d] a new political economy of modern personhood . . . in which differences among and between people (e.g., race, sex, gender, class, age, etc.) appear as attributes of bodies rather than the gradations of souls."[31]

This biopoliticized body, then, houses not only the person but also the person's specificity; all forms of difference emanate from and are located within this form. Sexuality, too, becomes moored in the human body, and thus in the human more generally. The climatic, architectural, and social environments, once understood to be critically definitive influences on sexual behavior, lose their explanatory power under this framework. Within the new architecture of the biopoliticized body, sexual behavior is no longer the result of the dynamic interaction between the human form and the many environments in which it circulates; instead, sexual behavior is captured by the body, becoming one of the attributes that distinguish it from other bodies and from its environment, rather than a set of behaviors that serves to tie the singular human form to the natural and social worlds. Sexual behavior, in a biopolitical framework, becomes *expressive,* telling a story about the human who possesses it, for it is now something possessed, rather than something situationally produced by the alchemy of incidental environmental exposure. The biopoliticized body thus serves an individuating function, and its sexual behavior thus appears individual in turn, something closer to property. And like property, sexual behavior becomes a vector of enfranchisement, something owned, something that must be read almost exclusively symptomatically, telling a story about the subject rather than a story about itself. In short, as the biopoliticized body emerged to dominance (and this is, of course, Foucault's account) sexual behavior became a technique of subjectification and what Foucault might have termed a "dispositif" of the subject. The emergence of the modern, biopolitical subject thus constitutes the occasion for modern sexuality—differently put, sexuality as we know it now—and is thus for all intents and purposes inseparable from it. And while certain forms of environmental thinking about sexuality have indeed persisted to overlap with and perhaps even contest the dominance of this modern framework that ties sexuality firmly to the subject, imagining sex outside of or at a distance from the subject has become increasingly difficult, if not impossible.

But is this necessarily a problem? While Foucault, in his work on biopolitics, as well as queer theorists from Eve Kosofsky Sedgwick to Michael Warner, Jasbir Puar to José Muñoz, have all expressed concern about sexuality's increasing potential to bind us to state power (or at least to compel us to enact its prerogatives), one could point to certain functionally useful effects that the bind between sexuality and subjectivity—especially as it coalesced into a sexual identity politics—precipitated.[32] Even some of the most avowedly radical of political workers have at times expressed an openness to the utility of what Gayatri Chakravorty Spivak once called "strategic essentialism" (although she has since retracted this idea), speaking to the calculated use of legible subject positions (e.g., woman, subaltern, queer) for the purpose of making political claims.[33] And this approach is not, of course, exclusive to queer political movements; in the United States, antiracist politics, including a broad range of liberation movements (civil rights movements, Black Power, etc.) have at times taken a similar approach. In other words, insofar as the mutual imbrication of sexuality in subjectivity has facilitated the articulation of political movements organized around sexual politics and practices, why should we mourn previous frameworks for making sense of sexual behavior? If previous, environmentalist frameworks for understanding sexuality have in fact been lost, then what, precisely, does that loss consist of? Is the subject so bad?

Queer theory has always been skeptical of the subject. From Leo Bersani's landmark 1987 essay "Is the Rectum a Grave?" to later proponents of the antisocial turn such as Lee Edelman, queer theory has been constructed upon critiques of the subject for being neither politically useful nor even entirely possible; indeed, what makes queer theory "queer" is that it presumes a fundamental instability at the level of the subject. And queer theory is not alone in this concern. Predating the emergence of queer theory, black feminist and later Afropessimist theory have made different, yet similarly concerned, calls for the reevaluation of the place of not only the subject but the human in scholarly investigations into political and metaphysical questions surrounding consciousness, liberation, and the political.[34] For queer theorists, the concern stems broadly from the disciplinary function of the subject, as outlined by Foucault, Butler, and others; for philosophers such as Wynter and Spillers, skepticism about the subject, in its rough coterminousness with liberal personhood and the human more generally, derives from both histories of the enslavement, subjugation, and deprioritization of black people and black life as a whole and the ongoing imbrication of the subject in a vision of the political that is still deeply invested in and structured by the prerogatives of antiblackness. If the subject, in all of its proliferating disciplinary power, girds a biopolitical state that differentially distributes life chances along the axes of race, class, and the potential for a specific vision of social reproduction, then it is no wonder that queer theory, emerging out of the devastation of the early years of the AIDS crisis, and black feminist philosophy, always already operating under the sign of its own putative impossibility, might have refused it. I want to be careful to make clear that neither the antisocial turn in queer theory (which arrived more or less twenty years after early antisubjective work in queer theory by Bersani and others) nor early queer theory on the whole was particularly interested in racial justice; the antisocial turn, in particular, was widely criticized for an uncritical lack of attention to racist violence in its refusal of a particular vision of the political.

I tarry on these intellectual genealogies because in their skepticism toward the subject, I perceive an articulation of what was lost by the emergence of modern, subjective visions of personhood, and of what is forced to remain lost or impossible by the persistent dominance of subjective modes of imagining freedom, operating to the exclusion of other ways of theorizing life. Thinkers in each of these fields have, at times, taken either agonistic or avowedly antagonistic positions in relation to the human writ large, coterminous as it has become with a disciplinary version of subjectivity that insists on individuation and inculcates the bearer into a politics designed to simply reproduce the limited potential of the status quo. They are not wrong to do so, of course, but I wonder to what degree older understandings of the environmental body might have changed, complicated, ameliorated, hindered, or revised the potential of the human not only as a productive site around which to politically organize, but as a different cartography of the political itself. The environmental body was ecological, understood as ineluctably imbricated in its community, and nothing without its situatedness. It was distinctly not individual, even though it could, at times, be singular. Was this environmental body pointing toward a vision of the human that never came to be?

Unexpectedly, I think we actually find mirage-like shimmerings of the twentieth-century descendants of this earlier body, this other human, throughout what we now consider some of the most canonical works in early queer theory. The same is true of black feminist theory such as the work of Sylvia Wynter, although Wynter had explicitly identified the human as a question for black thought as early as the late 1970s. Yet because early queer theory was not being consistently read with (nor was it reading) black feminist theory, the intellectual sympathies between the two are only rarely addressed. Just as scholars have demonstrated the degree to which early gender theory—Judith Butler's *Gender Trouble*, for example—did not initiate brand new ways of thinking about either gender or the subject but rather drew from and built upon a decade or so of scholarship in the women-of-color feminist tradition, here I return to Bersani's "Is the Rectum a Grave?" in particular to suggest its indebtedness not only to the feminist theory it directly cites (Catherine MacKinnon and Andrea Dworkin), but to black feminist theory as well.

In "Is the Rectum a Grave?" Bersani is notably interested in sex itself, over and above (but not to the exclusion of) the account of modern sexuality offered by Foucault. This is unusual in queer theory, with a few notable exceptions (Annamarie Jagose's *Orgasmology* comes most immediately to mind); thirty years later, the biggest secret about sex may in fact be that not even queer theorists like it. Indeed, in this piece we find Bersani unconvinced by the vision of sexuality as totally or primarily a subjectivating tactic of power. Instead he turns to a Freudian analytic tradition to effectively argue that to read sex as power—and specifically, to read the "top" and "bottom" social and sexual configuration of much of the gay sex he discusses as a reflection of power relations—is actually to misread sex itself. More curious about the way (gay) sex typifies an antirelational act than a power-relationship act, Bersani anatomizes the self-shattering *jouissance* that gay sex is capable of producing, which arrives precisely in the moment of assuming the pathic position, the role of the "bottom." For Freud, he argues, "the sexual emerges as the jouissance of exploded limits, as the ecstatic suffering into which the human organism momentarily plunges when it is 'pressed' beyond a certain threshold of endurance. Sexuality, at least in the mode in which it is constituted, may be a tautology for masochism."[35]

He does not mean "masochism," here, to refer to a sexual orientation (or exclusively to it), but rather to the death drive, that libidinal call toward self-erasure. This psychoanalytic vision of the relationship between sexuality and the subject, itself as modern as the sexuality of Foucault's account, is positioned in a contradictory tension with poststructuralist commonsense regarding the relationship between sexuality and subjectivity, an understanding that sees the former as a technique of the latter. The version of the relationship between sex and subjectivity that Bersani finds so compelling is one in which sex is poised to unsettle the self, refuse its prerogatives, foreclose its unfolding into the subject. And this unsettling is the site of Bersani's critique of Foucault; whereas Bersani sees the experience of assuming the pathic position in gay sex as one that points to sex's potential for self-shattering, he seems to interpret Foucault as reading the self-shattering potential of sex as evidence of sex's work as a form of power.[36] This, Bersani implies, is a fundamental misreading, one likely derived from Foucault's noted skepticism toward psychoanalysis; "the self which the sexual shatters," Bersani asserts, "provides the basis on which sexuality is associated with power."[37] While it is important to note that Bersani plays a bit fast and loose with terms, here (Freud's *self* is not coterminous with Foucault's *subject,* and Bersani's *sex* and *the sexual* is not Foucault's *sexuality*) in "Is the Rectum a Grave?" Bersani nonetheless uses an analytic model of the relationship between the self and the sexual to critique Foucault's understanding of the bind between sexuality and the subject, suggesting that certain sexual practices—the experience of bottoming in gay sex, for example, in its willful lean over the precipice of "suicidal ecstasy"—may be one place where the sexual refuses sexuality and where sex does not serve the subject, at least not exclusively. What he is making a claim for here is sex itself, sex as precisely not the same thing as sexuality, sex as another means by which the individuating, disciplinary work of sexuality might unexpectedly be undone.

Indeed, Bersani imagines the self-shattering potential of pathic sexual subjection as a drive with the potential to undo the human itself. This is perhaps the most understudied element of his essay and of the antisocial turn (which properly arrived to queer studies about thirty years after Bersani published "Is the Rectum a Grave?") more generally, and while Bersani is not interested in the problem of the human in the same terms as, for example, Wynter is, this essay does espouse a careful, specific vision of the human that understands it in privileged relationship to masochistic pleasure and the death drive. This should be important even to scholars unconvinced by or uninterested in Freud, because it points to a different genealogy of the human, one with radically different roots and goals than the critiques of the human emerging from black feminist theory, yet one that nonetheless refuses the exhaustive biopoliticization of the body and its commensurate merger into subjectivity and personhood. This biopoliticized body, even in its hermetic self-possession, is a fundamentally social body.

This is why Bersani takes such issue with any formulation in which sexuality is understood as a technique of power (which, to be honest, I find to be a rather ungenerous reading of poststructuralist theories of sexuality): he argues that "the ambition of performing sex as only power is a salvational project, one designed to preserve us from a nightmare of ontological obscenity, from the prospect of a breakdown of the human itself in sexual intensities, *from a kind of selfless communication with 'lower' orders of being.*"[38] To reiterate—because this is critical—Bersani critiques poststructuralist understandings of sexuality as a disciplinary tactic because these narratives refuse the potential of sex to let us lose ourselves (or simply be lost) in "ontological obscenity"; these narratives foreclose the possibility that sex might prompt "a breakdown of the human itself"; and, implicitly, these narratives create or enforce social order by preventing "selfless communication with 'lower' orders of being." Bersani's assessment of what people found and still find so threatening in the contemplation of gay sex, then, is a total loss of order, a loss of order of the self (the ontological), the species (the human), and the race ("lower" or "higher" orders of being). The "selflessness" referred to here is not about goodwill but about a total lack of distinction between self and other, inside and outside, you and me, body and world. In a sense, the structure of this psychic selflessness looks a bit like the structures of the kinds of environmental theories of the body, the human, and the social that were dominant before the arrival of a putative sexual modernity. That these older ways of making sense of people, bodies, and pleasures persisted quietly in psychoanalytic thinking in particular should surprise no one; Freud's rather mythological explanations of the etiology of psychic processes were based in ethnographies of distant, non-European peoples, and one need think only of

Totem and Taboo or *Moses and Monotheism* to see glimpses of possible *longue durée* sympathies or continuities between psychoanalysis and natural history.

I have no intention here of carrying a brief for the installation or restoration of psychoanalysis to some sort of pride of place in contemporary understandings of sexuality. Indeed, one of the reasons that psychoanalytic thinking has not earned that place on its own may be precisely the ease with which some analytic thinkers have been able to ignore the realities of social power and, with it, the enormous centrality of racialization to the meaning that is made of sexual behavior, and especially aberrant sexual behavior. I do believe that psychoanalysis, like so much of the sexological science from which it drew, is fundamentally indebted to the natural historical tradition, which in turn may explain the way some psychoanalytically inflected work in queer theory does seem to tell a radically different story about what sexuality is, how it is structured, and how it works, than Foucauldian accounts, which still claim something more or less like dominance in the US American academy. Detailing the ways in which this might be true, however, is not the work of this book. Suffice it to say that in early queer theory's deep reliance on psychoanalytic thought in the development of some of its most cherished questions and most entrenched premises, we might hear, I suggest, traces of older frameworks for thinking about sex outside of the subject, perhaps even reverberations of natural historical thought. This is not a celebration—natural history and its deep imbrication in the violence wrought by racisms structural, social, interpersonal, and casual is not a mode of thought I am prepared to uncritically revive—but it does index the way eighteenth-century, environmentalist frameworks for making sense of the human body and its morphological and behavioral diversity were not eradicated wholesale by the arrival of modern sexuality and its subjectivating project. It also suggests a new range of questions that contemporary sexuality studies scholars might gain from engaging primary sources from the presexological era, and it suggests innovative directions for scholarly work that promise to enliven eighteenth-century studies.

This book outlines an alternative genealogy of sexuality, one that began in a world without the modern subject and quietly entered into modernity alongside it. The vision of sexuality explored here might at first seem strange, since it is a vision of sexuality unmoored from the human body, an image of sexuality that quite literally shifts with the wind and weather, an image of sexuality that is the ugly issue of racist epistemological violence. This environmental vision of sexuality was not understood in its time as an iteration of freedom, choice, or enfranchisement and should not be understood as such now. Instead, I hope that the environmental theory of early sexuality that I propose here sounds a broad warning about—and perhaps offers an incitement to refuse—the perils of individuation, especially as I write in a moment in which the politics of sexuality, and queer politics specifically, have been so effectively co-opted and reorganized against affinity, against collectivity, and against justice. Because the best books, I think, leave readers with as many productive questions as definitive conclusions, I close this one in precisely such an aspirational pose, with the hope that it helps us reflect on two sets of inquiries. First, how might we imagine, today, an environmental theory of sexuality? Does one exist? If so, how might we recognize it? Do we actually see these histories of sexuality in present-day structures of racialization? The second is inspired by Joan Nestle's 1981 warning that "if we ask decorous questions of history, we will get a genteel history": how might we ask not only less decorous questions about history, but less decorous questions about the history of sexuality, in particular? What might a deep dive into sexuality's ugliest and least compelling pasts offer by way of new archives, new understandings, new ways forward? To imagine that the work of producing histories of sexuality is liberating, hopeful, or consolidating is to accede to an understanding of the relationship between sexuality and the political that seems, at best, profoundly misguided, at least to most of us who work on the eighteenth century. It is one way of producing a usable past, but another might be to take the hardest look that we can at what we know, and start from the assumption that these histories of sexuality are likely not charming, recuperable, or political in the way we want them to be. This is not to restate or even ascribe to any of the central tenets of the antisocial turn, but to simply acknowledge that most histories of sexuality are ugly—violent, unbidden, unusable. Let us start there.

NOTES

Introduction. Toward an Environmental Theory of Early Sexuality

1. W. Davis, "Travels and Captivity of Davis," 476. The narrative by Davis (also spelled Davies) was initially printed in London in 1614 under the title *A true relation of the travailes and most miserable captiuitie of William Dauies, barber-surgion of London*. It states that his travels begin in 1597; Elizabeth Baigent, in the entry on Davis in the *Oxford Dictionary of National Biography*, dates his travels to 1598. See Baigent, "Davies, William (*fl.* 1598–1614)," in *Oxford Dictionary of National Biography* (Oxford University Press, 2004); www.oxforddnb.com/view/article/7272, accessed November 23, 2015.

2. W. Davis, "Travels and Captivity of Davis," 476.

3. W. Davis, 477. Davis's 1597 narrative is reprinted in a collection titled *A Collection of Voyages and Travels* (1745). Although Davis's narrative is generally considered alongside other Barbary captivity narratives, the *Francis of Saltash* was neither captured nor its ship laborers sold into slavery by Barbary corsairs, but by ships belonging to the Duke of Florence, who mistook the *Francis of Saltash* for a Turkish ship. Because, as Davis explains, a significant minority of the sailors on board were "Turks," and because the ship was leaving from Tunis, "freighted with *Turkish* Goods by *Turks* . . . for we traded as well with the *Turk* as the Christian" (478), the ship was intercepted by a fleet belonging to the Duke of Florence on its way to "Syo" (Syo Syo, Cameroon), who "took [them] for a *Turkish* prize" (478). Upon being captured by the Duke of Florence's ships, the *Francis of Saltash* and its crew were taken to "Leghorn" (Livorno, Italy), and the crew members were sold as slaves. Davis spent "eight years and ten months" there (479).

4. Many historians of sexuality have noted that representations of sodomy appear frequently in discussions—or voiced disapproval—of cultural or religious difference, but as far as I have been able to determine, no one has tracked the pervasive appearance of sodomy in discussions of ethnic difference, other than Rudi Bleys in *Geography of Perversion*. See also Jonathan Goldberg's *Sodometries*, esp. xv; and his *Queering the Renaissance*.

5. See Linnaeus, *Systema naturæ, sive regna* (1735), 1–12; and Linnaeus, *Systema naturæ per regna* (Stockholm), [1–4], 1–824.

6. In *Systema Naturae*, Linnaeus defines humans as "mammalia," animals characterized by live birth and the feeding of young from milk produced in mammary glands. Humans are differentiated from other primates (such as apes and, curiously, sloths) by the presence of *mind* (*Homo sapiens*).

7. For considerations of the relationship between Linnaeus's taxonomies of the human and the development of modern racial categories, see W. Jordan, *White over Black*, 218–219; and Wheeler, *Complexion of Race*, 151–152, 169.

8. This is a translation of the Latin text that appears in Linnaeus, *Systema Naturae, Per Regna* (Holmiae: Impensis Direct), 20–24. The Latin descriptions under the "Afer" category—"women without shame" and "mammae lactate profusely"—are not fully translated into English when they appear in the 1802 translation of the Johann Friedrich Gmelin Latin edition (in Latin: "feminis sinus pudoris; mammae lactantes prolixae"), and furthermore, this is a contested translation. See chapter 1, note 47. The rest of the English description remains the same and simply reproduces and translates the Latin, but it omits "women without shame" and "mammae lactate profusely." Scholars who cite this passage in English are likely citing about forty years worth of citations of James Slotkin's *Readings in Early Anthropology*, which includes English translations of the 1758–59 Latin tenth edition.

9. Linnaeus's reference to the *mammae* of women of the *Afer* variety refers in a general way to the defining characteristics of the category of *mammalia*, which includes *Homo sapiens*. This reference to the *mammae* appears only in descriptions of women of the *Afer* class, not in the descriptions of any of the other taxa of *Homo sapiens*. This representational trope that associates black women with breastfeeding is well documented by eighteenth-century historians; see Morgan, "'Some Could Suckle.'"

10. Note that an English translation of Linnaeus's famous chart of the "sexual system" of botanical taxonomy was available as early as 1787; it was reproduced in the Lichfield Society's 1787 *Families of Plants*, which drew heavily from Linnaeus's 1737 *Genera plantarum*. The first English translation of the *Systema Naturae* seems to be one rendered by William Turton (Linnaeus, *General System of Nature*). Incidentally, the *Afer* category of *Homo sapiens* in the tenth edition is the only category to include characteristics specifically attributed to women. Beyond the two direct translations of Linnaeus's *Systema Naturae* in 1801–2 and 1802–6, however, the ideas behind Linnaeus's taxonomy circulated widely in print productions in North America and Europe well before the turn of the nineteenth century. Consider, for example, Erasmus Darwin's *Botanic Garden*, which I discuss in chapter 4; the second part of Darwin's poem *The Loves of the Plants*, published first in 1789, relies explicitly on Linnaeus's "sexual system" of taxonomy, also introduced in the tenth edition of *Systema Naturae*. Mavor's *Complete System of Natural History* also reproduces Linnaeus's division of *Homo sapiens* into four regional categories, but in prose form. For further examples of the circulation of Linnaeus's work in contemporary naturalist writing, see Stillingfleet, *Miscellaneous Tracts*, which is based on Linnaeus's *Oeconomia naturae* from *Amoenitates Academica* (1751); Lee, *Introduction to Botany*, which collates pieces of both Linnaeus's *Genera* and the *Systema*; Milne, *Institutes of Botany*, which also translates parts of the *Genera*; Yeats, *Institutions of Entomology*; the Lichfield Botanical Society's translation of part of Linnaeus's *Systema Vegetabilium*, titled *System of Vegetables*; Lichfield's translation of *Families of plants* from Linnaeus's *Genera plantarum*; and Shaw, *Speculum Linnaeanum*. Thanks to Devin Griffiths for his aid in tracing the translation history of Linnaeus's *Systema* and other works into English and the circulation of Linnaeus's taxonomic theories into Anglophone texts.

11. Some of the best examples of studies of these intersections include scholarship coming out of early modern studies, including the writings of Ann Laura Stoler, Valerie

Traub, and Carmen Nocentelli (see Nocentelli, *Empires of Love*). In eighteenth-century studies, Felicity Nussbaum's *Torrid Zones* is the only monograph I've encountered that considers at length the question of sexuality vis-à-vis colonial writings.

12. See, for example, Katy Chiles's *Transformable Race*, a study of the way natural historical thinking shaped eighteenth-century literary works, which also argues that these literary works, in turn, engaged and shaped natural historical thinking.

13. Historians of science have recently begun to consider the implications of "observation" as a specific methodology for the production of knowledge. See, for example, Daston and Lunbeck, *Histories of Scientific Observation;* and especially Daston, "Empire of Observation."

14. On environmental influence on humoral makeup and complexion, see Wheeler, who argues,

> Subject to fluctuation, the humoral body was porous and thus easily affected by what went in it and around it, and it was especially sensitive to heat and cold, the major forces understood to alter bodies externally and internally. In the Middle Ages and Renaissance understanding, complexion "was never an absolute but always a relative quality" (Siraisi 103), a characteristic that continues to inflect it throughout the eighteenth century, even as complexion undergoes certain changes in usage. Complexion, or humoral theory, was often called on to explain psychological, social, and physiological characteristics as well as to formulate stereotypes (103). A vast majority of eighteenth-century Europeans still believed that cultural, educational, or environmental change altered the humoral mix and thus affected both appearance and behavior. (*Complexion of Race*, 23)

On the transformability of race, see Chiles, *Transformable Race*.

15. To name only a few, consider Bersani, "Is the Rectum a Grave?"; John Boswell, *Christianity, Social Tolerance, and Homosexuality,* esp. section 2, "The Christian Tradition"; Bray, *Homosexuality in Renaissance England,* esp. "Words and Symbols"; J. Goldberg, *Sodometries* and *Reclaiming Sodom;* and M. Jordan, *Invention of Sodomy*.

16. "Turk" was one of a handful of relatively generic terms for Muslim peoples living in the Islamic city-states of the northern coast of Africa.

17. See Warner, "New English Sodom."

18. See chapter 4, which discusses Herman Mann's fictionalized memoir of Deborah Sampson, *The Female Review* (1797).

19. On mid-twentieth-century behavioralism, and efforts to use behavior modification to "cure" homosexuality, see Jagose, "Behaviorism's Queer Trace: Sexuality and Orgasmic Reconditioning," in *Orgasmology,* 106–134; on the twentieth-century use of behavior modification techniques to "treat" people convicted of sexual offenses, see Moster, Wnuk, and Jeglic, "Cognitive Behavioral Therapy Interventions."

20. In a 2003 issue of the *William and Mary Quarterly,* Anne Myles argues that "early Americans did not recognize the sexual as a discrete category the way people in the present do." "Queering the Study of Sexuality," 199.

21. Danforth, *Cry of Sodom Enquired Into*.

22. Danforth's fear of the widespread incidence of secret sins was consistent with contemporary Puritan cultural concerns over a decrease in church membership, and thus a general decrease in the vivacity of faith, in New England during the last half of the

seventeenth century. We see this concern in Danforth's *Cry of Sodom Enquired Into* (1674), published three years after his fairly widely circulated jeremiad *A brief recognition of New-Englands errand*. The *Cry of Sodom Enquired Into* was delivered nine years after the development of the "half-way covenant," a theological compromise promoted by Rev. Solomon Stoddard that allowed for a partial form of church membership, aimed at sustaining and increasing church participation. Importantly, a large percentage of early gallows literature that made it into print focused on crimes of what we would now define as sex; see, for example, the first compilation of execution narratives, *Pillars of Salt*, assembled by Cotton Mather in 1699. I address this text at greater length in chapter 3.

23. Danforth, *Cry of Sodom Enquired Into*, 17–18.

24. Onan, for example, appears in Genesis 38, and God denounces the Nicolaitans in Revelation 2.

25. References to the Menandrians and the "Gnosticks," for example, gesture toward religious sects as well as historical theological debates over the ideological origins of Christianity.

26. "Sodom," for example, is figured by Danforth as female ("There are your Filthy sisters, Sodom and her daughters"), and in this passage Sodom appears as more of a person than a city, although the only Sodom in biblical or Christian religious history is the city of the same name.

27. We should read Danforth's understanding of Goad's "lecherous Kindred" as not only retrospective, pertaining to sinners who had come before Goad, but also prospective, inclusive of sexual sinners who would come after him.

28. This genre had earlier roots in England and proliferated in the Atlantic world throughout the long eighteenth century; while Danforth's sermon was the first execution sermon written and printed in the North American colonies, it harked back to a generic tradition that had long been in existence. See D. Cohen, *Pillars of Salt*; Halttunen, *Murder Most Foul*; and DeLombard, *Shadow of the Gallows*.

29. DeLombard, *Shadow of the Gallows*.

30. An example of such demotic cultural narratives would be Herman Mann's *Female Review*, a late-eighteenth-century "warrior woman" tale that Mann writes into novel form but that critics identify as participating in a much longer, primarily ballad-based oral tradition of "warrior woman" stories (see Dugaw, *Warrior Women and Popular Balladry*). Chapter 2 of the current volume focuses on both print and manuscript texts. My definition of "popular" narrative here is quite capacious: while focusing on printed prose narratives, it acknowledges the longer and often more folkloric histories of these stories and contextualizes the incorporation of the stories into print culture within this longer historical framework. Burke's *Popular Culture in Early Modern Europe* informs this understanding. He defines popular culture as "a system of shared meanings, attitudes and values, and the symbolic forms (performances, artifacts) in which they are expressed or embodied" (xiii), understanding the "popularity" of popular culture as "perhaps best defined . . . in a negative way as unofficial culture, the culture of the non-elite, the 'subordinate classes' as the Italian Marxist Antonio Gramsci famously called them" (xiii). Burke builds his definition of popular culture and theorizations of the popular within early modern culture to 1800 upon the work of cultural historians (specifically, Jacob Burckhardt, Aby Warburg, and Johan Huizinga); *Popular Culture* proved to be a foundational text, one that subsequent literary critics and historians of popular culture

took as a point of departure for later studies of both early popular culture and literary and historical methodology.

I use "popular narrative" (rather than print culture, for example) to describe the texts that I consider in this study. I use "narrative" to indicate two characteristics of the texts that I study: first, my focus is popular writing that first appeared in print in prose form. Though these stories may appear in many genres—sermons, advice tracts, novels, or "true histories"—I use "narrative" in the descriptive mode to indicate a form of writing that appears in prose, one that tells a story that collates events and characters. My definition of narrative follows the eighteenth-century use of the term; the *OED* charts numerous uses of "narrative," ranging from the sixteenth to the nineteenth century, in which "narrative" indicates "a part of a legal document which contains a statement of alleged or relevant facts closely connected with the matter or purpose of the document" and an "account of a series of events, facts, etc., given in order and with the establishing of connections between them." "Narrative, n.," in *OED Online*, June 2011 (Oxford University Press); www.oed.com/view/Entry/125146?rskey=a3GRTK&result=1&isAdvanced=false.

For more on recent theorizations of methodologies for approaching the history of the popular, see Bourdieu, *Field of Cultural Production;* Scribner, "History of Popular Culture"; Harris, "Problematizing Popular Culture"; Shershow, "New Life"; and Storey, *Inventing Popular Culture.* The history of popular culture has been explored extensively within early modern studies, in particular; for meditations on specific sixteenth- and seventeenth-century popular histories, see Stallybrass and White, *Politics and Poetics of Transgression,* esp. the introduction; Bahktin, *Rabelais and His World;* Watt, *Cheap Print and Popular Piety;* and Chartier, *Cultural Uses of Print.* For work on the history of early American popular culture, see Cullen's collection *Popular Culture in American History;* David Hall, "The Atlantic Economy in the Eighteenth Century," in *A History of the Book in America:* vol. 1, *The Colonial Book in the Atlantic World,* ed. David Hall and Hugh Amory (Chapel Hill: University of North Carolina Press, 2010), 152–162; and Shields, "Eighteenth-Century Literary Culture."

31. Foucault, *"Society Must be Defended,"* 7.

32. Much eighteenth-century popular narrative poses as "true," and many of the most widely read narratives were based on the lives of people we know to have actually lived. Chapters 2 and 3 read narratives of actual sailors who were taken captive by Barbary pirates and narratives of actual people who were sentenced to death, respectively; and Deborah Sampson, the protagonist of Herman Mann's *Female Review* (the subject of chapter 4) was a real person who did indeed dress as a man and join the Continental Army (Sampson is also the state woman of Massachusetts).

33. Wheeler, *Complexion of Race,* 9.

34. Faderman, *Surpassing the Love of Men,* 20.

35. Arondekar, *For the Record,* 3.

36. See Godbeer, *Sexual Revolution in Early America;* Fisher, "Queer Money"; and J. Goldberg, *Sodometries.*

37. Myles, "Queering the Study of Sexuality."

38. Scholars date the emergence of sexology as a science to points as early as 1836, with the publication of Parent du Châtelet's *De la prostitution* and as late as 1905, when Freud's *Three Essays on a Theory of Sexuality* was first published.

39. I use *environmental* in much the same way as Foucault defines the "milieu": "an

ensemble of natural givens—rivers, marshes, hills—and an ensemble of artificial givens—an agglomeration of individuals, of houses etc. The milieu is a certain number of combined, overall effects bearing on all those who live in it." Foucault, *Security, Territory, Population*, 21.

40. Winthrop Jordan was one of the earliest Americanist scholars of racial difference to describe race as an environmental phenomenon; see his *White over Black,* esp. 307, 479. Thanks to Katy Chiles for pointing me toward this reference.

41. On work that has returned to the racial and colonial dimensions of biopower, see Mbembe, "Necropolitics"; Stoler, *Race and the Education of Desire;* and Weheliye, *Habeas Viscus.*

42. See D. Goldberg, *Threat of Race,* 67; and S. Browne, "Introduction, and Other Dark Matters," in *Dark Matters,* 16. Goldberg is thinking specifically about neoliberalism, a period much later than the eighteenth century, but my use of "racialization" in this project builds on each of these authors' insights.

43. See Allewaert, *Ariel's Ecology.*

44. Cronon. "Place for Stories," 1349.

45. S. Smith, *Essay,* 62–63.

46. Here I cite Foster's manuscript in progress "After Emancipation: Genres of the New Racial Ordinary."

47. See Gustafson's essay "Genders of Nationalism."

48. On the well-documented relationship between natural history and colonialism, see, for example, Meyers and Pritchard, *Empire's Nature;* Schiebinger, *Plants and Empire;* and *Nature's Body;* Schiebinger and Swan, *Colonial Botany.*

49. Hallock, "Male Pleasure and the Genders," 698.

50. This is certainly a well-established idea in twentieth- and twenty-first-century studies of racialization, emerging from fields as diverse as sociology, the history of science, black feminist theory, and studies of biopolitics; this note will thus only gesture to a fraction of the important scholarly work on this question. For studies of racialization that do not attend at length to sexual politics, see Said, *Orientalism;* Omi and Winant, *Racial Formation in United States.* For studies of racialization that focus on the twentieth and twenty-first centuries and attend to the sexual politics of racialization, see Anzaldua and Moraga, *This Bridge Called My Back;* Collins, *Black Sexual Politics;* Crenshaw et al., *Critical Race Theory;* A. Davis, *Women, Race, and Class;* Johnson and Henderson, *Black Queer Studies;* and Lorde, *Sister Outsider.* For studies of racialization and sexuality that focus on the nineteenth century, see Somerville, *Queering the Color Line;* Schuller, *Biopolitics of Feeling.*

51. "Dawn of a New Day for Marriage Equality," released November 27, 2012, by the Human Rights Campaign, Hrc.org., September 15, 2015.

52. I use the terms "gay and lesbian" both because these are the terms that the ad uses and because of the Human Rights Campaign's historically unsupportive relationship to transgender political lobbying, including its support for the 2007 bill H.R. 3685, an iteration of ENDA, the Employment Non-Discrimination Act, that included protection for gay and lesbian workers but not protection along the lines of gender identity.

53. By "prevalent cultural logics," I do not mean the views of scholars; while the logic of analogy does, of course, appear in some scholarly work (including Bersani's "Is the Rectum a Grave?"), there is no shortage of work in critical ethnic studies, American

studies, and queer studies that both critiques and refuses "like race" arguments. I am concurring with Siobhan Somerville's discussion of the wide cultural prevalence of this logic.

54. On analogy in eighteenth-century science, see Griffiths, "Intuitions of Analogy"; Porter, "Scientific Analogy and Literary Taxonomy"; and Allewaert, *Ariel's Ecology*, 63–70.

55. On the Curse of Ham as a narrative of the origins of human variety, see Curran, *Anatomy of Blackness*, 76–78, 82–83, 120; Chiles, *Transformable Race*, 206; Iyengar, *Shades of Difference*, 8–11, 67; W. Jordan, *White over Black*, 40–41, 200; Dain, *Hideous Monster of the Mind*, 126–127; Kidd, *Forging of Races*, 35; and Gossett, *Race*, 5.

56. Genesis 9:7.

57. Genesis 9:21–27.

58. Curran writes, "While there is no mention of the color of Noah's sons in the Old Testament, subsequent exegeses of this episode in both Christian and Jewish theological writings often claim that this was the moment that not only sentenced Africans to a life of base servitude but *darkened* the sinful descendants of Ham" (*Anatomy of Blackness*, 78); and Dain argues that "chronology traced human diversity back to the time of the Flood, to Noah's three sons, Ham, Shem, and Japheth, seen as the progenitors of Africans, Asians, and Europeans, respectively" (*Hideous Monster of the Mind*, 127).

59. Curran, *Anatomy of Blackness*, 76.

60. The story of the Curse of Ham, as its inclusion in Hakluyt's sixteenth-century *Principal Navigations* suggests, would have been very familiar to Christians long before the eighteenth century, and readings of this biblical story as an account of the emergence of human variety were well known in North America by the very beginning of the eighteenth century. See, for example, C. Mather, *Negro Christianized*, where Mather muses, "Suppose these Wretched *Negroes, to be the Offspring of Cham* (which yet is not so very certain,) yet let us make a Trial, Whether the Christ who *dwelt in the Tents of Shem*, have not some of His Chosen among them; Let us make a Trial, Whether they that have been Scorched and Blacken'd by the Sun of *Africa*, may not come to have their Minds healed by the more Benign *Beams* of the *Sun of Righteousness*" (2–3).

61. Quoted in W. Jordan, *White Over Black*, 40–41.

62. Iyengar, *Shades of Difference*, 9.

63. Dain, *Hideous Monster of the Mind*, 126.

64. Somerville, *Queering the Color Line*, 14.

Chapter 1. The Natural History of Sexuality

1. Crèvecœur, *Letters from an American Farmer*, 46.

2. The question of whether to describe what eighteenth-century intellectuals termed "human variety" as "racial difference" is a complicated one, and Wheeler, Tawil, and Chiles all take pains to define exactly how they use racialized terms ("black," "white," "*negre*") in their monographs and to specify how and whether they use the term "race." In this book I follow Chiles in using "race"; she argues that there are several benefits of using this term: "First, it follows how some eighteenth-century writers did use the term; second, it requires a defamiliarization of the term *race* to emphasize its specific eighteenth-century connotations; and third, it demands a similar kind of historical defamiliarization for *whenever* the term is used—the eighteenth, nineteenth, twentieth, or twenty-first

centuries.... I hope my troubling of this term ... will remind us to problematize it each and every time we employ it, even today." *Transformable Race*, 9–10.

3. "Contrapuntal reading" is a term and interpretive method first developed by Edward Said in *Culture and Imperialism*. In operatic vocabulary, "contrapuntal" means "employing combined or contrasting themes, structures, etc." By arguing that natural history contains a quiet or contrapuntal narrative about sexual behavior, I am suggesting that even though natural historical writing is not, for the most part "about sex" (although it is, at least in part, "about" race), eighteenth-century readers still consistently gleaned information about sex and human sexual diversity from these texts. Said uses "contrapuntal reading" to describe one approach to the archival problem of the putative absence of representation of the realities of colonialism from novels about the metropole (his classic example is Austen's references to Antigua in *Mansfield Park*); a contrapuntal reading of *Mansfield Park*, which does not extensively represent Antigua, would understand the novel to still be saying something about colonialism. Said argues that a "contrapuntal orientation in history ... sees Western and non-Western experiences as belonging together because they are connected by imperialism." *Culture and Imperialism* (New York: Vintage Books, 1993), 279. On "contrapuntal," see "contrapuntal, adj.," in *OED Online*, June 2016 (Oxford University Press); www.oed.com/view/Entry/40437?redirectedFrom=contrapuntal.

4. See, for example, Chiles, *Transformable Race;* Iannini, *Fatal Revolutions;* Rusert, *Fugitive Science;* Wheeler, *Complexion of Race;* Parrish, *American Curiosity;* Nussbaum, *Torrid Zones;* and Curran, *Anatomy of Blackness*.

5. I am also beholden to my colleague and friend Benjamin Kahan, whose *Book of Minor Perverts* (forthcoming from University of Chicago Press) includes a piece on what he terms "Queer Weather" that theorizes the afterlives of early modern environmentalist thought—specifically in terms of climate—within the context of late-nineteenth-century sexology. Kahan, "Toward a World-System."

6. "Sexuality," as a discrete field of knowledge about human sexual behavior, is typically thought to have arrived with the advent of sexology, a development that scholars date to moments as early as 1836 and as late as 1900. Foucault's 1976 *History of Sexuality* is probably the most authoritative source on the modernity of sexuality among sexuality studies scholars in the United States. He argues that modern sexuality constitutes a network of discursive power through which the subject both is disciplined and disciplines itself. For a longer discussion of the modernity of sexuality, see also Coviello, *Tomorrow's Parties*.

7. See Chiles, *Transformable Race;* Curran, *Anatomy of Blackness;* Wheeler, *Complexion of Race;* Nussbaum, *Torrid Zones*.

8. Susan Scott Parrish offers a clean synopsis of humoral theory in her 2006 *American Curiosity*, 78–79.

9. Hippocrates's writings were occasionally glossed, summarized, or copied and reprinted in North America, however; see, for example, Brevitt, *History of anatomy*.

10. See C. Mather, *Angel of Bethesda* (1722). Also see the much longer edition, *Angel of Bethesda* (1972).

11. Katy Chiles argues in *Transformable Race*, 2: "Drawing on natural historical thinking, early Americans largely considered race ... to be potentially mutable: it was thought to be an exterior bodily trait, incrementally produced by environmental factors (such as climate, food, and mode of living) and continuously subject to change."

12. Parrish, *American Curiosity*, 79.
13. Paster, *Body Embarrassed*, 12.
14. See Wheeler, *Complexion of Race*, 7.
15. Following Britt Rusert, I do not use the term "pseudo-science" to characterize early racial sciences, though I acknowledge their purposeful and explicit racism (and use toward the project of hierarchizing racial difference), for these bodies of knowledge were taken very seriously in their own time. See Rusert, *Fugitive Science*, 6.
16. Rusert, 22.
17. Wheeler argues that

> colors, especially embodied in black and white skin tones, functioned on several registers during the eighteenth century: Climate, humors, anatomy, Christianity, and neutral description were all available paradigms. The ancient Greeks and Romans believed that climate was responsible for the complexion of a nation's inhabitants, particularly air temperature and exposure to the sun. *Complexion* referred to inhabitants' temperament and disposition; it arose from the interaction of climate and the bodily humors (blood, bile, phlegm, and choler). Skin color, then, was only one component of complexion. Eighteenth-century Europeans maintained great faith in the strong effects of climate on the body. Their other traditional frame of reference for skin color derived from Christian semiotics, which combined moral and aesthetic meanings, primarily in the binary pair pure white and sinful black. This powerful color construction referred to internal turmoil, actions, spiritual states and external coloring. (*Complexion of Race*, 2)

18. Wheeler, 24.
19. Falconer, *Remarks on Climate*, iv–v.
20. Wheeler, *Complexion of Race*, 30.
21. S. Smith, *Essay*, 62–63.
22. Geographical theories of human difference also girded "conjectural history" (a *longue durée* theory of the history of human development advanced by Scottish Enlightenment thinkers and related to four-stages theory).
23. The definitions of which areas of the earth fell within specific zones, however, remained a live question throughout most of the eighteenth century, and philosophers actually "re-zoned" the world, such that over time, as Wheeler notes, "climate theory was reworked to suit the needs of a shift in power to the northern parts of Europe, especially in the seventeenth and eighteenth centuries. Two changes occurred in traditional climate/humoral theory. . . . At the hands of several writers, Britain was imaginatively reconceptualized as part of the temperate region, particularly in the seventeenth and eighteenth centuries. . . . Now Britain and Scandinavia—the former northern zones—became the center of eighteenth-century civilization, at least to many British minds, and white skin color no longer had the pejorative associations associated with classical climate/humoral theory." *Complexion of Race*, 23. This zoning was neatly aligned with increasingly sedimented racial hierarchies:

> in the eighteenth century, proximity to Europe and to temperate climates generated a theory of theoretical hierarchy—not a scale of horizontal differences—that placed Europeans, and a few groups from the Middle East and North Africa, at the

top and Africans and Laplanders at the bottom. Because of the excessive heat that was believed to enervate the body, mind, and morals, commonplaces about the torrid zone being the home of dark-skinned people who were indolent, lascivious, and subject to tyranny often seemed confirmed when Englishmen confronted social and political life as well as labor arrangements that were alien to them. (24–25)

24. A. Ferguson, *Essay on Civil Society*, 171.

25. Torrid zones refer to "the territory between the Tropic of Cancer and the Tropic of Capricorn." Nussbaum, *Torrid Zones*, 7.

26. See, for example, Wynter, "Unsettling Being/Power/Truth/Freedom." Also see Alexander Weheliye's work on Wynter and other woman-of-color critiques of the human, in his *Habeas Viscus*.

27. See Curran, *Anatomy of Blackness*. The possibility of racial transformation was a contentious hypothesis, and as Curran notes, many eighteenth-century philosophers took issue with Buffon and others' belief in the potential of racial transformation. Curran argues that the dominance of climate-based understandings of racial differentiation had begun to wane among natural philosophers by the late 1760s: "Twenty years after Buffon published the first volumes of his *Histoire Naturelle* (1749), a significant number of eighteenth-century thinkers began to distance themselves from the seemingly overoptimistic elements of his degeneration theory" (169). He continues:

> Several years later, the Italian economist and polygraph Abbé Ferdinando Galiani also refuted Buffon's theory of degeneration in the influential (but elite) *Correspondance litteraire*. . . . This was a three-step process. First, he flatly rejected the possibility of racial reversibility, stating unequivocally that "perfectibility is not a gift given to man in general, but only to the white and bearded race." Secondly, he scorned climate theory: "Everything that is said of climates is nonsense, a *no causa*, the most common error in our logic." And thirdly, and well ahead of his time, he affirmed that the aptitude and destiny of humankind were not products of environment: they were the result of biology-based logic. In a declaration that is strangely prescient of Arthur de Gobineau's belief that race explained the destiny of the human species, Galiani asserted: "Everything is connected to race." (169)

Finally, Curran reminds us that the Abbé de Raynal had "announced that the climate theory at the heart of the monogenetic enterprise was entirely misguided. While he readily acknowledged that there were certainly 'climates suited only for certain species,' he also refuted Buffon and de Paux (although not by name), asserting that 'climatic differences' did not 'change the same species from white to black' and that 'the sun does nothing to alter and modify the seeds of reproduction.' It was entirely wrongheaded, he asserted, to 'attribute the color of the *negres* to climate'" (194).

28. See Chiles, *Transformable Race*, which documents this strain of racial thinking throughout eighteenth-century colonial North American literature and science.

29. Buffon, *Natural History, General and Particular*, 201. Wheeler also reports, "It was commonly reckoned that it would take at least ten generations for Englishmen in the torrid zones to turn into Negroes or for Negroes in England to turn into northern Europeans." *Complexion of Race*, 4.

30. S. Smith, *Essay*, 20–22.

31. For discussion of Jefferson's defense of the climate of North America in *Notes on the State of Virginia*, see Iannini, *Fatal Revolutions*, 222–223.

32. Wheeler makes a point of "integrating the writings of the Scottish Enlightenment philosophers on civil and savage societies" into her discussion of Enlightenment racial ideology. She explains that, "with its emphasis on economic organization and social manners, covering everything from the protection of private property to men's treatment of women, four-stages theory arguably offered a more significant form of racialization of the body politics than the categories concerning the physical body found in natural history." *Complexion of Race*, 7

33. For scholarship on the emergence of stadial theory, see the introduction to Nussbaum, *Torrid Zones;* Wheeler, *Complexion of Race*, esp. 36; and Chiles, *Transformable Race*, esp. 155; for eighteenth-century examples of natural historical reliance on the principles of stadial theory, see J. Brown, *Estimate of the Manners*, esp. 77; S. Smith, *Essay*, esp. 9; and A. Ferguson, *Essay on Civil Society*, esp. 181–182.

34. For a longer discussion of the role of habit in eighteenth-century environmentalist accounts of sexual behavior, see chapter 3.

35. In John Wood Sweet's reading of Samuel Stanhope Smith, he argues that "Smith sought to rehabilitate the climatic theory by refining and extending it: he argued that complexion was governed not just by climate but also by what he called 'state of society' and 'mode of living.' Indians, for instance, would not turn white just because the landscape around them had been tamed unless they also started living in a civil manner. But if they did, they might." *Bodies Politic*, 279.

36. Chiles, *Transformable Race*, 34.

37. Chiles, 22.

38. For more on these colonial concerns about cultural degeneration in colonial locations, see Parrish, *American Curiosity*, chap. 2, esp. 82–84. For more on the circulation of British and Scottish texts in North America during the eighteenth and nineteenth centuries, see Rezek, *London and Provincial Literature*.

39. Crèvecœur, *Letters from an American Farmer*, 51.

40. While there is no shortage of scholarship on Crèvecœur's reliance upon and engagement with natural historical thinking, it is worth noting that in his 1782 *Letters from an American Farmer*, Crèvecœur echoes John Wood Sweet's assertion about the way writers like Samuel Stanhope Smith extended and amended theories of climatic determinism such that they could accommodate a plethora of different hypotheses about the sources of environmental influence, "state of society," or "manners" among them.

41. Crèvecœur, 53. On Crèvecœur's invocation of eighteenth-century naturalist philosophy in *Letters*, see Parrish, *American Curiosity*, chap. 2, esp. 95; Chiles, *Transformable Race*, 113–114; and Regis, *Describing Early America*, chap. 4.

42. Crèvecœur, *Letters from an American Farmer*, 52, 53.

43. "Venery, n. 2," in *OED Online*, June 2015 (Oxford University Press); www.oed.com/view/Entry/222133.

44. Parrish, *American Curiosity*, 79.

45. Carew, *Examination of Men's Wits*, 107. Carew is more or less reproducing Huarte, in English, in this text; for the relationship between Huarte and Carew, see the editor's introduction to this text: Rocio Sumillera, ed., *The Examination of Men's Wits* (London: Modern Humanities Research Association, 2014).

46. Wheeler refers to *Systema* as "a stellar example of humoral theory's lingering influence" in eighteenth-century naturalist thought. *Complexion of Race*, 25.

47. See the introduction, note 8; the tenth edition of Linnaeus's *Systema Naturae* (1758) was not translated into English until the first few years of the nineteenth century, and in that translation, the Latin "feminis sinus pudoris" ("women without shame") and "mammae lactantes prolixae" ("mammae lactate profusely") are omitted. Importantly, this translation of the Latin is controversial; while "women without shame" *is*, indeed, the English translation that appears most frequently in histories of racial thinking and in scholarship in the history of science that cites this text, both David Halperin and Adam Bishop have contested this translation with me during presentations of this work, arguing that "feminis sinus pudoris" should be read as "in women: stretched labia." In a personal correspondence between myself and Adam Bishop from February 24, 2018, he notes that "feminis sinus pudoris" is "translated literally as 'curtain of shame' or 'drape of shame.' 'Pudor' literally does mean 'shame' in classical Latin, but it became a euphemism for the genitals. You've probably heard the adjective form too, 'pudendus/pudenda.' I guess there is a lot to be said there about the moral judgements of early biologists, but I'm not sure they were even thinking of actual shame . . . in biological neo-Latin it's just a word for the genitals." Many thanks to both David and Adam for their insights.

48. Tawil warns us: "If we are careful to situate Linnaean thought in the context of eighteenth-century theories of difference, we immediately recognize that Linnaeus's four primary taxonomic categories—*Americanus, Europaeus, Asiaticus,* and *Afer*—are not functions of biology or morphology, but rather of geography. They are, moreover, geographical in a peculiarly eighteenth-century sense. The most influential eighteenth-century theorists of human variety, Linnaeus, Buffon, Blumenbach, and the American Samuel Stanhope Smith, all defined subvarieties of the human species as 'classes of inhabitants' of different regions." Tawil, *Making of Racial Sentiment*, 43. For more on the tradition of association between African women and breastfeeding, see Morgan, "Some Could Suckle."

49. Nussbaum, *Torrid Zones*, 8.

50. J. Mitchell, *Essay Upon the Causes*. Mitchell reports:

> This cutaneous Malady, of the Negroes, called the *Yaws,* laid the first Foundation of the *Lues Venerea;* which became to differ from it only by the Part affected, and the particular Manner of receiving the Infection, after being transplanted into another colder Clime, on People of a different Complexion; the virulent Acrimony of the cutaneous Contagion being inviscated, and consequently mitified, by the *Semen* which received it; the subtiler Parts of the Contagion being likewise exhaled in the white People, on account of the Perspirability of their Bodies, although the Distemper was drove more on the internal Organs, upon account of the Coldness of the Climate; and so appeared to partake less of a true cutaneous Malady, after this *Lues Venerea* was first propagated to Europe." (144–145)

51. Buffon, *Natural History, General and Particular,* 102.

52. Buffon, 161–162.

53. Buffon, 57.

54. See Curran, *Anatomy of Blackness.*

55. On the availability of Falconer's remarks in North America, see Library Company of Philadelphia, *Catalogue of the books*, 77, accessed via the online Readex database *Early American Imprints, Series I: Evans, 1639–1800*, 22066, Winans, R.B. Book cats., 131.

56. Falconer, *Remarks on Climate*, 6–7.

57. Falconer, 25.

58. Nussbaum glosses this relationship in *Torrid Zones:* "As the geographical terrain shifts from hot to cool, so do human desires and social relations. In Ferguson's representation, sexual passion varies according to the world's topography and climate, yet its unrestricted expression is predictably linked to the torrid zone's failure to nurture civilization or political freedom. With Ferguson and other early natural historians, sexuality as it is connected to civilization is zoned on a world scale" (9).

59. A. Ferguson, *Essay on Civil Society*, 177.

60. A. Ferguson, 176.

61. Buffon, *Natural History, General and Particular*, 190.

62. A. Ferguson, *Essay on Civil Society*, 176.

63. A temporal or historical narrative also inheres in four-stages theory; communities that European thinkers identified as occupying one of the first three stages (hunting and gathering, pasturage, and agriculture), such as we saw in Buffon's description of the Malagasy people, were implicitly defined as the living past of "civilized" European communities, which occupied the fourth and final stage of human development (commerce).

64. Millar, *Observations*.

65. The *Oxford English Dictionary* defines eighteenth-century uses of "passion" as "senses relating to emotional or mental states." See "passion, n.," in *OED Online*, June 2017 (Oxford University Press); www.oed.com/view/Entry/138504.

66. Millar, *Observations*, 1–2.

67. Nussbaum, *Torrid Zones*, 95.

68. Nussbaum, 147. She explains:

> The four-stages theory argues that commercial advancement creates still greater freedom between the sexes. Such freedom may at first increase friendship and companionship, but it ultimately diminishes women's dignity. At the same time, women in the highest form of civilization yet achieved by man paradoxically cultivate their domestic talents and take their greatest pleasure in the reclusive life of the family. Since the commercial society relies on the cult of domesticity, public display of sexual desire is perceived to be un-English and associated with primitive women of earlier stages of development. Their heightened sexuality and its public display encourage "the same free communication between the sexes as in the ages of rudeness and barbarism." (13)

69. Millar, *Observations*, 3.

70. Millar, 3–4.

71. Millar, 2–3.

72. Nussbaum, *Torrid Zones*, 18.

73. Katy Chiles insists on recognizing the importance of contributions made by Native people, Africans, and African Americans—frequently the *targets* of naturalist

theorizations of human development and variety—to the development of the intellectual range of eighteenth-century natural history. For more on this, see Chiles, *Transformable Race*.

74. J. Brown, *Estimate of the Manners*, 48.
75. J. Brown, 19.
76. J. Brown, 30.
77. J. Brown, 20.
78. On natural history and Jefferson's *Notes on the State of Virginia*, see Dain, *Hideous Monster of the Mind*, esp. chap. 2; Chiles, *Transformable Race*, esp. chap. 2; Thomson, *Passion for Nature*, esp. chap. 5; Regis, *Describing Early America*, esp. chap. 3; Farber, *Finding Order in Nature*, esp. the introduction; Parrish, *American Curiosity*, esp. the conclusion; Iannini, *Fatal Revolutions*, esp. part 2; and Tawil, *Making of Racial Sentiment*, esp. chap. 1, 64–67.
79. Pamela Regis, in *Describing Early America*, 80, states that *Notes*

> was written in response to a series of questions circulated in 1780 by François Marbois, the secretary of the French legation at Philadelphia. The diplomat sought information about each American state. Jefferson was deemed best able to answer for Virginia. Taken together, his responses define, delineate, number, list, and describe his "country," beginning with its boundaries and ending with its state papers. First circulated in manuscript, Jefferson's replies were corrected and enlarged, then finally printed in 1785. Seen as an instance of the literature of place, *Notes* can be appreciated for its comprehensive representation of Virginia, an accounting that Jefferson updated during the twenty years following its publication. The book's organization, method, and rhetoric are scientific. Natural history's primary activities—observation, description, and careful record keeping—were essential tools of Jefferson's wide-ranging genius, and natural history of the primary intellectual framework on which he built *Notes*.

80. Iannini writes:

> From its beginnings, *Notes on the State of Virginia* had always been a provisional and dialogic text. Written in response to a questionnaire circulated by François Marbois and in a polemic against the Comte de Buffon and Abbé Raynal, its claims were challenged during subsequent decades by figures including the Marquis de Chastellux, Benjamin Banneker, and, later, David Walker. In the 1790s, Caribbean interlocutors played a significant role in this dialogue because their writings placed pressure on the published version of 1787, a text engaged in perhaps surprising ways (much like Crèvecœur's *Letters* before it) in debates over the implications of West Indian commerce, slavery, and natural history for the future of enlightened Republicanism. (*Fatal Revolutions*, 222–223)

81. See Sweet, *Bodies Politic*, 272–273.
82. My citations of Jefferson are from the copy in the Huntington Library, *Notes on state of Virginia*, unless otherwise indicated. Many scholars, ranging from Thomas Jefferson and Ralph Waldo Emerson to Sacvan Bercovitch and Ezra Tawil, cite the following passage, attributing it to the Comte de Buffon's *Histoire Naturelle* (1749):

The savage is feeble and small in his organs of generation; he has neither body hair nor beard, and no ardor for the female of his kind. Although lighter than the European, on account of his habit of running more, he is nevertheless much less strong in body: he is also much less sensitive, and yet more fearful and more cowardly; he lacks vivacity, and is lifeless in his soul; the activity of his body is less exercise or voluntary movement than an automatic reaction to his needs; take from him hunger and thirst, and you will destroy at the same time the active cause of all his movements; he will remain either standing there stupidly or recumbent for days at a time. (Quoted in Tawil, *Making of Racial Sentiment*, 63)

None of the citations of this passage that I have encountered include a page or volume number, except in cases where the writer is quoting another author (secondary source) who has quoted it; this is significant, as the text they are citing—Buffon's *Histoire Naturelle*—is a forty-four-volume work. I have searched extensively for the passage in both Smellie's 1791 and 1812 English translations and in the 1749 French text. The passage does appear on page 271 of volume 3 of an 1838 edition of the *Histoire Naturelle* that appears in Pillot's 1837 *Oeuvres Complets de Buffon*, but Jefferson, who died in 1826, would not have had access to this text. I have not found it in any of the editions or translations I have read that were available during Jefferson's lifetime. As far as I can tell, Thomas Jefferson is the first person to either publish or quote this passage. He quotes it in French in *Notes on state of Virginia*, 102–104.

83. Jefferson, *Notes on state of Virginia*, 102–104. The translation of the French passage is mine; great thanks to Sara Phenix at Brigham Young University for her help and oversight.

84. Jefferson writes, "America, running through the torrid as well as the temperate zone, has more *heat,* collectively taken, than Europe. But Europe, according to our hypothesis, is the *dryest*" (*Notes on state of Virginia*, 80).

85. Jefferson, *Notes on state of Virginia*, 217.

86. See Jefferson, *Notes on State of Virginia* (London), 103.

87. Jefferson, *Notes on state of Virginia*, 110. Jefferson argues:

This belief [in the lack of veracity of Buffon's account above] is founded on what I have seen of man, white, red, and black, and what has been written of him by authors, enlightened themselves, and writing amidst an enlightened people. The Indian of North America being more within our reach, I can speak of him somewhat from my own knowledge, but more from the information of others better acquainted with him, and on whose truth and judgment I can rely. From these sources I am able to say, in contradiction to this representation, that he is neither more defective in ardor, no more impotent with his female, than the white reduced to the same diet and exercise: that he is brave, when an enterprize depends on bravery; education with him making the point of honor consist in the destruction of an enemy by stratagem, and in the preservation of his own person free from injury; or perhaps that is nature; while it is education which teaches us to honor force more than finesse: that he will defend himself against an host of enemies, always chusing to be killed, rather than to surrender, though it be to the whites, who he knows will treat him well: that in other situations also he meets death with more deliberation and endures tortures with a firmness unknown almost to religious enthusiasm with us: that he is affectionate to his children, careful of them, and indulgent in the extreme: that his affections comprehend his other connections, weakening, as with us, from circle to circle, as they recede from the center: that his friendships are strong and faithful to the uttermost extremity: that his sensibility is keen, even the warriors weeping most bitterly on the loss of their children, though in general they endeavor to appear superior to human events: that his vivacity and activity of mind is equal to ours in the same situation, hence his eagerness for hunting and for games of chance. The women are submitted to unjust drudgery. This I believe is the case with every barbarous people. With such, force is law. The stronger sex therefore imposes on the weaker. It is civilization alone which replaces women in the enjoyment of their natural equality. (105–110)

88. Jefferson, *Notes on state of Virginia*, 373.
89. Jefferson, *Notes on state of Virginia*, 374.
90. Jefferson, *Notes on state of Virginia*, 252–253. For a summary of eighteenth-century North American scientific theories of anatomical theories of racial difference, see J. Mitchell, *Essay Upon the Causes* (1744).
91. Jefferson, *Notes on state of Virginia*, 254–256.
92. Tawil, *Making of Racial Sentiment*, 65. He argues that

Jefferson cannot simply adopt the relevant categories of the Linnaean taxonomy, *Homo Europaeus* and *Homo Americanus*, because British-Americans do not fit comfortably in either category. And by extension, he cannot speak of the varieties of men as "nations" without running the risk of sanctioning all of the implications of the theory of degeneration and the particular global geography which it had generated, according to which "Europe" signified the perfection of nature in its original state and "America" its corruption and diminution. Yet by virtue of this terminological adjustment, Jefferson's language alone puts in circulation a way of thinking about difference that seemed to make intrinsic properties ultimately more important than geography, and inched toward a theory of variation that made human bodies less subject to degeneration in different climates.

93. Chiles, *Transformable Race*.
94. Coviello, *Tomorrow's Parties*, 7, 13.

Chapter 2. The Complexion of Sodomy

1. Dan, *Histoire de Barbarie*. The phrase "the Barbary states" or "the Barbary coast" in the early modern period through the early nineteenth century usually refers to the regencies of Algiers, Tunis, and Tripoli and to the (brief) Republic of Salé (which, by the late seventeenth century, fell under the control of the Sultan of Morocco), although there were other corsairing cities to which these monikers sometimes refer (especially Rabat). Dan's interest in the Barbary states derived from his very particular position: at the time Dan was the Superior of the Order of the Most Holy Trinity and of the Captives, a Catholic organization founded in 1193 and dedicated to the redemption of Christians from captivity under Muslims, a persistent problem throughout the Crusades and during other medieval and early modern religious conflicts in which Christians found themselves

at war with Islamic, often Ottoman, powers. Dan dedicated his life to a religious mission organized around the realities of captivity and attending efforts to restore Christian captives from the physical and especially spiritual dangers inherent in subjection to slavery under Muslims. *Histoire de Barbarie et de ses Corsaires* treats Barbary slavery at length and in detail, enlisting natural historical inquiry to anatomize the organization of the Barbary states to more effectively save Christians from suffering captivity therein.

2. On Pierre Dan and Trinitarian redemption campaigns, see Weiss, *Captives and Corsairs,* esp. 39–41. According to Joseph Wheelan, between 1575 and 1769, Redemptionist friars secured the freedom of more than fifteen thousand Christian captives over the course of eighty-two missions. *Jefferson's War,* 25. On the history of Barbary piracy, see Baepler, "Barbary Captivity Narrative," 106. Wheelan, in *Jefferson's War,* 25, cites a specific number (15,500) of redeemed captives, but in this chapter I have tended away from specific numbers, since across the growing amount of scholarship on Barbary captivity, both estimates and calculations of numbers of enslaved Christians and numbers of redeemed captives vary incredibly widely, with some scholars' estimates as low as 20,000 and others' as high as 150,000. Calculating a correct figure is less important for my purposes than acknowledging the impact of Barbary captivity on the representation of the Islamic world in North American Anglophone communities.

3. While certainly not all seventeenth-century natural historical texts contain illustrations, many do; Ogilby's *Africa* has more than a dozen, and copperplate engravings are commonly found in the most expensive editions of seventeenth-century geographies and ethnographies.

4. On the history of Barbary captivity, see Colley, *Captives.*

5. See W. Davis, "Travels and Captivity of Davis."

6. See Tyler, *Algerine captive.* For the staking scene, see 75–76.

7. *Description of nature of slavery,* 4–5.

8. A "pathic" is a man or boy who is penetrated (as opposed to penetrating) during homosexual anal intercourse; Parker writes that "Egypt and the East (including Persia, associated with the barbering of Macedonian Alexander as well as the broader reaches of the Islamic world) continued to be identified with effeminacy in ways that both iterated and renovated familiar classical and Roman descriptions. But at the same time, the Ottoman Turk as the new power of the East and the Barbary Coast was represented as not only bearded but as identifying beardlessness and shaving with pathics and slaves, including captives from Christian Europe." "Barbers and Barbary," 214. "Turning Turk" was a common early modern, seventeenth-century, and eighteenth-century expression used to describe the process by which a Christian might be voluntarily or coercively converted to Islam. On turning Turk, see Vitkus, *Turning Turk;* and Parker, "Barbers and Barbary." On pathic subjection, see Parker, 210–212.

9. Foucault muses:

> It is often said that the establishment of botanical gardens and zoological collections expressed a new curiosity about exotic plants and animals. In fact, these had already claimed men's interest for a long while. What had changed was the space in which it was possible to see them and from which it was possible to describe them. To the Renaissance, the strangeness of animals was a spectacle: it was featured in fairs, in tournaments, in fictitious or real combats, in reconstitutions of legends in

which the bestiary displayed its ageless fables. The natural history room and the garden, as created in the Classical period, replace the circular procession of the "show" with the arrangement of things in a "table." What came surreptitiously into being between the age of the theatre and that of the catalogue was not the desire for knowledge, but a new way of connecting things both to the eye and to discourse. A new way of making history. (*Order of Things,* 131)

10. See Cocks, *Visions of Sodom.* Thanks to Clare Lyons for pointing me toward this study.

11. The Levant has historically been home to multiple major world religions, including Christianity, Judaism, and Islam, but it has been a primarily Muslim region since the seventh century. On the geography of sodomy, see Warner, "New English Sodom." Warner notes that sodomy and lesbianism are among the only sexual behaviors understood in terms of *place.*

12. W. Davis, "Travels and Captivity of Davis," 47; Coke, "Of Buggery, or Sodomy," 58.

13. See Colley, *Captives,* 128–129, where she notes that "British captivity literature had traditionally been far more concerned to stress the sexual threat to male captives in Barbary. For every single reference to heterosexual sex I have seen in British discussions of Barbary and Ottoman captivity before 1750, there are at least five to sodomy: and this is true of polite as well as popular literature, public statements and the most private of writings." Early important scholarship in the history of sexuality that is organized around the representation of sodomy includes M. Jordan, *Invention of Sodomy;* Bray, "Word and Symbol," in *Homosexuality in Renaissance England;* and J. Goldberg, *Sodometries.* J. Goldberg's *Sodometries;* Bleys's *Geography of Perversion;* Traub's "Psychomorphology of the Clitoris"; Parker's "Barbers and Barbary"; and Godbeer's *Sexual Revolution in Early America* all explore to varying degrees the persistent belief that non-European peoples were sodomites, but none of these scholars explore the idea that sodomy might itself index racial difference, and only Traub and Parker address the Barbary states.

14. "Orientalism," here, refers specifically to Edward Said's theory of orientalist discourse, advanced in his landmark 1978 *Orientalism.* While Said discusses a period significantly later than the period I address in this book (Said is more interested in the period from the 1798 failed Napoleonic invasion of Egypt through the end of the nineteenth century, and specifically that period in England), many of the general insights he offers in *Orientalism* prove generative for a discussion of North American representation of the Barbary states in the seventeenth and eighteenth centuries as well. Scholars such as Malani Johar Schueller concur with this perception; see her *U.S. Orientalisms;* and Schueller and Watts, *Messy Beginnings.* Furthermore, Said's emphasis on the role of philology as an academic discipline weaponized by colonialism and militarism implicitly invokes the mutual imbrication of philology and natural history (Said, *Orientalism,* 233), a relationship that is well documented by scholars. See, for example, Ogilvie, *Science of Describing,* which argues that philology provided an important foundation for the kind of observational and descriptive methods central to natural historical inquiry; or Daston and Most, "History of Science and Philologies." Said himself directly names natural history as "the fourth element preparing the way for modern Orientalist structures." *Orientalism,* 119.

15. For scholarship on the significance of sodomy in the early modern and British colonial North American world, see Bleys, *Geography of Perversion;* Bray, *Homosexuality in Renaissance England;* Godbeer, *Sexual Revolution in Early America;* J. Goldberg, *Sodometries;* M. Jordan, *Invention of Sodomy;* Warner, "New English Sodom." Of these writers, only Bleys addresses the relationship between sodomy and racial difference and ethnographic writing; Goldberg gestures to the way non-Europeans were understood to be or accused of being sodomites; Godbeer discusses aberrant sociosexual organization in the southern colonies at length and addresses the assumption of the sexual availability of brown and black peoples.

16. On this effect—more or less what Jasbir Puar terms "homonationalism"—see J. Goldberg's introduction to *Sodometries;* and Puar, *Terrorist Assemblages.*

17. Following Merleau-Ponty, Husserl, and Marx, Ahmed, in *Queer Phenomenology,* 120, reminds her readers that orientations need to be understood as, as she puts it, "energetic," never neutral, incidental, or objectless.

18. On the mutual curiosity and intertwined existences of the "east" and the "west" during the eighteenth century, see Aravamudan, *Enlightenment Orientalism;* and Boone, *Homoerotics of Orientalism.* On travel narratives with an "east" to "west" directionality, see I'tesamuddin, *Wonders of Vilayet;* and the eighteenth-century travel narratives of Dean Mohammed and Ram Mohun Roy.

19. Colley argues that

> by 1616, Algiers alone was estimated to have seized over 450 English vessels, and this was just the beginning. Between the 1610s and 30s, Cornwall and Devon, both sea-going counties heavily involved in trade with southern Europe, lost a fifth of their shipping to North African corsairs. In just one year, 1625, nearly a thousand sailors and fishermen from the major West Country port of Plymouth were seized, most within thirty miles of its shore. Overall, David Hebb calculates, in the two decades before the outbreak of the Civil War in 1642, Barbary corsairs inflicted well over one million pounds of damage on English shipping, a sum that needs to be multiplied more than a hundredfold to gain any sense of its meaning in today's values. (*Captives,* 49–50)

20. For example, in June of 1631, Barbary corsairs arrived in Baltimore, Ireland, and took the entire town hostage; this attack is documented in Irish poet Thomas Davis's "Sack of Baltimore 1631." Also see Colley, *Captives,* 50.

21. Paul Baepler, whose work on Barbary captivity narratives constitutes the only recent scholarly engagement with this genre of writing, cites these figures from several pieces of historical research, namely Wolf, *Barbary Coast;* and Braudel, *Mediterranean and Mediterranean World.* As I followed their footnotes, however, it appears that this repeated estimate of the number of Christian slaves in Algiers during the 1620s and 1630s may have initially come from the 11th edition of the *Encyclopedia Britannica,* published in 1910, under the entry "Barbary Pirates."

22. See, for example, captivity narratives by James Leander Cathcart (1785), John Foss (1798), Mrs. Maria Martin (1807, spurious), and novels such as Tyler's *Algerine Captive* (1797).

23. For scholarship on Barbary captivity in the revolutionary and early national periods, see Bouanani, "Propaganda for Empire"; Berman, *American Arabesque,* esp. 7;

Egan, *Oriental Shadows;* P. Gould, *Barbaric Traffic;* Isani, "Cotton Mather and the Orient"; Sha'ban, *Islam and Arabs;* Weir, *American Orient.*

24. See Colley, *Captives,* 5, 17–18.

25. Colley, 65. Colley also argues that until 1800, "and so far as land warfare was concerned, there was no invariable gulf between the armaments of Western and non-Western powers" (68–69).

26. James Boswell, *Life of Samuel Johnson,* 1:442.

27. There are, however, at least a handful of extant manuscript accounts of Barbary captivity; manuscripts of this ilk that I have come across thus far have been written primarily in the late seventeenth century.

28. This was the case in both England and the British North American colonies; see Colley, *Captives,* 75–76; Baepler, "Barbary Captivity Narrative," 114–115; and C. Mather, *Pastoral Letter to English Captives.*

29. See C. Mather, *Pastoral Letter to English Captives;* and C. Mather, *Glory of Goodness.*

30. Hester Blum argues that "in Barbary captivity narratives, sailors directed their writing explicitly to an audience of their fellow mariners, providing navigational suggestions and concrete strategies for responding to the threat of captivity." *View from the Masthead,* 12.

31. Library Company of Philadelphia, *Catalogue of the books,* 89, accessed via the online Readex database *Early American Imprints, Series I: Evans, 1639–1800,* 22066, Winans, R.B. Book cats., 131.

32. Gee, *Narrative of Joshua Gee,* 15. My citations of Gee's narratives are all from the printed transcription of his account. I have been in touch multiple times with the Wadsworth Atheneum, in an attempt to access the manuscript itself, but to no avail.

33. Albert Carlos Bates, the editor and publisher of the 1943 sole printing of Gee's narrative, offers this estimate of the year of Gee's birth.

34. See Baepler, who argues: "Because Gee's narrative finally made it to press only fifty years ago, two centuries after his death, we can be relatively certain that it had no significant impact on his contemporaries, except on a few influential people such as Sewall, the Mathers, and Gee's own famous son. We might also imagine, however, that Gee spoke of his 'firy trials' informally. His intimate anecdotes of the ordeal would have confirmed for his listeners a familiar national story." "Barbary Captivity Narrative," 114.

35. Hester Blum argues in "Barbary Captivity and the Intra-Atlantic Print Culture," her chapter on Barbary captivity narratives in *View from the Masthead,* that among the specific purposes of Barbary narratives were communication and the distribution of professional advice among sailors, pertaining to where and how to sail to avoid capture. Though her chapter focuses primarily on Barbary captivity narratives published during the early national era, her argument nonetheless pertains to Joshua Gee's narrative; he is clearly familiar with the experience of Barbary captivity, and his narrative adheres so closely to the generic conventions of maritime captivity narratives that it seems likely that, as a literate sailor, he would have also encountered Barbary captivity narratives.

36. The printed edition, transcribed from a handwritten manuscript, practices a certain fidelity in its transposition of Gee's words into print, maintaining the page breaks of the original document. This means that, by the standards of a present-day reader, the

pages of Gee's printed manuscript are only half full. The manuscript itself, then, is quite short, much shorter than the fifteen pages that it spans in the transcription.

37. Gee, *Narrative of Joshua Gee*, 29.

38. The question of rationality is one that has characterized many Christian representations of Muslims from the early modern period through the present. Baepler, in "Barbary Captivity Narrative," points out that Muslims are alternately portrayed as the racial and cultural "equivalent" of American colonists in their careful cultivation of a developed culture and as the racial and cultural "equivalent" of American Indians, based on their skin color and "barbarousness." By the early national period, the issue of to what degree "Arabs," "Turks," "Mahometans," or "Saracens," could be assumed to possess a basic faculty for reason was a frequent question. In Abraham Browne's narrative, for example, he speaks of making the "Choyce Rather to dye then to fall into the hands of Those onreasonable men." Riley, "Abraham Browne's Captivity," 36.

39. Parker argues that "*Barber* in early modern English (related to *barbarius* and *barba* or beard) was repeatedly conflated with *barbarous* as well as Barbary, in ways enabled by both unstable orthographies and sound." "Barbers and Barbary," 205.

40. Parker, 207–208.

41. Parker, 220.

42. Gee, *Narrative of Joshua Gee*, 21–22.

43. I begin with Gee's narrative, rather than with the chronologically earlier narrative by Abraham Browne, because the extant transcription of Gee's narrative is much closer to a complete text than that of Browne. Many scholars who work on Barbary captivity cite Browne, but every one of these citations can be found in Riley's 1980 essay "Abraham Browne's Captivity." Riley transcribed several large pieces of Browne's diary (which is about 150 pages long) into his article; all of the transcribed segments describe the events leading up to and immediately following Browne's capture and sale. I made several trips to the Massachusetts Historical Society between 2011 and 2017 to read this diary; unfortunately, because of the tiny, faded script and the handwriting, I was not able to read or transcribe as much of it as I had hoped, even with digital enhancement technology. Thanks are due to Ashley Cataldo, assistant curator of manuscripts at the American Antiquarian Society, whose paleographic skills greatly aided my work on Browne's diary. I include this note because I want to register that despite the relative frequency with which Browne's narrative is cited and discussed by the small collection of scholars working on the history and literature of Barbary piracy and captivity, discussions of Browne, my own included, hinge entirely on readings of merely several paragraphs of his narrative, at most a few dozen pages out of more than 150. For more on Browne's narrative, see Riley's essay.

44. See Riley, 34. Browne's parents apprenticed him to local merchants as a young man, and he eventually worked on ships that brought goods to the North American colonies, Barbados, and ports along the northern coast of western Europe. In 1650 his master offered him the opportunity to move to New England and manage one of his businesses there, and on May 1 of that year, Browne set sail for Richmond's Island, in what is now Maine. He continued his career as a merchant's apprentice in New England, returning to England in 1654 to "settle accounts with his master" and see family and friends.

45. Riley, 38. References to "Sally" or "Sallee" in Barbary captivity narratives indicate what is now the port town of Salé in northwest Morocco, not far from Rabat.

46. Riley, 38–39.

47. As Parker, Vitkus, and Burton all note, it was a common practice to shave the head of any slave on the Islamic slave market, and it occurred so often that a shaved head became a cultural signifier of slave status.

48. Slaves were often tortured until they agreed to convert to Islam. The English sailor John Rawlins, for example, records how Christians were tortured into submission, writing in 1622:

> Secondly, concerning their enforcing them, either to turne *Turke,* or to attend their filthinesse & impieties, although it would make a Christians heart bleed to heare of the same, yet must the truth not be hid, nor the terror left vntold. They commonly lay them on their naked backs, or bellies, beating them so long, till they bleed at the nose and mouth, and if yet they continue constant, then they strike the teeth out of their heads, pinch them by their tongues, and vse many other sorts of tortures to conuert them; nay many times they lay them their whole length in the ground like a graue, and so couer them with boords, threatning to starue them, if they will not turne, and so many euen for feare of torment, and death, make their tongues betray their hearts to a most fearefull wickednesse, and so are circumcised with new names, and brought to confesse a new religion. Others againe, I must confesse, who neuer knew any God, but their owne sensuall lusts, and pleasures, thought that any religion would serue their turnes, and so for preferment or wealth very voluntarily renounced their faith, and became *Renegadoes* in despight of any counsell which seemed to intercept them: and this was the first newes we encountred with at our comming first to *Argier.*

Elsewhere in his narrative, Rawlins records a more explicit account of the other definition, within British colloquialism, of being forced to "turn Turk"; after being taken captive, he and his shipmates sought out and found "many *English* at worke in other ships, they spared not to tell vs the danger we were in, and the mischiefes we must needs incurre, as being sure if we were not vsed like slaues, to be sold as slaues: for there had beene 500. brought into the market for the same purpose, and aboue a 100. hansome youths compelled to turne *Turkes,* or made subiect to more viler prostitution, and all *English*" (cited in Baepler, "Barbary Captivity Narrative," 111). The more colloquial meaning of "turning Turk," as Patricia Parker points out in "Barbers and Barbary," 207–208, is to take the submissive role in sexual intercourse (she charts this expression within early modern drama). The "viler prostitution" to which the "100 hansome youths" were subject was likely the process of a different kind of "turning Turk," of sexual subjection to their masters.

49. Parker, 215.

50. Parker notes that "Barbary and North Africa continued to be associated with barbering into at least the early eighteenth-century: an account in 1704 reports a barber's shaving of a Christian slave in Algiers as an instance of the 'barbarous cruelty' of a master who forces him to 'turn Mussulman' or 'Turk.'" Parker, 228.

51. See Colley, *Captives,* 12–17.

52. Colley, 12–13.

53. Sewall, *Diaries of Samuel Sewall*, 20. I have not been able to locate this passage in any of the editions of the diaries of Samuel Sewall to which I have access, save an 1879 edition published by the Massachusetts Historical Society. Sewall's interest in Gee's account of "Turkish" temporal structures is cited in Bates's introduction to Gee, *Narrative of Joshua Gee*, 6–7, 10.

54. Pepys, *Diary of Samuel Pepys*, 2:33–34 (8 February 1661).

55. The *Oxford English Dictionary* lists seventeenth- and eighteenth-century historical uses of "bagnard" (a variation on *bagnio*) that include "a bath, a bathing-house, esp. one with hot baths, vapour-baths, and appliances for sweating, cupping, and other operations"; "an oriental prison"; and "a brothel, a house of prostitution." I suggest that Pepys is using this term in a way that invokes all three definitions. "Bagnio, n.," in *OED Online*, June 2017 (Oxford University Press); www.oed.com/view/Entry/14647.

56. To be clear, not all of the Barbary states were controlled by the Ottoman Empire; Morocco operated outside of Ottoman control for much of the seventeenth century.

57. See Colley, *Captives*, 35. She argues that "on paper, at least, the Ottoman armed forces—the janissaries, provincial militias and timariots—were well over 150,000 strong, many times as large again as the armies at the disposal of early modern England's monarchs." Anti-deism movements began to gain traction in North American print culture around the turn of the eighteenth century; these movements considered any non-Protestant, and most non-Calvinist, religions to be false. On antideism writings, see Prideaux, *True nature of imposture;* C. Mather, *Reason satisfied;* Leslie, *Religion of Jesus Christ;* Gillespie, *Treatise against the deists*. On the existence and engagement with the idea of Islamic republicanism in Enlightenment debates about governance, see Marr, *Cultural Roots of American Islamicism;* and Garcia, *Islam and the English Enlightenment*.

58. Colley, *Captives*, 34.

59. Brooks, *Barbarian cruelty*.

60. Brooks, 6.

61. Nabil Matar includes Francis Brooks in his list of Englishmen held captive in Barbary before 1693, but I think it is also worth considering whether "Brooks" is either a pseudonym or a spurious account. On Brooks, see Matar, *British Captives*, 134–135.

62. Matar, 3.

63. Matar, 66.

64. Matar, 6.

65. Traub, "Mapping the Global Body," 51.

66. Also see Colley, *Captives*, 52. Discussing Moulay Ismail's strong leadership, Colley notes that Morocco's corsairing force grew more powerful (and expensive) under his reign: "After 1680, [Morocco's] formidable sultan, Moulay Ismail, systematised corsairing as a weapon of state finance. All captives seized by Moroccan corsairs now became the sultan's property, and European states were no longer allowed to redeem nationals on an individual or group basis. Instead, they had to pay for all of their captives detained in Morocco at any given time" (52).

67. Brooks, *Barbarian cruelty*, 59.

68. Brooks, 48–49. On other discussions of Moulay Ismail ibn Sharif as being a "mulatto," see *Description of nature of slavery*, in which the narrator describes him as a mulatto (26) and "almost black" (27).

69. Brooks, *Barbarian cruelty*, 49.

70. Ogilby, *Africa*, 254.

71. Brooks often presents ethnographic information and descriptions of violence in the same breath; for example, he narrates that in the evening, Mully Ishmael "in one of his Rooms in the Castle, . . . lies down on a kind of Quilt on the Ground; and sleeping that Night, he rises early in the Morning, and falls to his old Tyrannous and Inhuman Practices, domineering over his poor Slaves, and sets the Negroes to whip, stone and beat them, to work harder than many times it's possible for them to think they can hold out or endure till Night." *Barbarian cruelty*, 57.

72. Brooks, 32–33.

73. Brooks, 35.

74. Brooks, 61.

75. The name of this man appears as both "Stuart" and "Stewart" in different accounts of these redemption efforts. See Matar, *British Captives*, 145–46.

76. See Windus, *Journey to Mequinez*.

77. The author describes sailing with the ship *Experiment*, under the command of Adam Rigdon, on a voyage to Lisbon. *Description of nature of slavery*, 1. This work is held by the British Library and has been digitized within the collections of *Eighteenth-Century Collections Online* (ECCO). The ECCO edition of this manuscript is missing page 1, however; many thanks are due to Emily Friedman, who was at the British Library during the summer of 2017 and photographed the manuscript for me, including the missing page.

78. *Description of nature of slavery*, 31.

79. *Description of nature of slavery*, 32.

80. *Description of nature of slavery*, 35.

81. *Description of nature of slavery*, 38.

82. *Description of nature of slavery*, 36.

83. *Description of nature of slavery*, 4–5.

84. The narrator alternately refers to Muley Ismael as "king" and as "prince."

85. *Description of nature of slavery*, 4–5.

86. See El-Rouayheb, *Before Homosexuality*, esp. 1–6.

87. *Description of nature of slavery*, 36–37.

88. Tyler, *Algerine captive*, 75–76, where the narrator continues:

> I will not wound the sensibility of my humane fellow citizens, by a minute detail of this fiend like punishment. Suffice it to say that, after they had stripped the sufferer naked, except a cloth around the loins, they inserted the iron pointed stake into the lower termination of the vertebrae, and thence forced it up near his back bone, until it appeared between his shoulders; with devilish ingenuity contriving to avoid the vital parts. The stake was then raised into the air, and the suffering wretch exposed to the view of the assembly, writhing in all the contortions of insupportable agony. How long he lived, I cannot tell, I never gave but one look at him: one was enough to appal my New England heart.

89. Boone, *Homoerotics of Orientalism*, xxi.

90. See note 31 in this chapter.

91. W. Davis, "Travels and Captivity of Davis," 476. As scholars such as Jonathan Goldberg, Valerie Traub, and Rudi Bleys have demonstrated, sixteenth- and seventeenth-

century increases in western European colonial expansion, exploration, and domination precipitated an attending rise in the writing and publication of travel and ethnographic narratives. As anyone with some familiarity with ethnographic works would attest (popular examples might include writing by Bartolomé de las Casas, Leo Africanus, Hernán Cortés, Bernal Díaz, and John Barbot), ethnographic travel writings frequently include accounts of the sexual behaviors of the peoples and cultures encountered during the voyage. Valerie Traub has summarized this convention in ethnographic travel writing by noting that "whether describing the New World, Africa, or the East, narrators obsessively remark upon those cultural practices that differentiate native inhabitants from Europeans, often employing rhetorics of gender and sexuality as explanatory tropes." Traub, "Psychomorphology of the Clitoris," 86.

92. "An Introductory Discourse Concerning Geography," in *Collection of Voyages and Travels*, i.

93. "An Introductory Discourse Concerning Geography," i.

94. W. Davis, "Travels and Captivity of Davis," 477.

95. W. Davis, 477.

96. Most scholars of narratives of Barbary piracy agree that the terms "Turk," "Moor," "Saracen," and "Arab" were often used interchangeably within Barbary captivity narratives from the late seventeenth through the late eighteenth century, although these terms do, technically, refer to different peoples. "Negro" was sometimes also used interchangeably with these terms. Moreover, all of these terms, in the context of Barbary captivity narratives, served to indicate the Islamic faith of the people they were applied to. The descriptor "Moor," for example, indicated someone of mixed Arab and Berber descent but would have automatically identified a person as a Muslim. Davis's turn to the discourse of complexion constitutes a very common idiom through which Europeans documented their encounters with other peoples in travel writing. As Roxann Wheeler has argued, "complexion" was a key term in early taxonomies of the differentiation of human kinds; it included but was not limited to skin color.

97. Racial differences are eclipsed by religious differences for Davis, whose understanding is that "Turks," as Muslims, constituted a fundamentally different class of humanity from himself as a Christian, or from even the citizens of Civitavecchia, who professed the "Romish" faith.

98. "Curious, adj.," in *OED Online*, March 2011 (Oxford University Press); www.oed.com/view/Entry/46040.

99. Berman, *American Arabesque*, 6.

100. W. Davis, "Travels and Captivity of Davis," 477. Interestingly, this language also characterizes Davis's descriptions of Catholics, whose churches and church officials he also identifies as lavishly decorated and adorned. See Cocks, *Visions of Sodom*, on the association between Catholicism and sodomy.

101. Boone, *Homoerotics of Orientalism*, xx.

102. W. Davis, "Travels and Captivity of Davis," 478.

103. W. Davis, 479.

104. Crewe, "Disorderly Love," 395.

105. W. Davis, "Travels and Captivity of Davis," 485.

106. The biblical city of Sodom is located in the land of Cham or Ham. Ham, whom I discuss in the introduction, is associated with the curse of Ham (the Old Testament

"punishment" of racialization). The Barbary Coast, including Tunis, Tripoli, Salé, and Algiers, the very powerful Barbary states, as a whole fell within the part of the earth inherited by the descendants of Ham, and this association appears persistently in representations of Barbary pirates and Muslims more generally.

107. Bleys, *Geography of Perversion*, 11, 47, 114.

108. See Darby, "Captivity and Captivation," 128. Darby, citing Daniel Vitkus's work on the imagination of the Turk within early modern British dramatic culture, writes: "Vitkus . . . suggests . . . that the fear of circumcision was the result of exaggerated representations in plays and other popular literature, and that 'for those who became Muslims in order to join the Barbary pirates, conversion was probably a mere formality,' involving little or no obligation of religious practice. Stories of forced circumcision were 'a scare tactic designed to discourage potential converts,' not an accurate description of what really happened" (Darby, 128). Vitkus asserts that "because Christians . . . did not practise circumcision . . . this feature of Islamic custom was viewed with horror as an atavistic ritual performed by barbarians." Here, Darby is citing Vitkus, *Three Turk Plays*, 5.

109. W. Davis, "Travels and Captivity of Davis," 477.

110. See Parker, "Barbers and Barbary," 213. It is important to read this "preposterous" position on horseback as a particular representational technology; Parker and Jonathan Goldberg have both elaborated on the early modern understanding of sodomy as "'preposterous' (reversed and unnatural)."

111. W. Davis, "Travels and Captivity of Davis," 483.

112. W. Davis, 486.

113. W. Davis, 479.

114. Crewe, "Disorderly Love," 395.

115. Warner, "New English Sodom," 21.

116. See Warner, 20–21; and Godbeer, *Sexual Revolution in Early America*, 3.

117. On Oriental "despotism," see Buffon, *Natural History, General and Particular*, 108; on climate and despotism, see Nussbaum, *Torrid Zones*, 12.

118. Brooks, *Barbarian cruelty*, 12.

119. While sex between women was often explicitly legally prohibited as well, historians of sexuality have argued that cases of sex between women were prosecuted only very, very rarely, whereas sodomy accusations—Richard Godbeer discusses the famous late-seventeenth-century case of Nicholas Sension at length in *Sexual Revolution in Early America*, 1–52, for example—appear more frequently in early American archives.

120. See Ariès, *Centuries of Childhood*, esp. 137–329.

Chapter 3. "Egyptian Lusts" at the Gallows

1. C. Mather, *Tremenda*, 34.
2. See Weiner, *Black Trials*, 36.
3. C. Mather, *Tremenda*, 3.
4. C. Mather, 4–5.
5. C. Mather, 6.
6. C. Mather, 7.
7. C. Mather, 8.
8. Weiner, *Black Trials*, 35.
9. C. Mather, *Tremenda*, 5.

10. C. Mather, 5.
11. C. Mather, 8–9.
12. "Habit" and "custom" were often used coterminously in the eighteenth century, and definitions of "custom" that appear in eighteenth-century citations in the *Oxford English Dictionary* all define it as a behavioral term, often as a synonym for "habit." For example, "a person's or animal's habitual practice or typical mode of behaviour; habit; an instance of this. Usually with possessive." "Custom, n. and adj.," in *OED Online,* June 2017 (Oxford University Press); www.oed.com/view/Entry/46306.
13. We know very little about Joseph Hanno, but we know more about his life than about the lives of other free black residents of Boston during the late seventeenth and early eighteenth centuries. Hanno was a former slave who had been brought to New England as a child sometime around 1677; he may have been born in Madagascar, he was baptized and raised in the Christian community in which he lived, and he was freed in 1707. The *Boston-News Letter* registered Hanno's execution:

> Boston, On Thursday last the 25th Currant was Executed here One Joseph Hono, Negro, for Murdering his Wife; he had been in the Country about 44 Years, and about free for himself, and by his Masters brought up in the Christian Faith; and he hoped that all Mankind would take warning by him to keep themselves from committing such Sin & Wickedness as he was guilty of, particularly, Sabbath-breaking and wilful Murder, the one being the Ringleader to the other, for which last he was justly Condemned, which had he not been guilty of the first he might probably have never committed the second. (May 22–29, 1721, 2 [1])

On his possible birth in Madagascar, see Weiner, *Black Trials,* 35.

14. See Slauter, "Neoclassical Culture." Slauter argues that "neo-roman political theorists in Europe and America were interested in understanding how people found themselves subjected to tyrannical and despotic governments and why such subjects stayed under them, so they described political slavery as a mental rather than a physical phenomenon, the effect not of chains and violence but of an irreversible intellectual debasement. Mental slavery made liberty impossible, radical writers argued, producing civilizations incapable of 'culture' and cultures incapable of 'civilization'" (84).
15. C. Mather, *Tremenda,* 26, 23.
16. C. Mather, 8. Mather is citing the King James Version of the Bible, Jeremiah 13:23: "Can the Ethiopian change his skin, or the leopard his spots? then may ye also do good, that are accustomed to do evil."
17. Traub, "Mapping the Global Body," 51.
18. Arondekar, *For the Record,* 3.
19. "To John Adams from Benjamin Rush, 16 September 1808," in *Founders Online,* National Archives, http://founders.archives.gov/documents/Adams/99-02-02-5254, last updated March 28, 2016.
20. The location through which modern sexuality has historically been thought is the body, or perhaps the psyche, which scholars frequently imply is itself located in the body (Judith Butler and Leo Bersani are the noted exceptions to this trend). The question that scholars of eighteenth-century sexuality have routinely failed to ask is whether the body is or should be the privileged site through which early sexuality should be thought

as well. See Butler, "Imitation and Gender Insubordination"; and Bersani, "Is the Rectum a Grave?"

21. See Chiles, *Transformable Race;* Wheeler, *Complexion of Race.*
22. Traub, "Mapping the Global Body," 59.
23. Traub, 51.
24. See, for example, the 1594 English translation of Juan Huarte's *Examen de ingenios,* where he cites Galen, Hippocrates, Plato, and Aristotle. This is an oft-cited passage among scholars of early modern natural philosophy; Huarte reports:

> Galen writ a booke, wherein he prooveth, That the maners of the soule, follow the temperature of the body [*Quid animi mores*], in which it keeps residence, and that by reason of the heat, the coldnesse, the moisture, and the drouth, of the territorie where men inhabit, of the meates which they feed on, of the waters which they drinke, and of the aire which they breath: some are blockish, and some wise: some of woorth, and some base: some cruel, and some merciful. . . . And to proove this, he cites many places of Hippocrates, Plato, and Aristotle, who affirme, that the differences of nations, as well in composition of the body, as in the conditions of the soule, springeth from the varieties of this temperature: and experience it selfe evidently sheweth this, how far are different Greeks from Tartarians: Frenchmen from Spaniards: Indians from Dutch: and Aethiopians from English. (21–22)

25. Traub makes an incredibly important argument about early modern maps:

> The cultural production of domestic heterosexuality also contributed to the hardening of both gender and racial lines. For Osborne's marital pairs, it must be noted, are segregated from one another by race. Precisely at the moment when erotic practices were more likely to take a cross-cultural form, when racialized bodies might cross borders, mingle, and touch in unprecedented ways, Osborne's engraving silently elides interracial sexuality from the global picture. Just as the instability of the gender binary fosters a disavowal of diverse erotic practices, so too the instability of racial divisions enables a disavowal of interracial sexuality. The naturalization of heterosexuality and the unnaturalness of miscegenation emerge simultaneously, as two sides of the same global body. Thus, the production of two ways of constructing race: the ethnographic, which depends on the notion of habit, and the biological, which depends on the concept of reproduction. On seventeenth-century maps, we can begin to glimpse these two modes of construction articulating themselves in terms of one another. It is almost as if domestic heterosexuality serves as an enabling condition of modern racialization, while race gives order and support to domestic heterosexuality. (Traub, "Mapping the Global Body," 83–84)

26. See Ariès, *Centuries of Childhood;* and Stone, *Family, Sex and Marriage.*
27. Witherspoon, for example, argues in 1789 that "the earliest, in general, is the fittest and best time for instruction." *Sermon on religious education,* 10. It should be noted here that whereas childhood was purportedly invented in the seventeenth century, Calvinist communities, by most accounts, tended to refer to the first twenty-one or twenty-eight years of a person's life as "youth." In Rush's "Mode of Education," for example, he relies on an understanding of education more proper to "youth" than to "childhood": "It

is well known that our strongest prejudices in favor of our country are formed in the first one and twenty years of our lives" (87).
28. A. Browne, *Religious Education of children recommended*, 7.
29. Witherspoon, *Series of Letters on Education*, 31–32.
30. Witherspoon, 19.
31. Witherspoon, *Sermon on religious education*, 10.
32. Cohen argues that

> ministers would continue to explain the criminal's progress to the scaffold in terms of innate depravity, God's withdrawn grace, and sin's natural progress well into the nineteenth century. But beginning in 1739 a number of execution sermons delivered by clergymen outside of the orthodox Congregational fold located the cause of crime less in the offender's nature than in his or her early social experience. Some ministers had begun to see deviance as the unfortunate product of a flawed upbringing or a bad education. They still held sinful individuals responsible for crime but now shifted much of the burden of guilt from the perpetrators themselves onto the negligent parents, teachers, or masters. (D. Cohen, *Pillars of Salt*, 89)

33. Williams, *Pillars of Salt*, introduction, 20.
34. The initial publication date of *Onania* is something of a problem, and I am unaware of the accessibility of any printings of the text before the fourth edition, printed in London around 1718, although the British Library does hold a copy with an estimated publication date of 1710. McKeon, in *Secret History of Domesticity*, 285, states that *Onania* was "purportedly first published in 1710." Foucault concurs, writing, "There are no known copies of the first three editions." *Abnormal*, 259n2. Authorship of the pamphlet (*Onania*) is sometimes attributed to the Dutch theologian Balthasar Bekker, who died in 1698 (which would make the initial writing of *Onania* much earlier than we currently imagine). Laqueur, however, contends that the author was John Marten. *Solitary Sex*, 32. Several other scholars cite Beverland's 1697 *De Fornicatione Cavenda Admonitio* as a possible source text for *Onania*.
35. See C. Mather, *Pure Nazarite*. Also see *Onania* (1724). Note that *The Pure Nazarite* was first published anonymously; Cotton Mather's name does not appear on the title page or in the front matter. This sermon was attributed to Mather by Holmes in *Cotton Mather*. Electronic databases such as *Early American Imprints* consistently list Mather as the author.
36. See Neuman, "Masturbation, Madness," 1, 6. Also see Hitchcock, "Redefining Sex," 82.
37. Stolberg, "Self-Pollution, Moral Reform," 47.
38. Stolberg, 58.
39. The nineteenth edition, published in London, was 343 pages long; see *Onania* (1759). The twentieth edition was printed in Glasgow and is slightly shorter: see *Onania* (1760). According to Stolberg, "we may question the exact figures, but clearly Onania was a great success, and its various editions and translations may well have sold almost fifty thousand copies" (40).
40. Tissot's *L'onanisme: Dissertation sur les maladies produites par la masturbation* was first published in Latin under the title *Tentamen de morbis ex manustupratione*, in the

1758 edition of Tissot's *Dissertatio de febribus biliosis*. The first French edition appeared in 1760, and it was quickly translated into English; the first English edition appeared no later than 1766, with *Treatise on the crime of Onan*. The work was widely reprinted in both French and English throughout the rest of the century; the latest edition I have found was printed in New York in 1832 and titled *A Treatise on the Diseases Produced by Onanism*. By the time an English edition of *L'onanisme* was released in the 1760s, many readers in the British North American colonies would have already been familiar with Tissot's medical advice literature, as his *Avis au peuple sur sa santé* (1761) was spectacularly popular and was printed in both Boston in 1767 and Philadelphia in 1771. See Tissot, *Advice to the people in general* (1767); *Advice to the people in general* (1771). All editions with English titles cited above are themselves English translations. See also Singy, "Popularization of Medicine."

41. See Rush, *Medical Inquiries and Observations*, 348–349.

42. *Onania* (1724), 22.

43. C. Mather, *Pure Nazarite*, 6. The sixth commandment referred to here is "Thou shalt not kill," and the seventh is "Thou shalt not commit adultery." In eighteenth-century sermons, the seventh commandment is sometimes invoked to prohibit lustful thoughts or behavior more broadly, based on Matthew 5:27–29.

44. Hume, *Enquiry concerning Human Understanding*, esp. under the heading "Of the Association of Ideas."

45. Witherspoon, *Series of Letters on Education*, 37.

46. See D. Cohen, *Pillars of Salt*; and Williams, *Pillars of Salt*.

47. C. Mather, preface to *Pillars of Salt*.

48. See Louis P. Masur, "The Design of Public Executions in the Early American Republic," in Masur, *Rites of Execution*, 25–50; and Towner, "True Confessions and Dying Warnings." Towner notes that "in the early years of the literary genre [gallows literature], through 1701, the eleven confessors were all members of the community. But between 1702 and 1776, while nineteen were members of the community, sixty-six were outsiders, among them sixteen Indians, ten blacks, eleven Irish, and twelve pirates" (79n51). Also see R. Slotkin, "Narrative of Negro Crime." Slotkin notes the same phenomenon, citing it as the difference between North American gallows literature and its British cousin. He argues, "Perhaps the most distinctive feature of the American crime narrative, as opposed to the British, is the role played by the racially alien and lower-class criminal: the Indian, the non-English immigrant and the black" (4). Slotkin similarly logs the frequency with which gallows literature features crimes of a sexual character: "In view of the lack (save in Rhode Island) of a slave population comparable in size and economic importance to that of New York, the symbolic value of the 'black example' as a deterrent to white crime or dissidence was far more important than the actuality of black crime, and it is this symbolism that chiefly concerns us. Here the association of black crime or rebellion with criminal sexuality becomes crucial, as a characterizing image of the black's relation to white society" (16).

49. For an example of an influential account of the relationship between popular print and secularization, see Williams, *Pillars of Salt*, 13–14; and D. Cohen, *Pillars of Salt*, 98–99.

50. Belcher, *Worst Enemy Conquered*, 6.

51. Belcher, 6–7, 29.

52. See, for example, Rousseau's Second Discourse, *Discours sur l'origine*, and his discussion of "l'homme naturel" (natural man) that appears therein. Richard Slotkin makes this connection more baldly, writing that "in addition, as an aboriginal the black criminal could also serve as the symbol of 'man in his natural state'—that is, the state of Original Sin unredeemed by Christian conversion. In this guise the black man might stand for 'natural man' in general, and for unconverted children as well." "Narrative of Negro Crime," 9.

53. See 1 Corinthians 2:14, King James Version: "But the natural man receiveth not the things of the Spirit of God: for they are foolishness unto him: neither can he know them, because they are spiritually discerned." On Mather's *"Barbarous Nations,"* see C. Mather, *Pillars of Salt*, 5. On "Sin *rules* and *reigns* in the natural man; he is its slave and vassal," see Belcher, *Worst Enemy Conquered*, 13.

54. Belcher, *Worst Enemy Conquered*, 27–28.

55. Belcher, 15.

56. See Godbeer, "Cry of Sodom." Discussing Samuel Danforth's *Cry of Sodom Enquired Into* (1674), Godbeer argues: "Just as pride, gluttony, drunkenness, sloth, disobedience, evil company, irreligion, and profanity constituted 'the very fodder and fuel of the sin of uncleanness,' Danforth taught, so unclean acts would in turn encourage all manner of sin by 'pollut[ing] the noble faculties of the soul, the mind, and the conscience.' Sinful acts were, then, best understood in association with each other." Godbeer, "Cry of Sodom," 86.

57. C. Mather, *Pillars of Salt*. While this was certainly not the first printing of an execution sermon in the North American colonies, the formal qualities of this text—in particular its status as a compilation of multiple execution narratives and the mixing of genres in the execution narratives themselves—distinguish it from previous forms of execution literature.

58. More of a compilation than a narrative, since the text assembles the stories of twelve men and women executed for murder, *Pillars of Salt* includes six previously published execution narratives, at least one of which was borrowed from a sermon that Cotton Mather's father, Increase Mather, had written in 1675. The collection of narratives was later included in the 1702 edition of *Magnalia Christi Americana*, and for its time, *Pillars of Salt* and the execution narratives published therein would have been widely read. Cohen also argues that Cotton Mather was a significant influence because his sermons were so widespread; he published "more than a dozen volumes of execution sermons over a period of forty years" and "by his own account, . . . would hand out 'half a dozen books, more or less' on a typical day of pastoral visits, thrusting them into the hands of young and old, male and female alike. Over the course of a year, he would distribute hundreds of pamphlets, mostly the products of his own pen." D. Cohen, *Pillars of Salt*, 5.

59. In *Pillars of Salt*, under subheading "II," the entire story of one nameless criminal is related by a minister in the third person. C. Mather, *Pillars of Salt*, 62–63.

60. *The Discourse of the Minister with James Morgan, on the way to his Execution*, a reprinting of a text by Increase Mather, is included in Cotton Mather's *Pillars of Salt*, 70–85. The piece was originally published by Increase Mather in 1685–66 as *A sermon occasioned by the execution of a man found guilty of murder preached at Boston in N.E.*

March 11th 1685[/]6. A second edition of it is held by the American Antiquarian Society and is catalogued digitally in *Early American Imprints: Evans*. See I. Mather, *Sermon occasioned by the execution.*

61. For a text that is presented as a first-person confession, see the account of the nameless female offender sentenced to death for murdering a child born out of wedlock, in C. Mather, *Pillars of Salt*, 99–102.

62. It should be noted that gallows literature was also quite popular. It had flourished at a historically earlier moment in England, and the popularity of the genre there certainly influenced its development in the North American colonies. Williams reminds us, for example, that "there was a sizeable market in eighteenth-century America for popular literature" but also that "English presses primarily were responsible for meeting the demands of this market." Williams, *Pillars of Salt*, xiii. Slotkin argues that "the American crime narratives differed from contemporary English works like the *Newgate Calendar* in a number of important respects. The central role played by the American clergy in the civil rite of execution and the subsequent literature was unique with the New England colonies." R. Slotkin, "Narrative of Negro Crime," 3–4. Caleb Smith insists that we read these narratives within the context of broader transatlantic Anglophone markets: "The history of crime literature in the region plays out, with unusual clarity, patterns that developed in an uneven way—sometimes sooner, sometimes later—in various jurisdictions around the Anglophone Atlantic. It should perhaps go without saying that the execution sermon emerged from, and positioned itself within, a transnational culture of letters, both before and after the American Revolution, and that its publics included people from many continents and groups." C. Smith, *Oracle and the Curse*, 68–69.

63. C. Mather, *Pillars of Salt*, 42.

64. C. Mather, 42–43.

65. C. Mather, 42–43, emphasis mine. "Nefandous" is a relatively archaic term—and a favorite of Mather's—meaning "not to be spoken of, unmentionable; abominable, atrocious." "Nefandous, adj.," in *OED Online*, June 2016 (Oxford University Press); www.oed.com/view/Entry/125821?redirectedFrom=nefandous.

66. C. Mather, *Pillars of Salt*, 42–43.

67. C. Mather, 5. The full passage, on 4–5, reads:

> It is a marvelous Thing, I have often marveled at it! That while the more *Barbarous Nations* have Worshipped the *Sun,* the *Moon,* the Stars, and the more specious and glittering Objects of Idolatry; such *Learned* Nations as the *Egyptians* worshipped a *Bull,* a *Dog,* and a *Rat;* The Learned *Grecians,* worshipped *Feavers,* as *Passions,* for their Gods; and our most Learned *Romans* worshopped the very *Furies* of Hell. Truly, there was the *Just Wrath* of God, in this Thing, That the Nations, who had most of *Light,* whereby to *Know God,* and yet *Glorified Him not as God,* should become thus *Vain in their Imaginations*. But stop'd they here? No; There is mention'd a further Depravation whereto God left them, in His *Just Wrath* against them, for their Idolatries: *For this cause God gave them up to Vile Affections;* which also imply and comprize *Vile Practices* in them. It is here indeed, Three Times over noted, That God punished their *Idolatries,* by leaving the wretched *Idolaters* to other and farther *Abominations;* Especially, He Punished the *Spiritual Uncleannes of Idolatry,* by leaving them to the *Corporal Uncleanness of Debauchery:* They who worshopped

God under the Shape, it may be, of *Bruits*, were left by God, unto the *Sun*, which made them worse than *Bruits*.

68. C. Mather, 5.

69. Mather's gesture to Sodom and Gomorrah is not explicit but rather implied, when he says, "There is mention'd a further Depravation whereto God left them, in His *Just Wrath* against them, for their Idolatries: *For this cause God gave them up to Vile Affections;* which also imply and comprize *Vile Practices* in them." C. Mather, 5.

70. C. Mather, 5.

71. C. Mather, 5.

72. See R. Slotkin, "Narrative of Negro Crime," 18; Williams, *Pillars of Salt*, 25; Towner, "True Confessions and Dying Warnings," 79n51. Slotkin offers an enumerated argument to support this claim and also notes that execution narratives detailing sex crimes by people of color also increase toward the end of the eighteenth century:

> This concentration of interest in crimes of murderous and sexual violence against whites distorts the historical picture of what "typical" black criminality must have been.... In fact, the vast majority of crimes committed by blacks quite naturally consisted of crime against property, since they themselves were chattels and their attempts to escape or commit suicide were a form of theft. The distortion is also incremental, in that there is a distinct *rise* in the interest in sexual crimes (rape, cohabitation with whites) as the 18th century wears to a close.... But of twenty editions appearing after 1765, only three do not concern rape or miscegenate cohabitation. From this data, and from the rhetoric of the execution sermons, one might well conclude that to the Puritans and their posterity all Negro crime partook of some element of "dark" sexuality. Of 33 editions concerning black crimes, 22 in some way involve sexual offenses, aberrations or promiscuity, even when such offenses are mere incidents to the crime for which the criminal is being pursued. (R. Slotkin, "Narrative of Negro Crime," 18)

73. I am including the execution of "W.C.," a man executed in Boston for raping a young girl, ostensibly a servant; I include this account because the narrative discusses the condemned's early habits of fornication. Williams, *Pillars of Salt*, 70. I am also including two narratives of women executed for infanticide, because the attending narrative focuses heavily on the sexual uncleanness that led the unmarried women to become pregnant. See C. Mather, *Pillars of Salt*, esp. 69–70; and Danforth, *Brief recognition of New-Englands errand*.

74. Scholars often attribute this growing market to a commensurate rise in literacy rates in colonial New England. See E. Jennifer Monaghan, *Learning to Read and Write in Colonial America* (Amherst: University of Massachusetts Press, 2005), 4.

75. See Williams, *Pillars of Salt*, xi, 5, 12–13, 27, 58–59; and C. Smith, *Oracle and the Curse*, 65.

76. See Mountain, *Sketches of Joseph Mountain*. Caleb Smith also reads the title page as notably sensationalizing Mountain's race: "The exotic sensationalism of the *Sketches* was intensified by the identification of the author as 'A Negro.'" C. Smith, *Oracle and the Curse*, 84.

77. By all scholarly accounts, Mountain's narrative was popular and widely reprinted.

In preparing my analysis of this text, I relied on the following five editions of the narrative, all of which are remarkably consistent, although there are small differences between them: [David Daggett], *Sketches of the Life of Joseph Mountain* (Norwich, CT: Trumbull, 1790), 16 pages, New York Historical Society, item 45856; David Daggett, *The Singular Adventures &c. of Joseph Mountain* (New Haven, CT[?], 1790[?]), Yale College, item 45763; [David Daggett], *Sketches of the Life of Joseph Mountain* (Hartford, CT: Patten, 1790), 14 pages, American Antiquarian Society, item 45855; Joseph Mountain, *Sketches of the Life of Joseph Mountain* (New Haven, CT: T&S Green, 1790), 19 pages, CHS (Connecticut Historical Society?), 22441; David Daggett, *The Life and Adventures of Joseph Mountain* (Bennington, VT: Anthony Haswell, 1791).

78. [Daggett], *Sketches of Joseph Mountain* (Hartford), 1–2. Much of the writing on Mountain reports him to be a slave, but the few facts of his life that we can glean from his narrative—including his master's agreement to let him go to sea—suggest he was under indenture rather than bondage.

79. [Daggett], *Sketches of Joseph Mountain* (Norwich), 15.

80. Williams, *Pillars of Salt*, 55.

81. Mountain, *Sketches of Joseph Mountain*, 9.

82. C. Smith, *Oracle and the Curse*, 78.

83. Reports of Mountain's death, as well as pieces of his execution narrative, and a document titled *Address to the Prisoner* appeared, for example, in *New-York Packet*, August 22, 1790; and in *Boston Herald of Freedom*, September 10, 1790; Mountain's execution narrative was advertised in newspapers as distant as Philadelphia, where the *Pennsylvania Packet and Daily Advertiser* ran an ad for it on November 17, 1790.

84. [Daggett], *Sketches of Joseph Mountain* (Hartford), 9n. In other editions, the note reads simply, "It may appear singular to many, that a woman of this description should be in the least interested in my favor; yet such was the fact, that she not only endured my society, but actually married me in about six months after our first acquaintance. Her mother and friends remonstrated against this connexion; but she quitted them all, and united herself to me. My whole residence with her was about three years; during which time I exhausted all the property which came into my possession by the marriage. We then separated, and she was received by her father." Mountain, *Sketches of Joseph Mountain*, 13.

85. It is worth mentioning that there is a long tradition of black men being executed, both under the law and extrajuridically at the hands of lynch mobs, for the rape of white women, and an enormous body of both scholarship and otherwise political writing contests the legitimacy of the rape claims in these cases. In Mountain's account, the Mountain character freely admits to attacking the girl but actually does not admit to the rape. The narrator recounts: "I have uniformly thought that the witnesses were mistaken in swearing to the commission of a *Rape:* That I abused her in a most brutal and savage manner—that her tender years and pitiable shrieks were unavailing—and that no exertion was wanting to ruin her, I frankly confess." Mountain, 17.

86. Mountain, 16–17.

87. [Daggett], *Sketches of Joseph Mountain* (Hartford), 8.

88. In all of Joseph Mountain's execution narratives, this white woman remains unnamed. Exactly one source—Andrews's *To Tell a Free Story*—that I have encountered gives this white woman a name. Andrews reports that the woman Joseph Mountain was

convicted of raping was his ex-wife, Nancy Allingame. Because no other source, including the narrative itself, corroborates this, I have left her nameless in the narrative of this chapter.

89. Mountain, *Sketches of Joseph Mountain*, 16–17.

90. See Masur, *Rites of Execution*, esp. chap. 3 (50–70). Also see Wilf, *Law's Imagined Republic*, 122.

91. The pardon petition is listed in the Connecticut State Archives as "Joseph Mountain Pardon Petition," Connecticut Archives, 2nd series, 4/131, August 15, 1790. Many thanks to Jeannie Sherman at the State Archives for volunteering to scan the petition and email it to me. Stephen Wilf insists that "castration was, on the other hand, most certainly a vestige of slave law." *Law's Imagined Republic*, 120.

92. Dana, *Intent of Capital Punishment*, 9. Dana explicitly cites Blackstone in his sermon; see Blackstone, *Commentaries on Laws of England*.

93. Dana, *Intent of Capital Punishment*, 26.

94. Dana, 20.

95. Dana, 27.

96. Dana, 26–27. He cites Jeremiah 13:23, "Can the Ethiopian change his skin, or the leopard his spots? then may ye also do good, that are accustomed to do evil."

97. See R. Slotkin, "Narrative of Negro Crime," 26, 54. Slotkin argues that "the image of the black man that emerges from the crime narratives is that of an oversexed brute, ungrateful for the blessings bestowed by his white masters, with a proclivity for crime and criminal conspiracies against the masters, and white society in general" (26).

98. C. Mather, *Negro Christianized*, 25.

99. See Chiles, *Transformable Race*, 25, where she argues that the eighteenth-century "use of *habit* links it to the 'modes of living' or social practices that many natural historians thought could affect one's racial status (in addition to indicating one's habitation)."

100. See Chiles, esp. chaps. 2 and 3.

101. Powers, *Narrative and confession of Powers*. Williams also suggests in *Pillars of Salt* that there are earlier versions of the Powers narrative than the one he and I both cite.

102. Powers, *Narrative and confession of Powers*, 4–5.

103. Williams concurs, arguing that "by referring to his natural inclinations, Powers (the character) encouraged readers to use their prejudices not only to attribute the causes of his crimes to his nature, both ethnic and personal, but also more generally to reinforce the link between criminality and iniquity. Lacking a moral awareness, Powers was evil, and the courses of his evilness belonged to his race. This narrative mixture of racial and literary character was reinforced by the sensationalized connection between lewdness and lawlessness." *Pillars of Salt*, 56.

104. Worcester, *Sermon delivered at Haverhill*. Please note that N. Coverly, the printer of Worcester's sermon, also printed the *Female Marine* series in the early nineteenth century.

105. Powers, *Narrative and confession of Powers*, 3–4.

106. Powers, 5–6.

107. As with any first-person "true narrative" or true account, we have little to no information about how much of the information offered in Powers's narrative is authentic. Similarly, we do not know to what degree his story was mediated by the writer or transcriber. I understand this text as performative and as having an effect on its readers, whether or not it is in any way an accurate reflection of Powers's own experience.

108. Powers, *Narrative and confession of Powers*, 6–7.
109. Williams, *Pillars of Salt*, 56.
110. Powers, *Narrative and confession of Powers*, 11.
111. Three of the broadsides exist, each one very slightly different from the others, but all contain three items: a woodcut image of the scaffold; a first-person account of the murder (supposedly transcribed from Frost's words), and a third-person "Account of Samuel Frost," that addresses his "rational faculties." Two of the broadsides also contain the "Poem on the Solemn Occasion." The three printers responsible for these broadsides were Henry Blake & Co., in Keene, New Hampshire; Isaiah Thomas, in Worcester, Massachusetts; and E. Russell, in Boston (but the latter broadside was apparently sold by one Jonathan Plummer in Newbury). The edition I cite here is that of Isaiah Thomas, who also printed Bancroft's *Importance of a religious education*.
112. [Frost], *Confession and Dying Words*, bottom of fourth column.
113. Bancroft, *Importance of a religious education*, 12.
114. Bancroft, 11–12.
115. Bancroft, 12–13.
116. Bancroft, 17, 18.
117. Bancroft, 17.
118. Bancroft, 18–19.
119. Bancroft, 22.
120. Bancroft, 22, includes a footnote that specifically defines the crime in question as patricide, not just murder.
121. Bancroft, 22.
122. Forbes, "Between Human and Brutal Creation," esp. 40–41.
123. Bancroft, *Importance of a religious education*, 17. I put "past" in quotes to acknowledge that Native ownership of North America was anything but a past-tense concern, in 1793 or today; I use "past" to document an increasingly prevalent trope in writing by natural historians and statesmen such as Thomas Jefferson, who write about Native peoples and their concerns as if they were a fully prior or historical set of political issues.
124. Forbes, "Between Human and Brutal Creation," 42.
125. [Frost], *Confession and Dying Words*, bottom of fourth column.
126. Bancroft, *Importance of a religious education*, 21.
127. See Williams, *Pillars of Salt*, 21; and Foucault, *Discipline and Punish*, esp. 135.
128. C. Mather, *Tremenda*, 3
129. See DeLombard, *Shadow of the Gallows*.

Chapter 4. Botanical Sexuality and the Colonial Landscape

1. See Polwhele, *The Unsex'd Females*, 7. Polwhele's quite explicit parody of Darwin's *Loves of the Plants* within his *Unsex'd Females* might strike readers as unexpected since Polwhele penned one of the handful of dedications to Darwin that appeared in the prefatory matter of the first London edition of *The Botanic Garden*. This dedication is also reprinted along with the original "Advertisement" in the first American edition, which explicitly states that it is reproducing the London edition's dedications. Written in the very year that Wollstonecraft's *Vindication* appeared, the dedication serves as a reminder that it was not "botanizing" across the board so much as coeducational botanizing (and high-quality education for women in general) for which Polwhele feels

such deep disdain. Neither was he alone in his association of botanical science with radical sociosexual reorganization; Mary Wollstonecraft, in a 1767 letter to her sister, famously used a taxonomic metaphor to refer to herself, calling herself "the first of a new genus." Mary Wollstonecraft to Everina Wollstonecraft, London, November 7, [1787], in *Wollstonecraft, Collected Letters*, 163–165.

2. In a review of Darwin's *Botanic Garden* that appeared in the New York–based *Loudon's Register*, July 25, 1792, an anonymous writer identifies Linnaeus as "the author of the sexual system of botany." In Ireland, John Walker, in the third edition of his *Elements of Geography* (1797), refers to his use of "the Linnaean system, which is founded on the sexuality of plants" (125). Erasmus Darwin himself explicitly refers to Linnaean taxonomy as the "Sexual System of Botany" in the preface to the 1795 third edition of *The Botanic Garden*. The organization of Linnaeus's "sexual system" (as well as Darwin's *Loves of the Plants*) relied on an analogical logic that was built upon extant understandings of human anatomy and dominant discourse surrounding sexual practices. This analogical epistemology was older than either Linnaeus or Darwin; Nehemiah Grew, in *Anatomy of Plants* (1682), for example, wrote of a plant that "the blade (or stamen) does not unaptly resemble a small penis, with the sheath upon it, as its praeputium" (170).

3. Hallock, "Male Pleasure and the Genders," 698. Hallock attributes this to Linnaeus, arguing that "the taxonomy developed by Carolus Linnaeus, to which Collinson and the self-taught Bartram converted in the 1740s, used pistils and stamens for classification, and commentary on Linnaeus's system ran the gamut from moral censure to liberated delight. William Smellie blasted the 'alluring seductions' of botany in the 1760 *Encyclopedia Britannica*, whereas Erasmus Darwin (Charles's grandfather) set Linnaeus to verse in a botanic paean to sexual freedom, *The Loves of the Plants* (1789)" (698). Hallock also asserts that even the garden itself became a means of sexualized sociality, as plant exchange and experimentation became a culturally institutionalized form of eroticism between men in the 1790s on the eastern seaboard of the United States. Lisa Moore, in her article "Queer Gardens," makes a similar case for the role of gardens in the sexualizing of Mary Delany's (1700–1788) friendships with other women.

4. Polwhele, *Unsex'd Females*, 10–11.

5. Polwhele, 10n.

6. I am using the term "sexuality" to indicate the way plants' sexual organs were assumed to bear a truth-telling function about the nature of the organism itself. The popular deployment of Linnaean taxonomic principles asserted that the characteristic essence of the plant—that which determined its place in relation to other plants—was determined by its sexual organs and its role in patterns of sexual reproduction.

7. Sampson poses a historical problem because her sexual behavior looks so dramatically like modern definitions of lesbianism to present-day readers, even though no such vocabulary was widely used at the time (although "lesbian" did exist as a term, as of 1736). I thus use the term "lesbian-like," drawn from Martha Vicinus's article on recovery historiography and lesbian history; see Vicinus, "Lesbian History." She uses this term to designate same-sex sexual behavior that present-day critics "recognize" as lesbianism but that, because of its place in a distinct cultural paradigm that did not regularly use the word "lesbian" to describe sexual behaviors (or think of sexuality as a category of identity), should not be called "lesbian" behavior. Michael J. Dukakis named Deborah Sampson "the Official Heroine of Massachusetts" on May 23, 1983.

8. Mann's prose is famously convoluted. Readers and editors of *The Female Review*, from John Adams Vinton in 1866 through Jonathan Ned Katz in 2009, consistently comment on how difficult it is to read Mann's novel. The passages in which Mann describes Sampson's relationships with her "sister sex" are especially difficult to parse. Many of Mann's other critics, in notes to their articles and books, thank colleagues for helping them parse Mann's convoluted prose.

9. On eighteenth-century sapphism, see Binhammer, "Sex Panic of the 1790s"; Fairchild-Craft, "Sexual and Textual Indeterminacy"; Gonda and Beynon, *Lesbian Dames;* Lanser, *Sexuality of History;* Moore, *Dangerous Intimacies;* and Traub, *Renaissance of Lesbianism*. Both "sapphist" and "lesbian" were terms used in the eighteenth century.

10. Mann, *Female Review,* 225.

11. Mann, 55. On Rush's theorization of *anarchia,* see his *Medical Observations and Inquiries,* 293. On Rush's theorization of anarchia and revolutiana, see Miller, "Body Politic and Body Somatic," esp. 67.

12. Mann, *Female Review,* 105.

13. The cultural history of botany, as a vehicle for popularizing scientific notions of sexual difference, provided readers of *The Female Review* with a vocabulary and taxonomy for accounting for human variation at the level of biological sex and sexual behavior.

14. I want to emphasize that this argument pertains to sex and sexual behavior, rather than gender; historians of science have put much work into demonstrating the degree to which the botanical science of the late seventeenth and early eighteenth centuries relied on "the implicit use of gender to structure [its] taxonomy." Schiebinger, *Nature's Body,* 22.

15. This form of botanical empiricism or classification—characterized as it was by anthropomorphic descriptions of plant sexual organization—was digested by popular culture as a taxonomy that could also include the human and pertain to human sexuality and sexual difference. Darwin's *Loves of the Plants* did the important work of linking sexuality to the human interior, as an internal characteristic of the human organism: "By portraying all of life's forms as organized around sexual reproduction, Darwin imagined plants and animals along with humans, as partaking of common instincts, emotions, and manners arising from their sexual impulses." Teute, "Loves of the Plants," 328. Darwin's insistence that we shift "the basic principle of natural existence away from Lockean notions of sentient intellect to sexuality" (Teute, 328) had far-reaching ramifications that we can observe in Mann's description of the observable sexual differences that characterize Deborah Sampson in *The Female Review*. Because of the popularization of botany as a means of understanding *human* sexuality within culture, "sexual characteristics as biological 'facts' were naturalized." Teute, 328.

16. Hallock, "Male Pleasure and the Genders," 698.

17. Schiebinger almost paradoxically suggests that despite the interest in the "new specimen" that characterized the eighteenth-century craze for naturalism (and also justified colonial expansion and the perpetuation of slavery), taxonomists "could not or would not recognize an unfamiliar sexual type" (*Nature's Body,* 22). I am skeptical of this assertion, but I think Schiebinger makes this argument to illustrate many scientists' resistance to the "discovery" of plants that demonstrate a "hermaphroditic" gender (e.g., plants that have both "male" and "female" sexual organs). For my purposes, however, it is important to note that a debate *did* exist over the "hermaphroditic" status of plants, and although many famous botanists (William Smellie, Sebastien Vaillant) disagreed with this desig-

nation and refused to use it in their taxonomic projects, the "hermaphroditic" was a recorded and debated category of gender or nongender as early as the mid-eighteenth century.

18. Readers should note that Mann's depiction of Sampson's sexual difference is, for the most part, apologetic or taxonomic, rather than judgmental. If anything, the narratological strategy of indigenization that I discuss in this chapter speaks to Mann's interest in assimilating Sampson into, rather than rejecting her from, social understandings of sexuality.

19. See Young, *Masquerade;* and Alex Myers, *Revolutionary* (New York: Simon and Schuster, 2014).

20. See my subsequent discussion of New York printer Philip Freneau, who referred to Sampson as "eccentric."

21. Note that Sampson was born in Plympton, Massachusetts; she arrived in Middleborough (about eight miles from Plympton) at the age of eight after being indentured out by her mother. She had been indentured for at least three years before coming to live with the Thatcher family in Middleborough.

22. "Conclusion" may be a misleading description of the final pages of the novel, which are followed by an appendix. Mann hastily ends *The Female Review* with a slapdash paragraph that reads, "Spring having once more wafted its fragrance from the South, our Heroine leaped from the masculine, to the feminine sphere. Throwing off her martial attire, she once more hid her form with the *dishabille* of Flora, recommenced her former occupation; and I know not, that she found difficulty in its performance. Whether this was done voluntarily, or compulsively, it is to me an enigma. But she continues a phenomenon among the revolutions of her sex." Mann, *Female Review,* 238. Before this paragraph, however, Mann lingers on Sampson's time posing as her brother and working on her uncle's farm, including a lengthy and syntactically confusing lamentation pertaining to Sampson's either intentionally or unwittingly seducing young, local women in her uncle's village. Sampson's uncle cautions her against consorting too carelessly with these young women, and Mann uses this event to discuss seduction more generally and to transition into the seduction story of Fatima and Philander, in which Sampson appears only momentarily as a character. The twenty-seven-page appendix concludes as Mann defends Sampson from accusations of compromising her chastity while in the Continental Army and accusations that she is not a proper wife to her husband (i.e., that she would not consummate the marriage, an assertion that Mann repudiates on Sampson's behalf by pointing out that she and her husband, Benjamin Gannett, have children).

23. I collated this publication history from Young's *Masquerade;* Hiltner's 1999 article "'She Bled in Secret'"; and editions of *The Female Review* that are digitally archived on *Early American Imprints: Evans.*

24. See Young, *Masquerade;* and Myers, *Revolutionary.*

25. Dugaw identifies the beginnings of the "warrior woman" as a motif in popular song and balladry, originating as a Restoration-era hero on broadsides and in chapbooks alongside other popular folk heroes such as Robin Hood. *Warrior Women and Popular Balladry,* 15–16. Gustafson, in her article "Genders of Nationalism," supplements Dugaw's research on the history of the warrior woman figure, reminding readers that not only was Mann participating in a tradition drawn from popular balladry, but he was also engaging in a Revolution-era conversation over the changing performances of republican woman-

hood. Alongside Charles Brockden Brown and John Daly Burk (whose 1798 play, *Female Patriotism*, offered a dramatic retelling of the story of Joan of Arc), Gustafson argues that representations of Sampson's life (including *The Female Review* as well as Sampson's subsequent speaking tour in the first years of the nineteenth century) pose serious questions about how emergent nationalisms shaped possibilities for new forms of womanhood as the eighteenth century came to a close. In turn, Gustafson highlights the mutable and performative nature of womanhood more generally. Describing Sampson's speaking tour, at the conclusion of which Sampson donned her military uniform and performed drills (a performance likely emulating that of Hannah Snell in London half a century earlier), Gustafson writes: "The performance description of her tour—the costume, the arms drill, Gannett's substantial physical presence—both aestheticized [Deborah Sampson Gannett's] own role and invited members of her audience to recognize the republican woman as a figure requiring a similar, if less dramatic, kind of role playing." "Genders of Nationalism," 398.

26. Both Dugaw and Daniel Cohen trace the imagery of the "warrior woman" to popular ballads and putatively (auto)biographical narratives that circulated in England and western Europe more generally from the second half of the seventeenth century through the mid-1800s. Dugaw's *Warrior Women and Popular Balladry* (1989) documents the earlier instantiations of this image, noting that readers of Herman Mann's *Female Review* would have also been familiar with the widely popular "warrior woman" narratives that circulated in England in the mid-eighteenth century, especially *The Female Soldier or The Surprising Life and Adventures of Hannah Snell* (London, 1750). For more on this tradition, see chapter 5 of this book.

27. I am referring to Lanser's "Tory Lesbians" and "Mapping Sapphic Modernity" and Binhammer's "Sex Panic of the 1790s."

28. While it would be difficult to determine precisely to what degree residents of the early Republic would have been familiar with Snell and her story, at least three American publications between 1797 and 1809 either refer to Snell's narrative explicitly or reproduce it directly.

29. Tabitha Tenny's 1801 parody of the romantic novel, *Female Quixotism*, also features an episode in which the female protagonist briefly lives as a man.

30. Mann, *Female Review*, 33.

31. Mann, 29.

32. Mann, 26, emphasis mine.

33. See John Adams Vinton, ed., *The Female Review: Life of Deborah Sampson, Female Solider in the War of Revolution*, by Herman Mann (Boston: J. K. Wiggin and Wm. Parsons Lunt, 1866), 9.

34. See Schiebinger, *Nature's Body*, 14, where she notes that the fad for taxonomic systems—"system madness"—by the end of the eighteenth century was "truly 'epidemical.'"

35. Technically, Linnaeus's taxonomic system was later replaced by the binomial taxonomic system (developed by Gaspard and Johann Bauhin two centuries after Linnaeus), in which all life forms have a two-word name (e.g., *Homo sapiens*—although this was actually a Linnaean binomial). Linnaeus's system was nonetheless the origin and inspiration behind the later binomial system, and he is widely considered the inventor of popular scientific taxonomy.

36. Mann, *Female Review*, 39.

37. Mann, 42–43.

38. Critics such as Karen Sanchez-Eppler, Elizabeth Susan Wahl, Susan Branson, and Nancy Cott have all written extensively on the widely documented epistemologies of republican womanhood and motherhood at the end of the eighteenth and the beginning of the nineteenth century in America. Between 1795 and 1820, Benjamin Rush also wrote extensively on the importance of republican motherhood in the making of good citizens. Two of the most extensive studies of this ideological construction are Kerber, *Women of the Republic;* and Norton, *Liberty's Daughters*.

39. As Cathy Davidson notes, Tabitha Tenney's *Female Quixotism* (1801) responded to the presumedly female or feminine culture of novel-reading (and its presumed dangers) by presenting a story that provides a "subtle how-not-to-read-a-novel novel." *Revolution and the Word*, 275. In William Brown's *Power of Sympathy* (1789), Mrs. Holmes famously warns against the dangers of "immoderate reading," because too much reading could result in a distasteful asociality. There is a wealth of scholarship on the history of the novel that links concerns about excessive novel reading to concerns about other materials that induce solitude, thoughtfulness, and masturbation. Brown writes, in *The Power of Sympathy*:

> "There is another medium," said Mrs. Holmes, assenting to my observation, "to be noticed in the study of a lady—she takes up a book, either for instruction or entertainment; the medium lies in knowing when to put it down. Constant application becomes labour—it sours the temper—gives an air of thoughtfulness, and frequently of absence. By *immoderate reading* we hoard up opinions and become insensibly attached to them; this miserly conduct sinks us to affectation, and disgustful pedantry; *conversation* only can remedy this dangerous evil, strengthen the judgment, and make reading really useful. They mutually depend upon, and assist each other. (W. Brown, *Power of Sympathy*, 27)

40. Proper parenting and the effective management of children's passions were a crucial cultural issue at the end of the eighteenth century. The issue was marked and largely directed by the vastly influential antimasturbation campaigns that extended from the early eighteenth century through the mid-twentieth. As a result, the management and cultivation of children's passions must be understood as one instantiation of contemporary discussions about immoderation, which we now consider either part of or related to the eighteenth-century history of sexuality.

41. This understanding of "sex" comprehends both what we think of as sexuality—sexual desire, "deinstrumentalized," as Stein would have it, from reproductive function—and what we think of as gender, the way biological sex is performed in a social sense. Stein, "Mary Rowlandson's Hunger," 476.

42. This rhetoric of revolutionary newness—"sensations and emotions different from whatever they had before experienced" (Mann, *Female Review*, 55), or "sensations, to which she had hitherto been unaccustomed" (60)—permeates the text. In light of my argument that Sampson's culture understood her aberrant sexual behaviors through a scientific lens of epistemological precipitousness, this rhetoric is remarkably consistent with the representation of Sampson as a "new specimen," subject to feelings "different from whatever [the population] had before experienced."

43. Mann, 60.

44. Mann, 122.

45. Cross-dressing and "woman warrior" tales often featured scenes in which their cross-dressed protagonists accidentally or intentionally find themselves the object of romantic attentions from other women. Scholars generally understand these scenes as intended to both titillate and amuse, although in the case of Deborah Sampson, Mann also made explicit efforts to emphasize her normative womanhood in the name of her campaign to receive an invalid's pension for her service to the Continental Army. Mann's defense of Sampson's normative marriage and family life appears in his appendix to *The Female Review* (also see Gustafson, "Genders of Nationalism," 393n23).

46. The first American edition of *The Botanic Garden* is Darwin, *Botanic garden* (1798).

47. See Schiebinger, *Nature's Body*, 22. Moore understands Darwin's poetically eroticized flowers as part of "a long tradition that viewed flowers as strongly suggestive of human eroticism. The later eighteenth-century English reception of Linnaeus capitalized on this history of sexual connotation for flowers. The notion of plants as possessors of a sexual life comparable to that of humans was extensively, even obsessively, imagined in *The Loves of the Plants*." "Queer Gardens," 65.

48. Linnaeus, *Families of Plants*, 1:77–78. Teute offers a clear and fairly concise overview of Darwin's extremely complicated system of organization in *Loves of the Plants*:

> The first classes contained both sexes and were arranged according to the number of males in the blossom; XII had twenty males; XIII, "Polyandria," had "Many Males." The next classes depended on the "Powers," that is, the different heights, of the stamens. The ways in which males adhered in "their union" to the females determined succeeding classes: "One Brotherhood," "Two Brotherhoods," "Many Brotherhoods," "Confederate Males," and "Feminine Males." Two classes of plants had flowers containing only one sex, "One House" and "Two Houses"; in "Polygamy," there was a mixture of single-sex and both-sex flowers. Finally, in "Clandestine Marriages," flowers were not discernible in the plants. (Teute, "Loves of the Plants," 324–325)

49. J. Browne, "Botany in Boudoir and Garden," 159.

50. The Friendly Club of New York City, founded by Dr. Elihu Hubbard Smith and active in the 1790s, was an all-male intellectual club that considered "science . . . a crucial collaborative medium for intellectual and social exchange." Teute, "Loves of the Plants," 321. This particular intellectual milieu was also part of a larger movement of groups of young, intellectual men and women who found in scientific learning and practice the opportunity to explore new forms of heterosexual sociality. In such circles, "the women botanized together; the men instructed the women in Lavoisier's chemical principles." Teute, 321. Teute notes: "For many young American intellectuals, among the published texts signaling the possibilities of a new age dawning were Erasmus Darwin's *The Botanic Garden* (1789) and Mary Wollstonecraft's *A Vindication of the Rights of Woman* (1792)." Teute, 319–320.

51. Another Massachusetts resident, Ann Bailey, had famously enlisted in the Continental Army in 1777, several years before Sampson, under the name of Samuel Gay. Bailey was exposed, arrested, and prosecuted. Weyler, "Actor in the Drama of Revolu-

tion," 183. Revolutionary-era war records from Elizabeth, New Jersey, refer to yet another soldier found to be a woman, who was later strip-searched and marched out of the camp. Weyler, 184; Hiltner, "'Like a bewildered star,'" 9–10.

52. Judith Hiltner argues that the widespread social belief that women in army camps were prostitutes accounted, at least in part, for Sampson's decision to enlist as Shurtliff. Hiltner, "'Like a bewildered star,'" 9–10.

53. While Freneau was initially hesitant to adopt her cause, privately calling Sampson an "eccentric," he later wrote the following poem of support, which appeared in the December 4, 1797, edition of the *Time Piece* (a New York newspaper that ran from March 1797 through August 1798) "bearing the enigmatic inscription, 'A Soldier Should be Made of Sterner Stuff!'"

> Ye Congress-men, and men of weight,
> Who fill the public chairs,
> Who many a favour have conferr'd,
> On men unknown to Mars—
> And ye, that on that lofty bench
> Decide by vote our great affairs,
> Ah, turn a calm attentive ear
> To Her, who never did fear—
> —Relieve this gallent Wench. (Weyler, "Actor in Drama of Revolution," 187)

54. Davidson, *Revolution and the Word*, 149.
55. Mann, *Female Review*, 225–227.
56. Mann, 248.
57. Comment's article "Charles Brockden Brown's *Ormond* and Lesbian Possibility in the Early Republic" offers a comprehensive list of literary and other print productions that appeared during the eighteenth century in England and the North American colonies that represented sapphists, tribades, and female husbands:

> Representations of graphic lesbian behavior pervade the era's medical, juridical, and literary texts. For example, Robert James's *Medicinal Dictionary*, published in London in 1745, identifies "tribades" as "those women who abuse the clitoris by using it to engage in sex with other women" . . . ; Swiss doctor Samuel Tissot's *Onania: A Treatise Upon the Disorders Produced by Masturbation*, translated into English in 1766, discusses mutual masturbation as a sexual practice between women and warns that it might diminish a woman's desire for heterosexual intercourse; William King's *The Toast* (1736) contains a satiric depiction of a lascivious hermaphrodite who has sex with both women and men; John Cleland's infamous *Fanny Hill, or Memoirs of a Woman of Pleasure* (1747), which enjoyed underground popularity in both Britain and America, describes in graphic detail the sexual initiation of the protagonist by a female prostitute; Mathieu Francois Marobert's French novel *L'Espion Anglois* (1777–78) portrays the initiation of the narrator, a young girl named "Sapho," into a brothel of lesbians who profit by servicing "tribades"; and Denis Diderot's 1796 publication of *La Religieuse* (The Nun) depicts a lecherous Mother Superior who preys on her young initiates. The eighteenth century also saw a number of so-called female husbands publicly prosecuted

in Germany, England, and the Netherlands. . . . Henry Fielding's *The Female Husband: or the Surprising History of Mrs. Mary, Alias George Hamilton, Convicted for Marrying a Young Woman of Wells* (1746) records the real life case in which Mary (alias George) Hamilton was whipped and imprisoned for marrying several women while a impersonating a man. And one of the most famous women of the era, Marie Antoinette, was graphically accused by her political opponents of having sexual relations with other women. (Comment, "Brown's *Ormond* and Lesbian Possibility," 62)

58. Vicinus, "Lesbian History," 58.
59. Mann, *The Female Review*, 247.
60. Mann, 150.
61. Morgensen, "Biopolitics of Settler Colonialism," 60.
62. Mann, *The Female Review*, 219.
63. Mann, 215.
64. Mann, 216. "Begging quarters" means asking to be exempted from being put immediately to death in exchange for total surrender.
65. Many Native studies scholars, scholars of empire and colonialism, and scholars of the history of sexuality have extensively documented the way colonial projects forcibly ensured that Native peoples conform to European sexual and gender standards. See, for example, Schneider, "Oklahobo"; J. Goldberg's chapter "Discovering America" in *Sodometries*, 179–232; Morgensen, "Biopolitics of Settler Colonialism"; Miranda, "Extermination of the Joyas"; Rifkin's introduction and "Reproducing the Indian" in *When Did Indians Become Straight?*; and Spear, "Indian Women, French Women."
66. See Branson, *These Fiery Frenchified Dames;* and Hunt, "Many Bodies of Marie Antoinette."
67. On the changing meaning and significance of whiteness during this period, see Jacobson, *Whiteness of a Different Color;* and Roediger, *Wages of Whiteness.*
68. Morgensen, "Biopolitics of Settler Colonialism," 60.
69. Mann, *Female Review*, 199.
70. Young's statement that "we are in the dark" refers specifically to Sampson's relationship—rumored to be sexual—with a local free black domestic worker in her hometown of Middleborough, Massachusetts.
71. "White feminism" is a term used both diagnostically and derisively to refer to a form of feminist politics that tends to envision empowerment in a way that benefits wealthy and white women over poor women or women of color. During Hilary Clinton's presidential campaign in 2016, actress Meryl Streep, in a speech at the Democratic National Convention (where Clinton was officially named and endorsed as the Democratic Party's candidate), referred to Deborah Sampson as "the first woman to fight for our country." Streep demanded:

> What does it take to be the first female anything? It takes grit. And it takes grace. Deborah Sampson was the first woman to take a bullet for our country. She served, disguised as a man, in George Washington's Continental Army. And she fought to defend a document that didn't fully defend her. 'All men are created equal,' it read. No mention of women. And when she took a blast in battle to her leg, she was afraid to reveal her secret. So she took out a penknife, she dug out the musket ball,

and she sewed herself back up again. That's grit. (https://www.bustle.com/articles/175088-transcript-of-meryl-streeps-dnc-speech-calls-for-the-grit-and-grace-that-hillary-clinton-embodies, accessed September 25, 2017)

72. The BBC historical drama *Tipping the Velvet* (2002) is based on a historical novel of the same name (by Sarah Waters, 1998) that follows its protagonist as she falls in love with a cross-dressing woman in the Victorian era in England. *Patience and Sarah* (1971; first published under the title *A Place for Us* in 1969) is a historical novel by Isabel Miller based on the real-life story of two women who built a life together in Connecticut in 1816.

73. Love's essay "'Spoiled Identity'" assesses Prosser as "argu[ing] that every effort to read *The Well* as a lesbian novel has been 'a case of trying to fit a square peg into a round hole' . . . Prosser imagines gender inversion as a precursor to transsexual subjectivity, a longing for biological, psychic, and sexual transformation not yet medically available to the subjects in the case histories." Love, 506.

74. Love, "Modernism at Night," 746. Love argues that "opacity, for Martinez, is not an absolute quality or the negation of knowledge; it has its characteristic climates and textures and is part of the experience of living with (if uncomfortably or partially) social identities" (746). She suggests that the writers whose essays she introduces "may suggest a new scene for queer modernism: not the epistemology of the closet but an encounter with the illegible. Queer lives and queer feelings scribbled over but still just visible—you can half make them out in the dark" (747).

75. Gannett, *Address, Delivered with Applause*, 24.

76. Mann, *Female Review*, 219.

Chapter 5. "Negro Hill" and the Sexuality of Space

1. Whitehill, *Boston*. Whitehill notes on 7–8 that while Mount Whoredom was eventually renamed "Mount Vernon," "there was always a certain veiled sniggering among local antiquaries about the name of this peak, which appeared quite unequivocally on Lieutenant Page's 1776 map as Mount Whoredom." See also the map *A Plan of Boston, and its environs. Shewing the true situation of His Majesty's army. And also those of the rebels*, drawn by an engineer at Boston, October 1775 (London: Andrew Dury, [1776]), which is housed in the collections of the American Antiquarian Society in Worcester, MA.

2. D. Cohen, *The Female Marine*, 29. "Negro Hill" described a very specific area of the larger neighborhood of Beacon Hill. The distinction between "Negro Hill" and "Beacon Hill" more generally is important: the area called "Negro Hill" is in Beacon Hill, but "Negro Hill" or "the Hill" was only one section of the Beacon Hill neighborhood, an area in that neighborhood that might be referred to as a "vice district" today. Beacon Hill also housed a large community of working-class free black people and a growing population of wealthy white people. The distinction between these different areas of Boston's West End was the subject of an ongoing tension throughout the first three decades of the nineteenth century, a tension characterized by what one might call a certain politics of respectability surrounding the differing socioeconomic communities within the population of Boston's West End: the burgeoning population of white, wealthier residents (especially merchants and artisans); free, working- and middle-class black residents; and the "vice district" ("Negro Hill"). Barbara Meil Hobson writes, "For

West End residents, disorderly houses represented more than a nuisance; they were a threat to property values, a point made in a petition signed by over a hundred persons in 1820." *Uneasy Virtue*, 13. Daniel Cohen notes:

> Although many hundreds of African Americans had been living in the capital of Massachusetts since the colonial period (most held in bondage), it was not until after the American Revolution that slavery was abolished in Massachusetts and that Boston blacks were able to establish a somewhat autonomous communal and institutional presence. But unfortunately most African Americans were forced to live in the city's poorest and least reputable residential neighborhoods, particularly in the West End. As early as 1810, more than half of Boston's black population resided on or near "Negro Hill," living by necessity amidst the area's numerous prostitutes and other entrepreneurs of urban vice. (D. Cohen, *Female Marine and Related Works*, 29).

3. While there is some debate about when sexology developed as a discrete arm of the human sciences, most scholars more or less date the earliest instantiations of sexological research to Alexandre Parent du Châtelet's 1836 *De la prostitution*, an eight-year study of sex workers in Paris. See also Bullough, *Science in the Bedroom;* and Terry, *American Obsession*. Importantly, Benjamin Kahan also just finished an edition of Heinrich Kaan's *Psychopathia Sexualis*, the 1844 text that Foucault identifies as one of the earliest works of sexology. See Kahan, *Heinrich Kaan's Psychopathia Sexualis*.

4. Daniel Cohen argues that Nathaniel Coverly, a Boston-based printer, printed all of the editions of *The Adventures of Louisa Baker* and the related works and that said works were written by Nathaniel Hill Wright. D. Cohen, *Female Marine and Related Works*, 5.

5. Cohen hypothesizes in *The Female Marine and Related Works* about the month of publication of these texts, but he does so with an acknowledgment of his own uncertainty. See D. Cohen, 47–49.

6. See, for example, popular seduction fiction that had appeared in the recent past, such as William Hill Brown, *Power of Sympathy* (1789); Susanna Rowson, *Charlotte Temple* (1791); and Hannah Webster Foster, *The Coquette* (1797). Stories of what Cathy Davidson terms "the female picaresque"—featuring women traveling, often in men's garb—also appeared frequently in the popular fiction of this era. See Cathy N. Davidson, "The Picaresque and the Margins of Political Discourse: The Female Picaresque," in *Revolution and the Word: The Rise of the Novel in America* (New York: Oxford University Press, 1986), 179–192.

7. The count of 21 editions of *The Adventures of Lucy Brewer* includes all the editions of the related works (many of which overlap with, and at times reproduce verbatim, the original story).

8. Important work in both the history of sexuality and queer studies has carefully considered the relationship between the emergence of sexuality as a component dimension of human experience and changes in the social and professional architecture of cities following industrialization. Works such as Chauncey's *Gay New York;* Walkowitz's *City of Dreadful Delight;* and Rubin's landmark essay "Thinking Sex" have all plumbed the relationship between urbanity and sexuality, especially socially aberrant sexual practices, but none of these texts have read sexuality as an environmental phenomenon.

9. *Adventures of Lucy Brewer*, 30–31; *Affecting Narrative of Louisa Baker*, 11.

10. See D. Cohen, *Female Marine and Related Works;* and Davidson, *Revolution and the Word.*

11. D. Cohen, *Female Marine and Related Works,* 5.

12. See DeLombard, *Slavery on Trial;* and D. Cohen, *Pillars of Salt.*

13. The first (1815) edition of *The Adventures of Louisa Baker* refers specifically to Deborah Sampson at numerous points, and one of the 1816 editions reproduces text verbatim from Rowson's 1791 *Charlotte Temple:* "If what I have exposed to public view is sufficient to induce youths of my sex never to listen to the voice of love, unless sanctioned by paternal approbation: and to resist the impulse of inclination when it runs counter to the precepts of Religion and Virtue, then, indeed, have I not written in vain." Quoted in *Affecting Narrative of Louisa Baker,* 24.

14. This pronouncement appears on the title page of *The Awful Beacon* (Boston: N. Coverly Jr., 1816).

15. D. Cohen, *Female Marine and Related Works,* 5, 3.

16. See, for example, Malthus, *Essay on Principle of Population* (1798), a treatise on political economy responding to writing on population theory by William Godwin (the 1793 *Enquiry Concerning Political Justice and its Influence on General Virtue and Happiness*) and the Marquis de Condorcet (the 1794 *Sketch for a Historical Picture of the Progress of the Human Mind*).

17. Hopkins, *Evils and Remedy of Lewdness.* The writer of this sermon is ostensibly citing the infamous 1831 Magdalen Report: New York Magdalen Society, *Magdalen Report,* 19–20.

18. Whitehill, *Boston,* 47. Also see Hobson, *Uneasy Virtue,* 12–13.

19. Hobson writes: "Two sections of the city, the North End and the West End, became the most densely populated. They had the most transients and the highest concentrations of taverns and brothels. The North End, the oldest part of the city, had once been the home of the illustrious Mathers, Paul Revere, and some of the richest colonial Bostonians, but by the late eighteenth century it was already in a state of decline." *Uneasy Virtue,* 13.

20. Hobson, 13.

21. Cited in Whitehill, *Boston,* 7–8.

22. Cohen writes:

Another related source of popular anxiety reflected in Coverly's cross-dressing pamphlets stemmed from the increasing visibility of Boston's African American community during the early decades of the nineteenth century—and its perceived association with urban vice. Although many hundreds of African Americans had been living in the capital of Massachusetts since the colonial period (most held in bondage), it was not until after the American Revolution that slavery was abolished in Massachusetts and that Boston blacks were able to establish a somewhat autonomous communal and institutional presence. But unfortunately most African Americans were forced to live in the city's poorest and least reputable residential neighborhoods, particularly in the West End. As early as 1810, more than half of Boston's black population resided on or near "Negro Hill," living by necessity amidst the area's numerous prostitutes and other entrepreneurs of urban vice." (D. Cohen, *Female Marine and Related Works,* 29)

23. Whitehill, *Boston*, 70.

24. Whitehill, 70. It is worth noting that even historical work that seeks to offer a more complicated or textured account of this neighborhood frequently resorts to language that, to contemporary readers, relies on subtly racist idioms that are still frequently used today to describe "bad," "dangerous," or simply poor neighborhoods. Walter Whitehill, for example, refers to "Negro Hill" as "mangy" (70) and Hobson's book-length study of sex workers at times reads sex work as ontologically (rather than intellectually, historically, or culturally) problematic. This idiom for the discussion of "bad" neighborhoods—language that frequently indexes the racial makeup of those that live within them—persists today as a mode of representation similar to, if not necessarily continuous from, that which was used in the early nineteenth century to describe urban crime, poverty, and vice.

25. Hobson, *Uneasy Virtue*, 13.

26. What is now the area of Boston termed "Back Bay" was, until 1857, a tidal bay covered by both water and marsh. In 1857 the city began a large-scale project, completed in the 1880s, to fill in the back bay so that it could be built upon.

27. As Cohen details, all of these developments were coeval with the publication, circulation, and growing popularity of the *Female Marine* pamphlets:

> Indeed, it was at precisely the period when Coverly was marketing his pamphlets on the female marine that Boston residents of differing social ranks first mobilized to combat prostitution by a variety of means. In September 1815, just a month after the publication of *The Adventures of Louisa Baker*, a mob of over twenty local laborers and others pulled down a "disorderly house" in Boston, the first of several documented anti-brothel riots to take place in that city between the mid-1810s and mid-1820s. Less than two years later, in 1817, the Boston Female Society for Missionary Purposes, an organization of respectable evangelical women, hired two preachers to spread the gospel in Boston's vice districts of the West End ("Negro Hill") and the North End. At about the same time, civic and evangelical reformers also launched successful campaigns to establish a Penitent Female's Refuge and a House of Industry for the reform and confinement of prostitutes. Finally, in 1823, Mayor Josiah Quincy of Boston personally spearheaded urban America's first sustained police crackdown on prostitution." (D. Cohen, *Female Marine and Related Works*, 28–29)

Furthermore, the anonymous *Big Dick* (1849) contains a passage describing the history of mob violence through which brothels and disorderly houses on the Hill had been attacked:

> "Nigger Hill"—as it was once termed in derision—though every shade of human color assembled there—was once well known to the inhabitants of Boston, and as to its character, the "Five Points" of New York bear the nearest resemblance to it of any section that has yet gained a notoriety in any of the Atlantic cities; and as this latter place has been so often described we will not shock our readers with a relation of the disgusting scenes which nightly occurred on "Nigger Hill." That the younger portion of our readers may know where this famous place had a locality, they will find it on the summit of Beacon Hill, in the vicinity of Pinckney, Myrtle and

Belknap streets, an elevation which commands a fine view of West Boston and all those lovely villages that skirt the Back Bay from Roxbury around to Charlestown.

So many and so frequent were the riotous and outrageous scenes that transpired on "the Hill," and finding the laws of the City inadequate to suppress them, the people took the matter into their own hands, resolving no longer to submit to such an extended nuisance right in their midst, hit upon a bold expedient to effect their wishes." (*Big Dick*, 57)

Thanks to Paul Erickson for pointing me in the direction of *Big Dick* during a month-long fellowship at the American Antiquarian Society.

28. For example, the Magdalen Society of Philadelphia (a largely Episcopalian organization inspired by British Magdalene's Asylums, the first of which appeared in Whitechapel, England, in 1758) was founded in 1800, the Boston Society for the Moral and Religious Instruction of the Poor in 1816, the Boston Female Society for Missionary Purposes in 1817, and the Penitent Female's Refuge in 1822; New York's Madgalen Society (founded 1830) circulated its important *Magdalen Report* in 1831.

29. See, for example, the directory of sex workers in New York authored by "Butt Ender" in 1839; the author believes his *Prostitution Exposed* will be useful for "those residing at a distance from the city, not only because it displays the amount of evil practiced therein, but if they ever visit our busy Gotham, it may be used as a guide to direct them how to shun the dangers." Butt Ender, *Prostitution Exposed*, 1.

30. For more on the sexual politics of vice and virtue—or what Carroll Smith-Rosenberg terms "the feminization of virtue"—see Dimock, *Residues of Justice*, esp. 57–60.

31. Dimock, 49. On the sexual politics of virtue, see Smith-Rosenberg, "Dis-covering the Subject." Also see Hirshman and Larson, *Hard Bargains*, esp. 90–91.

32. Hopkins, *Evils and Remedy of Lewdness*. The writer of this sermon is ostensibly citing the infamous 1831 Magdalen Report: New York Magdalen Society, *Magdalen Report*, 19–20.

33. See Andrew, *Philanthropy and Police*.

34. "Vice, n. 1," in *OED Online,* December 2015 (Oxford University Press); www.oed.com/view/Entry/223113.

35. Hobson, *Uneasy Virtue*, 17. Here Hobson is referring to Josiah Quincy's *Remarks on Some of the Provisions of the Laws of Massachusetts Affecting Poverty, Vice and Crime* (1822), in which Quincy both equates poverty with vice and crime and distinguishes the "virtuous poor" from the vicious poor. He writes, "Among all the general relations of man, the most interesting to the individual, and the most important to society, are those of poverty, vice, and crime. They are, in truth, often little else than modifications of each other; and, though the class of virtuous poor form an honorable exception to the fact, yet in the more depressed classes, they are so frequently found together, that in every general survey, they may be considered, for the purpose of analysis and remark, in some measure as inseparable." Quincy, *Remarks*, 3. Importantly, Quincy is echoing earlier notions, dating from at least the late sixteenth century, about the distinction between forms of poverty. On this, see Friedlander, "Rogue Sexuality."

36. "Butt Ender's" directory of sex workers operating in New York in 1839 includes a heading "On the influence of occupations" (*Prostitution Exposed*, 2), in which he offers a chart of known sex workers that includes information on their past, ostensibly legitimate,

professions. He prefaces this chart by arguing that "it is a matter of vital importance, not only to parents, but to young men, . . . to become acquainted with the chances of obtaining a proper companion from the working classes of females. It is with a desire to inform the minds of the young, as well as to show to parents the most suitable occupations for their daughters, that we subjoin the following tablet as a correct statement of the victims of debasing passions that are attached to the various branches of female labor, all of which have been obtained, after a diligent research by the 'Reform Society'" (2).

37. Hopkins, *Evils and Remedy of Lewdness*, 24.

38. Hobson notes that in the 1816 writings on sexual immorality of the Boston Society for the Moral and Religious Instruction of the Poor, "they wove together moral and economic consequences. For example, the society estimated in one report that if two thousand prostitutes were each paid $1.25 a day, then wage earners were spending $912,500 a year for licentiousness." Furthermore, "Dudley Rosseter and James Davis, the missionaries hired by the Female Missionary Society, suggested that there were incredible numbers of prostitutes in Boston—over 2000 in 1817. That would be one out of every fourteen women between the ages of twenty and forty-five. However, this figure included nonmarital sexual relationships and today that would not be considered prostitution." Hobson, *Uneasy Virtue*, 20–21. And finally, in a sermon that borrowed heavily from McDowall's *Magdalen Report*, Asa Hopkins speculated: "The probable annual expense of this sin [lewdness] in that single city [New York] is shown, by particulars, into which I may not need not enter, to exceed *six millions of dollars*." Hopkins, *Evils and Remedy of Lewdness*, 19.

39. Sharp, *Report of the Union Committee*.

40. D. Cohen, *Female Marine and Related Works*, 28.

41. Hobson, *Uneasy Virtue*, 17–18.

42. Hobson, 17.

43. Foucault, *History of Sexuality*.

44. See, further, Lyons, *Sex among the Rabble;* Gilfoyle, *City of Eros;* Shah, *Contagious Divides*.

45. Hopkins, *Evils and Remedy of Lewdness*, 14.

46. Hopkins, 9.

47. Hopkins, 19.

48. Sharp, *Report of the Union Committee*, 4.

49. Sharp, 18.

50. The Boston Female Society for Missionary Purposes (Baptist and Congregational) commissioned two preachers to minister specifically to Boston's West End and North End neighborhoods. D. Cohen, *Female Marine and Related Works*, 28.

51. Hobson, *Uneasy Virtue*, 22.

52. *Adventures of Lucy Brewer*, 23.

53. *Affecting Narrative of Louisa Baker*, 17.

54. *Affecting Narrative of Louisa Baker*, 15.

55. *Adventures of Lucy Brewer*, 23.

56. *Adventures of Lucy Brewer*, 30–31; *Affecting Narrative of Louisa Baker*, 11.

57. *Affecting Narrative of Louisa Baker*, 14.

58. Hobson, Whitehill, and Cohen are all careful to note that the West End of Boston housed a wide range of communities of different classes, from the very impoverished to

what today might be called working- or middle-class people. And while the West End did include the largest African American population in Boston for many years, it was always a racially mixed neighborhood.

59. *Affecting Narrative of Louisa Baker*, 16.
60. *Case of Young Robinson*, 4.
61. *Affecting Narrative of Louisa Baker*, 15.
62. *Adventures of Lucy Brewer*, 33.
63. *Affecting Narrative of Louisa Baker*, 15–16.
64. *Adventures of Lucy Brewer*, 26–27.
65. On the "Pompey" character in the minstrel tradition, see Mahar, *Behind the Burnt Cork Mask*, 136. The minstrel character "Pompey" also sees an afterlife in William Wells Brown's *Clotel; or, the President's Daughter* (1853).
66. Hervey, *Meditations and contemplations*, 7.
67. *Adventures of Lucy Brewer*, 30–31.
68. *Affecting Narrative of Louisa Baker*, 11.
69. Bennett, *Vibrant Matter*, xiii, describes her project in the following terms: "My aim, again, is to theorize a vitality intrinsic to materiality as such, and to detach materiality from the figures of passive, mechanistic, or divinely infused substance. This vibrant matter is *not* the raw material for the creative activity of humans or God. It is my body, but also the bodies of Baltimore litter, Prometheus' chains, and Darwin's worms, as well as the not-quite-bodies of electricity, ingested food, and stem cells."
70. Boston Female Society, *Brief Account*, 1.
71. Boston Female Society, 8.
72. [Report of the] *Penitent Female's Refuge*, 9–12.
73. See the broadside *Proceedings in the town of Worcester, on the subject of moral reform: At a large and very respectable meeting of citizens holden in the Union Church, on Sunday, July 31st* [1836] (American Antiquarian Society BDSDS, 1836).
74. [Report of the] *Penitent Female's Refuge*, 1–2.
75. Hobson, *Uneasy Virtue*, 18.
76. Hobson, 18.
77. *Constitution of Penitent Female's Refuge*.
78. *Constitution of Penitent Female's Refuge*, 7–8.
79. *Affecting Narrative of Louisa Baker*, 17.
80. Boston Female Society, *Brief Account*, 5.
81. It should also be noted that this rhetorical rendering-foreign of black Bostonians works in tandem with the indigenization of white, ostensibly Euro-American settlers as "natives" of the area, a trend I discuss at length in chapter 4. Indeed, the full title of the 1816 edition of *An Affecting Narrative of Louisa Baker* is actually *An Affecting Narrative of Louisa Baker, a Native of Massachusetts*; the *Female Marine* texts work in step with a series of white literary nativisms that were characteristic of popular writing in this era. Yet, at the beginning of the nineteenth century, both Boston and the Massachusetts cape were home to a large population of Wampanoag people. We can thus read these missionary efforts on the part of the Boston Female Society for Missionary Purposes as also deploying white nativisms to tighten the boundary between white and nonwhite peoples in the city. See O'Brien, *Firsting and Lasting*. Thanks to J. Kēhaulani Kauanui for helping me parse this representative tension.

82. Boston Female Society, *Brief Account*, 6.
83. Boston Female Society, 8.
84. Sweet, *Bodies Politic*, 287–288.

Epilogue. Thinking Sex—without the Subject

1. Here I am primarily addressing the work of Giorgio Agamben, his *Homo Sacer: Sovereign Power and Bare Life*, trans. Daniel Heller-Roazan (Palo Alto, CA: Stanford University Press, 1998); and *State of Exception*, trans. Kevin Attell (Chicago: University of Chicago Press, 2005); and the work of Michel Foucault, in his *"Society Must Be Defended"*; *Birth of Biopolitics*; and *History of Sexuality*, vol. 1. Additionally, Mbembe's "Necropolitics" is increasingly canonical among scholars considering the impact of the emergence of biopolitical governance, specifically as a corrective to Foucault. For a succinct account of Foucault's argument about the emergence of biopolitics, see E. Cohen, *Body Worth Defending*; Cohen argues:

> Appearing after the disciplines, biopolitics emerges in the eighteenth century as a regulatory ensemble that both constitutes and conditions a new aggregate form of life: population. One of a series of modern abstractions that hypostatize the regularities of collective living and discern quasi-natural laws within them (e.g., the economy, society, human nature), population conceives the individual lives of national subjects as units belonging to a more encompassing vital domain which the state now recognizes as a valuable resource for its own ends. The overinvestment in life at the levels of the natural body (discipline) and of the population (biopolitics) constitutes the new regime that Foucault names "biopower." (20)

2. E. Cohen, *Body Worth Defending*, 22.
3. See E. Cohen, introduction to *Body Worth Defending*.
4. To be clear, there are endless examples of scholarship in queer studies that think carefully about racialization and endless examples of scholarship in critical ethnic studies that think carefully about gender and sexuality. But it is simultaneously very easy to find critiques of each of these fields for their lack of attention to the intersections of race, gender, and sexuality. See, for example, the special issue of *Social Text* 84–85 (2005), "What's Queer about Queer Studies Now?," especially the critique in the introduction of the "Gay Shame" conference, hosted by the University of Michigan in 2003. Also see Sharon Holland, *The Erotic Life of Racism* (Durham, NC: Duke University Press, 2012); and R. Ferguson, *Aberrations in Black*.
5. See Eve Kosofsky Sedgwick, "Axiom 5" of the essay "Axiomatic," in *Epistemology of the Closet* (Berkeley: University of California Press, 2008), 44; and Peter Coviello, "Introduction: The Unspeakable Past," in *Tomorrow's Parties*.
6. On blackness and the down low, see C. Riley Snorton, *Nobody Is Supposed to Know: Black Sexuality on the Down Low* (Minneapolis: University of Minnesota Press, 2014).
7. E. Cohen, *Body Worth Defending*, 20.
8. There have since been a vast range of efforts aimed at correcting the lack of attention to race and colonialism in studies of biopower, among them Mbembe, "Necropolitics"; Weheliye, *Habeas Viscus*; and Puar, *Terrorist Assemblages*, to name only a very few.
9. That this is so is the subject of a recent book, Bashford and Chaplin. *New Worlds of Malthus*.

10. As of this writing, scholarship on the racial politics of sexology in the nineteenth century is rather limited but growing fast. Somerville's *Queering the Color Line* is the only monograph dedicated to this question, but work by scholars such as Omnia El-Shakry, Howard Chiang, Kadji Amin, Durba Mitra, and Zohar Weiss-Berman is expanding on the relationship between sexology, racialization, and colonialism.

11. Godwin, *Enquiry Concerning Political Justice;* and Condorcet, *Sketch for a Historical Picture.* See also Franklin, *Observations concerning increase of mankind.*

12. See Foucault, *History of Sexuality;* "Society Must Be Defended"; *Security, Territory, Population;* and *Birth of Biopolitics.* Also see Engelmann, "Queer Utilitarianism."

13. McLane, "Malthus Our Contemporary?"

14. Klyve, "Darwin, Malthus, Sussmilch, and Euler."

15. Malthus, *Essay on Principle of Population,* 12.

16. Malthus, 17.

17. See chapter 1 for a discussion of stadial theory or four-stages theory, especially Curran, *Anatomy of Blackness;* Nussbaum, *Torrid Zones;* Buffon, *Histoire Naturelle* (1750); J. Brown, *Estimate of the Manners;* Millar, *Observations* (1773).

18. Malthus, *Essay on Principle of Population,* 12.

19. Jefferson, *Notes on state of Virginia,* 80. This copy, in the Huntington Library, is missing the title page and the attending copyright information. The translation of Jefferson's reproduction of the Buffon passage is mine, and I offer great thanks to Sara Phenix at Brigham Young University for her help and oversight.

20. Malthus, *Essay on Principle of Population,* 12.

21. Jefferson, *Notes on state of Virginia,* 80.

22. On January 29, 1804, Jefferson wrote to his friend Dr. Joseph Priestley Washington that he had recently read Malthus's *Essay:*

> Have you seen the new work of Malthus on population? It is one of the ablest I have ever seen. Altho' his main object is to delineate the effects of redundancy of population, and to test the poor laws of England, & other palliations for that evil, several important questions in political economy, allied to his subject incidentally, are treated with a masterly hand. It is a single 4'to. volume, and I have been only able to read a borrowed copy, the only one I have yet heard of. Probably our friends in England will think of you, & give you an opportunity of reading it. (www.let.rug.nl/usa/presidents/thomas-jefferson/letters-of-thomas-jefferson/jefl161.php, accessed March 18, 2018)

See also Hoff, *State and the Stork;* Hodgson, "Malthus' Essay on Population"; and McCoy, "Jefferson and Madison on Malthus."

23. See Somerville, *Queering the Color Line;* Terry, *American Obsession;* and Duggan, *Sapphic Slashers.*

24. Franklin writes:

> The negroes brought into the English Sugar Islands have greatly diminished the whites there; the poor are by this means depriv'd of employment, while a few families acquire vast Estates, which they spend on foreign luxuries, and educating their children in the habit of those luxuries, the same Income is needed for the support of one that might have maintain'd one hundred. The Whites who have slaves, not

labouring, are enfeebled, and therefore not so generally prolific; the slaves being work'd too hard, and ill fed, their constitutions are broken, and the deaths among them are more than the births; so that a continual supply is needed from Africa. The Northern Colonies having few slaves increase in Whites. Slaves also pejorate the Families that use them; the white children become proud, disgusted with labour, and being educated in idleness, are rendered unfit to get a Living by industry. (*Observations concerning increase of mankind*, 7)

25. Nelson, "Making Men," 1393.
26. Malthus states this in no uncertain terms: "I think I may fairly make two postulates. First, that food is necessary to the existence of man. Secondly, That the passion between the sexes is necessary and will remain nearly in its present state" (*Essay on Principle of Population*, 4). See also McLane, "Malthus Our Contemporary?," 342–343, who refers to Malthus as "a sexual theorist worth taking seriously" (338).
27. See Burgett, "Between Speculation and Population." See also Finger, *Contagious City*.
28. Foucault, *History of Sexuality*, 139.
29. E. Cohen, *Body Worth Defending*, 18, emphasis mine.
30. E. Cohen, 10, 9.
31. E. Cohen, 9–10.
32. See Sedgwick, "Axiomatic"; Michael Warner's coinage of "heteronormative," in "Introduction: Fear of a Queer Planet," *Social Text* 29 (1991): 3–17; Lisa Duggan's coinage of "homonormative" in *The Twilight of Equality? Neoliberalism, Cultural Politics, and the Attack on Democracy* (New York: Beacon Press, 2003); Lisa Duggan and José Esteban Muñoz, "Hope and Hopelessness: A Dialogue," *Women & Performance* 19, no. 2 (2009): 275–283; and Puar, *Terrorist Assemblages*.
33. Gayatri Chakravorty Spivak, "Can the Subaltern Speak?"
34. On black feminist and Afro-pessimist critiques of both the subject and the human, see Spillers, "Mama's Baby, Papa's Maybe"; N. Mitchell, "On Afropessimism"; Weheliye, *Habeas Viscus;* Wynter, "Unsettling Being/Power/Truth/Freedom"; and Moten, "Case of Blackness." On the antisocial turn in queer theory, see Edelman, *No Future;* and Caserio et al., "Antisocial Thesis in Queer Theory."
35. Bersani, "Is the Rectum a Grave?"
36. Importantly, Foucault may not believe in the self-shattering potential of sex at all.
37. Bersani, 218
38. Bersani, 221, emphasis mine.

BIBLIOGRAPHY

The Adventures of Lucy Brewer (Alias) Louisa Baker. Boston: Printed by H. Trumbull, 1815. Available in the collections of the American Antiquarian Society, Worcester, MA; Shaw and Shoemaker 36515.
An Affecting Narrative of Louisa Baker, a Native of Massachusetts. Boston: Printed by Nathaniel Coverly; reprint, New York: John Low, no. 62 Vesey-Street, and Sold by Him Wholesale and Retail, 1816.
Agamben, Giorgio. *Homo Sacer: Sovereign Power and Bare Life.* Palo Alto: Stanford University Press, 1998.
———. *State of Exception.* Chicago: University of Chicago Press, 2005.
Ahmed, Sara. *Queer Phenomenology: Orientations, Objects, Others.* Durham, NC: Duke University Press, 2006.
Allewaert, Monique. *Ariel's Ecology: Plantations, Personhood, and Colonialism in the American Tropics.* Minneapolis: University of Minnesota Press, 2013.
Andrew, Donna T. *Philanthropy and Police: London Charity in the Eighteenth Century.* Princeton, NJ: Princeton University Press, 1991.
Andrews, William L. *To Tell a Free Story: The First Century of Afro-American Autobiography, 1760–1865.* Urbana: University of Illinois Press, 1986.
Anzaldua, Gloria, and Cherrie Moraga, eds. *This Bridge Called My Back: Writings by Radical Women of Color.* New York: Kitchen Table / Women of Color Press, 1983.
Aravamudan, Srinivas. *Enlightenment Orientalism: Resisting the Rise of the Novel.* Chicago: University of Chicago Press, 2011.
Ariès, Philippe. *Centuries of Childhood: A Social History of Family Life.* Translated by Robert Baldick. New York: Vintage, 1965.
Arondekar, Anjali. *For the Record: On Sexuality and the Colonial Archive in India.* Durham, NC: Duke University Press, 2009.
Baepler, Paul. "The Barbary Captivity Narrative in Early America." *Early American Literature* 30, no. 2 (1995): 95–120.
———. *White Slaves, African Masters: An Anthology of American Barbary Captivity Narratives.* Chicago: University of Chicago Press, 1999.
Bahktin, Mikhail. *Rabelais and His World.* Translated by Hélène Iswolsky. Bloomington: Indiana University Press, 2009.
Bancroft, Aaron. *The importance of a religious education illustrated and enforced.* A sermon printed at Worcester, MA: Isaiah Thomas; sold at the Worcester bookstore, 1793.

Bashford, Alison, and Joyce E. Chaplin. *The New Worlds of Thomas Robert Malthus: Rereading "The Principle of Population."* Princeton, NJ: Princeton University Press, 2016.
Belcher, Joseph. *The Worst Enemy Conquered. A Brief Discourse On the Methods and Motives to pursue A Victory Over those Habits of Sin, Which War against the Soul.* Delivered on June 6, 1698, at Boston. Boston: Bartholomew Green and John Allen, 1698.
Bennett, Jane. *Vibrant Matter: A Political Ecology of Things.* Durham, NC: Duke University Press, 2010.
Berman, Jacob Rama. *American Arabesque: Arabs and Islam in the Nineteenth-Century Imaginary.* New York: New York University Press, 2012.
Bersani, Leo. "Is the Rectum a Grave?" *October* 43, special issue: "AIDS: Cultural Analysis/Cultural Activism" (Winter 1987): 197–222.
Beverland, Hadriaan. *De Fornicatione Cavenda Admonitio.* Augusta, Italy, 1697.
Big Dick, the King of the Negroes; or, Virtue and Vice Contrasted: A Romance of High & Low Life in Boston. Boston: N.p., 1849.
Binhammer, Katherine. "The Sex Panic of the 1790s." *Journal of the History of Sexuality* 6, no. 3 (January 1996): 409–434.
Blackstone, William. *Commentaries on the Laws of England.* Oxford: Printed at the Clarendon Press, 1765.
Bleys, Rudi. *Geography of Perversion: Male-to-Male Sexual Behavior outside the West and the Ethnographic Imagination, 1750–1918.* New York: New York University Press, 1996.
Blum, Hester. *The View from the Masthead: Maritime Imagination and Antebellum American Sea Narratives.* Chapel Hill: University of North Carolina Press, 2008.
Boone, Joseph Allen. *The Homoerotics of Orientalism.* New York: Columbia University Press, 2014.
Boston Female Society for Missionary Purposes. *A Brief Account of the Origin and Progress of the Boston Female Society for Missionary Purposes: With Extracts from the Reports of the Society, in May, 1817 and 1818, and Extracts from the Reports of Their Missionaries, Rev. James Davis, and Rev. Dudley D. Rosseter.* Boston: Printed by Lincoln & Edmands, no. 53 Cornhill, [1818].
Boswell, James. *The Life of Samuel Johnson.* 2 vols. London: Printed by Henry Baldwin, for Charles Dilly, in the Poultry, 1791.
Boswell, John. *Christianity, Social Tolerance, and Homosexuality: Gay People in Western Europe from the Beginning of the Christian Era to the Fourteenth Century.* Chicago: University of Chicago Press, 1981.
Bouanani, Ali. "Propaganda for Empire: Barbary Captivity Literature in the US." *Journal of Transatlantic Studies* 7, no. 4 (2009): 399–412.
Bourdieu, Pierre. *The Field of Cultural Production.* New York: Columbia University Press, 1993.
Branson, Susan. *These Fiery Frenchified Dames: Women and Political Culture in Early Philadelphia.* Philadelphia: University of Pennsylvania Press, 2001.
Braudel, Fernand. *The Mediterranean and the Mediterranean World in the Age of Philip II.* New York: Harper & Row, 1972.
Bray, Alan. *Homosexuality in Renaissance England.* New York: Columbia University Press, 1996.
Brevitt, Joseph. *The History of anatomy, from Hippocrates, who lived four hundred years before Christ.* Baltimore: Printed by Samuel Sower, in Fayette-Street, [1799].

Brooks, Francis. *Barbarian cruelty. Being a true history of the distressed condition of the Christian capitol under the tyranny of Mully Ishmael Emperor of Morocco, and King of Fez and Macqueness in Barbary.* . . . Reprint, Boston: S. Phillips, at the brick shop, 1700.

Brown, John. *An Estimate of the Manners and Principles of the Times.* London, Printed; Boston, New-England: Reprinted and sold by Green and Russell, at their Printing-Office in Queen Street, 1758.

Brown, William Hill. *The Power of Sympathy.* New York: Penguin Classics, 1996. Original publication 1789.

Browne, Arthur. *Religious Education of children recommended . . . Being the day appointed for the execution of Penelope Kenny.* Boston: Printed and sold by S. Kneeland and T. Green, in Queenstreet over against the Prison, 1739.

Browne, Janet. "Botany in the Boudoir and Garden: The Banksian Context." In *Visions of Empire: Voyages, Botany, and Representations of Nature*, edited by David Philip Miller and Peter Hanns Reill, 153–172. Cambridge: Cambridge University Press, 2011.

Browne, Simone. *Dark Matters: On the Surveillance of Blackness.* Durham, NC: Duke University Press, 2015.

Buffon, Georges-Louis Leclerc, le Comte de. *Natural History, General and Particular.* Translated by William Smellie. Vol. 3. London: Printed for A. Strathan and T. Cadell in the Strand, 1791.

Bullough, Vern. *Science in the Bedroom: A History of Sex Research.* New York: Basic Books, 1994.

Burgett, Bruce. "Between Speculation and Population: The Problem of 'Sex' in Our Long Eighteenth Century." *Early American Literature* 37, no. 1 (2002): 119–153.

———. "The History of X in Early America." *Early American Literature* 44, no. 1 (2009): 215–225.

Burke, Peter. *Popular Culture in Early Modern Europe.* 1978. Reprint, Farnham, UK: Ashgate, 2009.

Butler, Judith. "Imitation and Gender Insubordination." In *The Judith Butler Reader*, edited by Sarah Salih and Judith Butler, 119–137. Malden, MA: Blackwell, 2004.

Butt Ender [pseud.]. *Prostitution Exposed; or, a Moral Reform Directory Laying Bare the Lives, Histories, Residences, Seductions, &c. of the Most Celebrated Courtezans and Ladies of Pleasure in the City of New York.* New York: Public Convenience, 1839.

Carew, Richard. *The Examination of Men's Wits.* Edited by Rocio Sumillera. Cambridge: Modern Humanities Research Association, 2014.

Case of Young Robinson. Boston: Printed at the Office of the Mercantile Journal, 1836. Original publication in the *Boston Mercantile Journal*.

Caserio, Robert L., Lee Edelman, Judith Halberstam, José Esteban Muñoz, and Tim Dean. "The Antisocial Thesis in Queer Theory." *PMLA* 121, no. 3 (2006): 819–828.

Chartier, Roger. *The Cultural Uses of Print in Early Modern France.* Translated by L. C. Cochrane. Princeton, NJ: Princeton University Press, 1987.

Chauncey, George. *Gay New York: Gender, Urban Culture, and the Making of the Gay Male World, 1890–1940.* Reprint, New York: Basic Books, 1995.

Chiles, Katy. *Transformable Race: Surprising Metamorphoses in the Literature of Early America.* Oxford: Oxford University Press, 2014.

Cocks, H. G. *Visions of Sodom: Religion, Homoerotic Desire, and the End of the World in England, c. 1550–1850.* Chicago: University of Chicago Press, 2017.

Cohen, Daniel, ed. *The Female Marine and Related Works: Narratives of Cross-Dressing and Urban Vice in America's Early Republic*. Amherst: University of Massachusetts Press, 1997.

———. *Pillars of Salt, Monuments of Grace: New England Crime Literature and the Origins of American Popular Culture, 1674–1860*. Amherst: University of Massachusetts Press, 2006.

Cohen, Ed. *A Body Worth Defending: Immunity, Biopolitics, and the Apotheosis of the Modern Body*. Durham, NC: Duke University Press, 2009.

Coke, Edward. "Of Buggery, or Sodomy." In *The Third Part of the Institutes of the Laws of England*. London: M. Flesher, for W. Lee and D. Pakeman, 1644.

A Collection of Voyages and Travels . . . Compiled from the curious and valuable Library of the late Earl of Oxford, vol. 1. London: Printed for and sold by Thomas Osborne of Gray's-Inn, 1745.

Colley, Linda. *Captives: Britain, Empire, and the World, 1600–1850*. New York: Anchor Books, 2004.

Collins, Patricia Hill. *Black Sexual Politics: African Americans, Gender, and the New Racism*. New York: Routledge, 2005

Comment, Kristin. "Charles Brockden Brown's *Ormond* and Lesbian Possibility in the Early Republic." *American Literature* 40, no. 1 (2005): 57–78.

Condorcet, Marquis de. *Sketch for a Historical Picture of the Progress of the Human Mind*. London: Printed for J. Johnson, [1795].

Constitution of the Society and Directors of the Penitent Female's Refuge: Adopted April 14, 1819. [Boston] Yankee Office. Boston: Printed by True & Weston, 1819.

Coviello, Peter. *Tomorrow's Parties: Sex and the Untimely in Nineteenth Century America*. New York: New York University Press, 2013.

Crenshaw, Kimberlé, Neil Gotanda, Gary Pellar, and Kendall Thomas, eds. *Critical Race Theory*. New York: New Press, 1995.

Crèvecœur, J. Hector St. John de. *Letters from an American Farmer*. Oxford: Oxford World's Classics, 2009.

Crewe, Jonathan. "Disorderly Love: Sodomy Revisited in Marlowe's *Edward II*." *Criticism* 51, no. 3 (2009): 385–399.

Cronon, William. "A Place for Stories: Nature, History, and Narrative." *Journal of American History* 78, no. 4 (March 1992): 1347–1376.

Cullen, Jim, ed. *Popular Culture in American History*. New York: Blackwell, 2000.

Curran, Andrew. *Anatomy of Blackness: Science and Slavery in an Age of Enlightenment*. Baltimore: Johns Hopkins University Press, 2011.

[Daggett, David]. *Sketches of the Life of Joseph Mountain*. Hartford, CT: Patten, 1790. Available at the American Antiquarian Society, Worcester, MA; item 45855.

[———]. *Sketches of the Life of Joseph Mountain*. Norwich, CT: Trumbull, 1790. Available at the New York Historical Society, New York; item 45856.

Dain, Bruce R. *A Hideous Monster of the Mind: American Race Theory in the Early Republic*. Cambridge, MA: Harvard University Press, 2002.

Dan, Pierre. *Histoire de Barbarie, et de ses Corsaires*. Paris: Chez P. Rocolet, 1649.

Dana, James. *The Intent of Capital Punishment. A Discourse Delivered in the City of New-Haven, October 20, 1790. Being the day of Execution of Joseph Mountain for a RAPE.*

New Haven, CT: T & S Green, 1790. Available at the American Antiquarian Society, Worcester, MA; 22446.

Danforth, Samuel. *A brief recognition of New-Englands errand into the wilderness; made in the audience of the General Assembly of the Massachusets Colony, at Boston in N.E. on the 11th of the third month, 1670* [i.e., 1671]. *Being the day of election there.* Cambridge, MA: Samuel Green and Marmaduke Johnson, 1671.

——. *The Cry of Sodom Enquired Into.* Cambridge, MA: Marmaduke Johnson, 1674.

Darby, Robert. "Captivity and Captivation: Gullivers in Brobdingnag." *Eighteenth-Century Life* 27, no. 3 (2003): 124–139.

Darwin, Erasmus. *The Botanic Garden.* Lichfield, England: J. Jackson for J. Jonson, 1789.

——. *The botanic garden. A poem, in two parts. Part I. Containing the economy of vegetation. Part II. The loves of the plants. With philosophical notes.* New York: T. & J. Swords, printers to the faculty of physic of Columbia College, no. 99 Pearl-Street, 1798.

Daston, Lorraine. "The Empire of Observation, 1600–1800." In Daston and Lunbeck, *Histories of Scientific Observation*, 81–114.

Daston, Lorraine, and Elizabeth Lunbeck, eds. *Histories of Scientific Observation.* Chicago: University of Chicago Press, 2011.

Daston, Lorraine, and Glenn Most. "History of Science and History of Philologies." *Isis* 106, no. 2 (2015): 378–390.

Davidson, Cathy. *Revolution and the Word: The Rise of the Novel in America.* Oxford: Oxford University Press, 2004.

Davis, Angela. *Women, Race, and Class.* New York: Vintage, 1983

Davis, Thomas. "The Sack of Baltimore 1631," 1844. In *The Poems of Thomas Davis.* Dublin: James Duffy, 1846.

Davis, William. "The Travels and miserable Captivity of William Davis, Barber-Surgeon of London." In *Collection of Voyages and Travels*, 1:476–488. London: Printed for and sold by Thomas Osborne of Gray's-Inn, 1745. Originally published as *A true relation of the travailes and most miserable captiuitie of William Dauies, barber-surgion of London.* London: Thomas Snodham, 1614.

DeLombard, Jeannine Marie. *In the Shadow of the Gallows: Race, Crime, and American Civic Identity.* Philadelphia: University of Pennsylvania Press, 2012.

——. *Slavery on Trial: Law, Abolitionism, and Print Culture.* Chapel Hill: University of North Carolina Press, 2007.

A Description of the nature of slavery among the Moors, and the cruel sufferings of those that fall into it. London: J. Peele, at Lock's Head in Pater-noster Row, 1721.

Dimock, Wai Chee. *Residues of Justice: Literature, Law, Philosophy.* Berkeley: University of California Press, 1997.

Dugaw, Dianne. *Warrior Women and Popular Balladry, 1650–1850.* Chicago, University of Chicago Press, 1989.

Duggan, Lisa. *Sapphic Slashers: Sex, Violence, and American Modernity.* Durham, NC: Duke University Press, 2001.

——. *The Twilight of Equality: Neoliberalism, Cultural Politics, and the Attack on Democracy.* Boston: Beacon Press, 2004.

Duggan, Lisa, and José Esteban Muñoz. "Hope and Hopelessness: A Dialogue." *Women & Performance: a journal of feminist theory* 19, no. 2 (2010): 275–283.

Edelman, Lee. *No Future: Queer Theory and the Death Drive*. Durham, NC: Duke University Press, 2004.
Egan, Jim. *Oriental Shadows: The Presence of the East in Early American Literature*. Columbus: Ohio State University Press, 2011.
El-Rouayheb, Khaled. *Before Homosexuality in the Arab-Islamic World, 1500–1800*. Chicago: University of Chicago Press, 2009.
Eng, David, J. Halberstam, and José Esteban Muñoz. "Introduction" to "What's Queer about Queer Studies Now?" *Social Text* 23.3–4 (2005): 1–17.
Engelmann, Stephen G. "Queer Utilitarianism: Bentham and Malthus on the Threshold of Biopolitics." *Theory & Event* 17, no. 4 (2014). Available at Project Muse, muse.jhu.edu/article/562822.
Faderman, Lillian. *Surpassing the Love of Men: Romantic Friendship and Love between Women from the Renaissance to the Present*. New York: HarperCollins, 2001.
Fairchild-Craft, Catherine. "Sexual and Textual Indeterminacy: Eighteenth-Century English Representations of Sapphism." *Journal of the History of Sexuality* 15, no. 3 (2006): 408–431.
Falconer, William. *Remarks on the Influence of Climate*. London: Printed for C. Dilly, in the Poultry, 1781.
Farber, Paul Lawrence. *Finding Order in Nature: The Naturalist Tradition from Linnaeus to E. O. Wilson*. Baltimore: Johns Hopkins University Press, 2000.
Ferguson, Adam. *An Essay on the History of Civil Society*. Edinburgh: A. Millar & T. Caddel in the Strand, London; A. Kincard & J. Bell, 1767.
Ferguson, Roderick. *Aberrations in Black: Toward a Queer of Color Critique*. Minneapolis: University of Minnesota Press, 2003.
Finger, Simon. *The Contagious City: The Politics of Public Health in Early Philadelphia*. Ithaca, NY: Cornell University Press, 2012.
Fisher, Will. "Queer Money." *ELH* 66, no. 1 (1999): 1–23.
Forbes, Erin. "'Between the Human and Brutal Creation': Posthuman Agency and the Samuel Frost Corpus." *Eighteenth-Century: Theory and Interpretation* 57, no. 1 (2016): 39–69.
Foster, Travis. "After Emancipation: Genres of the New Racial Ordinary." Unpublished manuscript, revised November 2017. Microsoft Word files.
Foucault, Michel. *Abnormal: Lectures at the Collège de France, 1974–75*. Translated by Graham Burchell. Reprint, New York: Picador, 2003.
———. *The Birth of Biopolitics: Lectures at the Collège de France, 1978–1979*. Edited by Michel Senellart, François Ewald, and Allessandro Fontana. Translated by Graham Burchell. Reprint, New York: Picador, 2010.
———. *Discipline and Punish: The Birth of the Prison*. Translated by Alan Sheridan. New York: Vintage Books, 1978.
———. *The History of Sexuality*. Vol. 1. Reprint, New York: Vintage 1978.
———. *The Order of Things: An Archaeology of the Human Sciences*. New York: Vintage, 1994.
———. *Security, Territory, Population: Lectures at the Collège de France, 1977–1978*. Edited by Michel Senellart. Translated by Graham Burchell. Reprint, New York: Picador, 2007.
———. *"Society Must Be Defended": Lectures at the Collège de France, 1975–76*, edited

by Mauro Bertani, Allessandro Fontana, and François Ewald; translated by David Macey. Reprint, New York: Picador, 2003.

Franklin, Benjamin. *Observations concerning the increase of mankind, peopling of countries, &c*. Boston: Printed and sold by S. Kneeland in Queen Street, 1755.

Freud, Sigmund. *Three Essays on a Theory of Sexuality*. New York: Basic Books, 1962. Original publication 1905.

Friedlander, Ari. "Rogue Sexuality: The Erotics of Social Status in Early Modern England." Unpublished manuscript, November 2017. Microsoft Word files.

[Frost, Samuel.] *The Confession and Dying Words of Samuel Frost*. Broadside. Worcester, MA: Printed and sold at Mr. Thomas's Printing Office, in Worcester, 1793.

Gannett, Deborah Sampson. *Address, Delivered with Applause*. Dedham, MA: Mann, 1802.

Garcia, Humberto. *Islam and the English Enlightenment, 1670–1840*. Baltimore: Johns Hopkins University Press, 2011.

Gee, Joshua. *Narrative of Joshua Gee of Boston, Mass. While He Was Captive in Algeria of the Barbary Pirates, 1680–1687*. With an introduction by Albert Carlos Bates. Hartford, CT: Wadsworth Atheneum, 1943.

Gilfoyle, Timothy. *City of Eros: New York City, Prostitution, and the Commercialization of Sex, 1790–1920*. New York: Norton, 1994.

Gillespie, George. *A treatise against the deists or free-thinkers: proving the necesity of revealed religion*. Philadelphia. Printed for the author by A. Bradford, at the Sign of the Bible in Second-Street, 1735.

Godbeer, Richard. "The Cry of Sodom: Discourse, Intercourse, and Desire in Colonial New England." In *Long before Stonewall: Histories of Same-Sex Sexuality in Early America*, edited by Thomas A. Foster, 81–113. New York: New York University Press, 2007.

———. *Sexual Revolution in Early America*. Baltimore: Johns Hopkins University Press, 2002.

Godwin, William. *An Enquiry Concerning Political Justice and its Influence on General Virtue and Happiness*. Vol. 1. London: G. G. J. and J. Robinson, 1793.

Goldberg, David Theo. *The Threat of Race: Reflections on Racial Neoliberalism*. Oxford, UK: Blackwell, 2009.

Goldberg, Jonathan. *Queering the Renaissance*. Durham, NC: Duke University Press, 1993.

———, ed. *Reclaiming Sodom*. New York: Routledge, 1994.

———. *Sodometries: Renaissance Texts, Modern Sexualities*. Stanford, CA: Stanford University Press, 1992.

Gonda, Caroline, and Beynon, John. *Lesbian Dames: Sapphism in the Long Eighteenth Century*. Surrey: Ashgate, 2010.

Gossett, Thomas. *Race: The History of an Idea in America*. Dallas: Southern Methodist University Press, 1963.

Gould, Phillip. *Barbaric Traffic: Commerce and Antislavery in the Eighteenth-Century Atlantic World*. Cambridge, MA: Harvard University Press, 2009.

Gould, Stephen J. "The Geometer of Race." *Discover* 15 (1994). Available at Discover, discovermagazine.com/1994/nov/thegeometerofrac/441.

Grew, Nehemiah. *Anatomy of Plants*. London: Printed by W. Rawlins, for the author, 1682.

Griffiths, Devin. "The Intuitions of Analogy in Erasmus Darwin's Poetics." *Studies in English Literature, 1500–1900* 51, no. 3 (2011): 645–665.

Gustafson, Sandra. "The Genders of Nationalism: Patriotic Violence, Patriotic Sentiment in the Performances of Deborah Sampson Gannett." In *Possible Pasts: Becoming Colonial in Early America*, edited by R. Blair St. George, 380–399. Ithaca, NY: Cornell University Press, 2000.

Hallock, Thomas. "Male Pleasure and the Genders of Eighteenth Century Botanic Exchange: A Garden Tour." *William and Mary Quarterly* 62, no. 4 (2005): 697–718.

Halttunen, Karen. *Murder Most Foul: The Killer and the American Gothic Imagination.* Cambridge, MA: Harvard University Press, 2000.

Harris, Tim. "Problematizing Popular Culture." In *Popular Culture in England c. 1500–1850*, edited by Tim Harris, 1–27. New York: St. Martin's Press, 1995.

Hervey, James. *Meditations and contemplations. In two volumes. Containing, Vol. I. Meditations among the tombs. Reflections on a flower-garden; and, A descant on creation. Vol. II. Contemplations on the night. Contemplations on the starry heavens; and, A winter-piece.* Boston: New-England, Printed and sold by Daniel Fowle in Queen-Street and by Daniel Gookin in Marlborough-Street, 1750.

Hiltner, Judith. "'Like a bewildered star': Deborah Sampson, Herman Mann, and *Address, Delivered with Applause*." *Rhetoric Society Quarterly* 29, no. 2 (2000): 5–24.

———. "'She Bled in Secret': Deborah Sampson, Herman Mann, and 'The Female Review,'" *Early American Literature* 34, no. 2 (1999): 190–220.

Hirshman, Linda, and Jane Larson. *Hard Bargains: The Politics of Sex.* Oxford: Oxford University Press, 1998.

Hitchcock, Tim. "Redefining Sex in Eighteenth-Century England." *History Workshop Journal* 41 (Spring 1996): 72–90.

Hobson, Barbara Meil. *Uneasy Virtue: The Politics of Prostitution and the American Reform Tradition.* New York: Basic Books, 1987.

Hodgson, Dennis. "Malthus' Essay on Population and the American Debate over Slavery." *Comparative Studies in Society and History* 51, no. 4 (October 2009): 742–770.

Hoff, Derek. *The State and the Stork: The Population Debate and Population Making in US History.* Chicago: University of Chicago Press, 2012.

Holland, Sharon. *The Erotic Life of Racism.* Durham: Duke University Press, 2012.

Holmes, Thomas James. *Cotton Mather: A Bibliography of His Works.* 3 vols. Cambridge, MA: Harvard University Press, 1940.

Hopkins, Asa. *The Evils and Remedy of Lewdness: A Sermon, Delivered in the Bleeker-Street Church, Utica; and in the Second Presbyterian Church, Rome. March, 1834.* Utica, [NY]: Gardiner Tracy, printer, 1834.

Huarte, Juan. *Examen de ingenios. The Examination of Men's Wits . . . Translated out of the Spanish tongue by M Camillo Camilis. Englished out of the Italian by R.C. Esquire* [Richard Carew, Esq.]. London: Printed by Adam Islip, for Richard Watkins, 1594.

Hume, David. *An Enquiry concerning Human Understanding.* Harvard Classics, vol. 37, part 3. New York: P. F. Collier & Son, 1909–14.

Hunt, Lynn. "The Many Bodies of Marie Antoinette: Political Pornography and the Problem of the Feminine in the French Revolution." In *Eroticism and the Body Politic*, edited by Lynn Hunt, 108–130. Baltimore: Johns Hopkins University Press, 1991.

Iannini, Christopher. *Fatal Revolutions: Natural History, West Indian Slavery, and the Routes of American Literature*. Chapel Hill: University of North Carolina Press, 2012.

Isani, Mukhtar Ali. "Cotton Mather and the Orient." *New England Quarterly* 43, no. 1 (1970): 46–58.

I'tesamuddin, Mirza Sheikh. *The Wonders of Vilayet, Being the Memoir, Originally in Persian, of a Visit to France and Britain in 1765*. Translated by Kaiser Haq. Leeds: People Tree Press, 2002.

Iyengar, Sujata. *Shades of Difference: Mythologies of Skin Color in Early Modern England*. Philadelphia: University of Pennsylvania Press, 2005.

Jacobson, Matt. *Whiteness of a Different Color: European Immigrants and the Alchemy of Race*. Cambridge: Harvard University Press, 1999.

Jagose, Annamarie. *Orgasmology*. Durham, NC: Duke University Press, 2012.

Jefferson, Thomas. *Notes on the State of Virginia*. London: Printed for John Stockdale, Opposite Burlington-House, Piccadilly, 1782.

———. *Notes on the state of Virginia; : written in the year 1781, somewhat corrected and enlarged in the winter of 1782, for the use of a foreigner of distinction, in answer to certain queries proposed by him respecting 1. Its boundaries. 2. Rivers. 3. Sea port*. [Paris], 1782 [i.e., 1785]. Available at the Huntington Library, San Marino, CA, RB 492811.

Johnson, E. Patrick, and Mae Henderson, eds. *Black Queer Studies: A Critical Anthology*. Durham, NC: Duke University Press, 2005.

Jordan, Mark. *The Invention of Sodomy in Christian Theology*. Chicago: University of Chicago Press, 1997.

Jordan, Winthrop. *White over Black: American Attitudes toward the Negro, 1550–1812*. Chapel Hill: Published for the Omohundro Institute for Early American History and Culture by the University of North Carolina Press, 1968.

Kahan, Benjamin, ed. *Heinrich Kaan's Psychopathia Sexualis (1844): A Classic Text in the History of Sexuality*. Ithaca, NY: Cornell University Press, 2016.

———. "Toward a World-System of Sexuality: Queer Weather and the Macro Environments of Sex." Invited speech, Humanities Center, University of Pittsburgh, January 2012.

Kerber, Linda. *Women of the Republic: Intellect and Ideology in Revolutionary America*. Chapel Hill: Published for the Omohundro Institute for Early American History and Culture by the University of North Carolina Press, 1997.

Kidd, Colin. *The Forging of Races: Race and Scripture in the Protestant Atlantic World, 1600–2000*. Cambridge: Cambridge University Press, 2006.

Klyve, Dominic. "Darwin, Malthus, Sussmilch, and Euler: The Ultimate Origin of the Motivation for the Theory of Natural Selection." *Journal of the History of Biology* 47 (2012): 189–212.

Lanser, Susan. "Mapping Sapphic Modernity." In *Comparatively Queer: Interrogating Identities across Time and Cultures*, edited by Jarrod Hayes, Margaret R. Higonnet, and William J. Spurlin, 69–89. London: Palgrave Macmillan, 2010.

———. *The Sexuality of History: Modernity and the Sapphic, 1565–1830*. Chicago: University of Chicago Press, 2014.

———. "Tory Lesbians: Economies of Intimacy and the Status of Desire." In *Lesbian*

Dames: Sapphism in the Long Eighteenth Century, edited by Caroline Gonda and John C. Beynon, 173–190. London: Routledge, 2010.

Laqueur, Thomas. *Solitary Sex: A Cultural History of Masturbation.* New York: Zone Books, 2004.

Lee, James. *An Introduction to Botany . . . Extracted from the Works of Dr. Linnaeus.* London: Printed for J. and R. Tonson, 1760.

Leslie, Charles. *The religion of Jesus Christ the only true religion, or, A short and easie method with the deists.* Boston: Printed by T. Fleet, and to be sold by John Checkley, at the Sign of the Crown and Blue Gate over against the west end of the Town-House, 1719.

Library Company of Philadelphia. *A catalogue of the books, belonging to the Library Company of Philadelphia; to which is prefixed, a short account of the institution, with the charter, laws and regulations.* Philadelphia: Printed by Zachariah Poulson, Junior, in Fourth-Street, between Market-Street and Arch-Street, [1789].

Lichfield Botanical Society, trans. *The families of plants.* Lichfield: printed by John Jackson. Sold by J. Johnson, St. Paul's Church-yard, London; T Byrne, Dublin, and J Balfour, Edinburgh, 1787.

———, trans. *A System of Vegetables.* Lichfield: printed by John Jackson, for Leigh and Sotheby, York Street, Covent Garden, London, [1783].

Linnaeus, Carolus [Carl von Linné]. *The Families of Plants, with their natural characters.* 2 vols. Lichfield: printed by John Jackson. Sold by J. Johnson, St. Paul's Church-yard, London; T. Byrne, Dublin, and J. Balfour, Edenburgh [sic], 1787.

——— [C. Linnæus]. *Genera plantarum.* Leiden, Netherlands: Printed by [Conrad Wishoff], 1737.

——— [Carl von Linné]. *A General System of Nature.* Translated by William Turton. London: Printed for Lackington, Allen, and Co., Temple of the Muses, Finsbury Square, 1801–2.

——— [Caroli Linnaei]. *Systema Naturae, Per Regna Tria Naturae. . . .* 10th ed., revised. Holmiae: Impensis Direct, Laurentii Salvi, 1758.

——— [C. Linnæus]. *Systema naturæ per regna tria naturæ, secundum classes, ordines, genera, species, cum characteribus, differentiis, synonymis, locis.* Vol. 1. 10th ed., revised. Holmiæ: (Salvius) Stockholm, 1758.

——— [C. Linnæus]. *Systema naturæ, sive regna tria naturæ systematice proposita per classes, ordines, genera, & species.* Lugduni Batavorum (Haak). Leiden, Netherlands, 1735.

Lorde, Audre. *Sister Outsider: Essays and Speeches.* New York: Crossing Press, 1984.

Love, Heather. "Introduction: Modernism at Night." *PMLA* 124, no. 3 (May 2009): 744–748.

———. "'Spoiled Identity': Stephen Gordon's Loneliness and the Difficulties of Queer History." *Gay and Lesbian Quarterly* 7, no. 4 (2001): 487–519.

Lyons, Clare. *Sex among the Rabble: An Intimate History of Gender and Power in the Age of Revolution, Philadelphia, 1730–1830.* Chapel Hill: Published for the Omohundro Institute for Early American History and Culture by the University of North Carolina Press, 2006.

Mahar, William John. *Behind the Burnt Cork Mask: Early Blackface Minstrelsy and Antebellum American Popular Culture.* Urbana: University of Illinois Press, 1999.

Malthus, Thomas Robert. *An Essay on the Principle of Population.* First printed for J. Johnson, in St. Paul's Church-Yard, London, 1798.

Mann, Herman. *The Columbian Primer.* Dedham [MA]: Printed & sold by H. Mann, 1802. Published according to act of Congress. Sold also by the booksellers in Boston, Salem, Worcester, Portsmouth, (NH) Hartford, (CT), Providence (RI), and at other principal places in the United States, [1802]

———. *The Female Review, or Memoirs of an American Young Lady.* Dedham [MA]: Printed by Nathaniel and Benjamin Heaton, for the author, 1797.

———. *The Material Creation: Being a Compendious System of Universal Geography and Popular Astronomy.* Dedham, MA: Herman Mann, 1818.

Marr, Timothy. *The Cultural Roots of American Islamicism.* Cambridge: Cambridge University Press, 2006.

Masur, Louis P. *Rites of Execution: Capital Punishment and the Transformation of American Culture, 1776–1865.* Oxford: Oxford University Press, 1991.

Matar, Nabil. *British Captives from the Mediterranean to the Atlantic, 1563–1760.* Leiden, Netherlands: Koninklijke Brill, 2014.

Mather, Cotton. *The Angel of Bethesda.* Edited by Gordon W. Gones. Barre, MA: American Antiquarian Society; Barre, 1972.

———. *The angel of Bethesda, visiting the invalids of a miserable world.* New London, CT: Printed and sold by Timothy Green, 1722.

———. *The Glory of Goodness. The Goodness of God Celebrated.* Boston: Printed by Timothy Green, for Benjamin Eliot, 1703.

———. *The Negro Christianized: An Essay to Excite and Assist that Good Work, the Instruction of Negro-Servants in Christianity.* Boston: Printed by B. Green, 1706.

———. *A Pastoral Letter to the English Captives in Africa. From New-England.* Boston: Printed by B. Green and J. Allen, 1698.

———. *Pillars of Salt. An History of Some Criminals Executed in this Land, for Capital Crimes.* Boston: Printed by B. Green and J. Allen, for Samuel Phillips at the Brick Shop near the Old Meeting House, 1699.

———. *The pure Nazarite. Advice to a young man. . . .* Boston: Printed by T. Fleet, for John Phillips, at his shop on the south side of the Town-House, 1723.

———. *Reason satisfied: and faith established. The resurrection of a glorious Jesus demonstrated by many infallible proofs.* Boston: Printed by J. Allen, for N. Boone, at the sign of the Bible, 1712.

———. *Tremenda. The dreadful sound with which the wicked are to be thunderstruck.* Boston: Printed by B. Green, for B. Gray & J. Edwards, & sold at their shops, 1721.

Mather, Increase. *A sermon occasioned by the execution of a man found guilty of murder preached at Boston in N.E. March 11th 1685[]6 The second Edition.* Boston: Printed by R.P. [i.e., Richard Pierce]; sold by J. Brunning book-seller, at his shop at the corner of the Prison-Lane next the Exchange, Anno, 1687.

Mavor, William Fordyce. *A Complete System of Natural History, Founded on the Linnæan Arrangement of Animals; with Popular Descriptions in the Manner of Goldsmith and Buffon. Illustrated by fourty-four Copper-Plates, Representing upwards of One Hundred and Seventy of the Most Curious Objects.* Philadelphia: Printed by P. Byrne, 72, Chesnut-Street, 1802.

Mbembe, Achille. "Necropolitics." Translated by Libby Mjientes. *Public Culture* 15, no. 1 (2003): 11–40.
McCoy, Drew. "Jefferson and Madison on Malthus: Population Growth in Jeffersonian Political Economy." *Virginia Magazine of History and Biography* 88, no. 3 (July 1980): 259–276.
McKeon, Michael. *The Secret History of Domesticity: Public, Private, and the Division of Knowledge*. Baltimore: Johns Hopkins University Press, 2009.
McLane, M. N. "Malthus Our Contemporary?: Toward a Political Economy of Sex." *Studies in Romanticism* 52, no. 3 (2013): 337–362.
Meyers, Amy, and Margaret Pritchard, eds. *Empire's Nature: Mark Catesby's New World Vision*. Chapel Hill: Published for the Omohundro Institute of early American History and Culture by the University of North Carolina Press, 1998.
Millar, John. *Observations Concerning the Distinction of Ranks in Society*. 2nd ed. London: Printed for J. Murray, 1773. Original publication 1767.
Miller, Jaclyn. "The Body Politic and the Body Somatic." In *Centre of Wonders: The Body in Early America*, edited by Janet Moore Lindman and Michele Lise Tarter, 61–76. Ithaca, NY: Cornell University Press, 2001.
Milne, Colin. *Institutes of Botany . . . translated from the Latin of the celebrated Charles von Linné. . . .* London: sold by W. Griffin, Bookseller, Catharine-Street; J Nourse, Bookseller to His Majesty; P. Elmsley, opposite Southampton-Street; Messrs. Richardson and Urquart, under the Royal Exchange; F. Noble, opposite Gray's-Inn Gate, Holborn; and J. Robson, New-Bond-Street, 1771.
Miranda, Deborah. "The Extermination of the Joyas: Gendercide in Spanish California." *Gay and Lesbian Quarterly* 16, no. 1 (2010): 253–284.
Mitchell, John. *An Essay Upon the Causes of the different Colours of People in Different Climates. Philosophical Transactions* 44 (1744–1745): 223–271. *Giving Some Account of the Present Undertakings, Studies, and Labours of the Ingenious in Many Considerable Parts of the World, Vol 43 for the years 1744 and 1745*. London: Printed for C. Davis, Printer to the Royal Society, Over-against *Gray's-Inn-Gate* in Holburn, 1744.
Mitchell, Nick. "On Afropessimism; or, The People Critique Makes." Unpublished manuscript, December 2017. Microsoft Word file.
Moore, Lisa. *Dangerous Intimacies: Toward a Sapphic History of the British Novel*. Durham, NC: Duke University Press, 1997.
——. "Queer Gardens: Mary Delany's Flowers and Friendships." *Eighteenth-Century Studies* 39, no. 1 (Fall 2005): 49–70.
Morgan, Jennifer L. "'Some Could Suckle over Their Shoulder': Male Travelers, Female Bodies, and the Gendering of Racial Ideology, 1500–1770." *William and Mary Quarterly* 54, no. 1 (January 1997): 167–192.
Morgensen, Scott Lauria. "The Biopolitics of Settler Colonialism: Right Here, Right Now." *Settler Colonial Studies* 1, no. 1 (2011): 52–76.
Moster, Aviva, Dorota Wnuk, and Elizabeth Jeglic. "Cognitive Behavioral Therapy Interventions with Sex Offenders." *Journal of Correctional Health Care* 14, no. 2 (2008): 109–121.
Moten, Fred. "The Case of Blackness." *Criticism* 50, no. 2 (2008): 177–218.
Mountain, Joseph. *Sketches of the Life of Joseph Mountain*. New Haven, CT: Printed and

sold by T&S Green, 1790. Available from the Connecticut Historical Society, Hartford, CT; 22441, 19 pages.

Myles, Anne. "Queering the Study of Early American Sexuality." *William and Mary Quarterly* 60, no. 1 (2003): 199–202.

Nelson, William Max. "Making Men: Enlightenment Ideas of Racial Engineering." *American Historical Review* 115, no. 5 (December 2010): 1364–1394.

Neuman, R. P. "Masturbation, Madness, and the Modern Concepts of Childhood and Adolescence." *Journal of Social History* 8, no. 3 (1975): 1–27.

New York Magdalen Society. *Magdalen Report: Annual Report of the Executive Committee of the N.Y. Magdalen Society*. New York: New York Magdalen Society, 1831.

Nocentelli, Carmen. *Empires of Love: Europe, Asia, and the Making of Early Modern Identity*. Philadelphia: University of Pennsylvania Press, 2013.

Norton, Mary Beth. *Liberty's Daughters: The Revolutionary Experience of American Women, 1750–1900*. Ithaca, NY: Cornell University Press, 1996.

Nussbaum, Felicity. *Torrid Zones: Maternity, Sexuality, and Empire in Eighteenth-Century English Narratives*. Baltimore: Johns Hopkins University Press, 1995.

O'Brien, Jean M. *Firsting and Lasting: Writing Indians out of Existence in New England*. Minneapolis: University of Minnesota Press, 2010.

Ogilby, John. *Africa: Being an Accurate Description of the Regions of AEgypt, Barbary, Lybia, and Billedugerid*. London: Printed by Tho. Johnson for the Author, 1670.

Ogilvie, Brian. *The Science of Describing: Natural History in Renaissance Europe*. Chicago: University of Chicago Press, 2008.

Omi, Michael, and Howard Winant. *Racial Formation in the United States*. 3rd ed. New York: Routledge, 2014.

Onania, or the Heinous Sin of Self-Pollution. London: Printed for, and sold by, C. Corbette at the Correct State Lottery-Office, opposite St. Dunstan's Church, in Fleet-Street; and by T. Cooke, under the Royal Exchange, Cornhill, 1759.

Onania, or the Heinous Sin of Self-Pollution . . . The Twentieth Edition. Glasgow: Printed for A. MacIntosh, [1760]. Accessed through *Eighteenth Century Collections Online*, range 15340.

Onania; or, The heinous sin of self-pollution, and all its frightful consequences, in both sexes, considered. Boston: London printed; reprinted at Boston, for John Phillips, and sold at his shop on the south side of the town-house, 1724.

Parent du Châtelet, Alexandre. *De la prostitution dans la ville de Paris, considérée sous le rapport de l'hygiène publique, de la morale et de l'administration*. Paris: J. B. Bailliere, 1836.

Parker, Patricia. "Barbers and Barbary: Early Modern Cultural Semantics." *Renaissance Drama* 33 (2004): 201–243.

Parrish, Susan Scott. *American Curiosity: Cultures of Natural History in the Colonial British Atlantic World*. Chapel Hill: Published for the Omohundro Institute for Early American History and Culture by the University of North Carolina Press, 2006.

Paster, Gail Kern. *The Body Embarrassed: Drama and the Discipline of Shame in Early Modern England*. Ithaca, NY: Cornell University Press, 1993.

Pepys, Samuel. *The Diary of Samuel Pepys*. 11 vols. Edited by Robert Lathan and William Matthews. Berkeley: University of California Press, 1970.

Peskin, Lawrence. *Captives and Countrymen: Barbary Slavery and the American Public, 1785–1816.* Baltimore: Johns Hopkins University Press, 2009.

Pillot, F. D., ed. *Oeuvres Complets de Buffon.* Paris: Chez F. D. Pillot, Rue de Seine-Saint-Germaine, 49, 1837.

Polwhele, Richard. *The Unsex'd Females.* New York: W. M. Cobbett, 1800.

Porter, Dahlia. "Scientific Analogy and Literary Taxonomy in Darwin's *Loves of the Plants.*" *European Romantic Review* 18, no. 2 (2007): 213–221.

Powers, Thomas. *The narrative and confession of Thomas Powers, a Negro. . . .* Norwich, CT: Printed by John Trumbull, August 19, 1796.

Prideaux, Humphrey. *The true nature of imposture fully displayed in the life of Mahomet.* London: Printed for William Rogers, at the Sun against St. Dunstan's Church, in Fleetstreet, 1697.

Puar, Jasbir. *Terrorist Assemblages: Homonationalism in Queer Times.* Durham, NC: Duke University Press, 2007.

Quincy, Josiah. *Remarks on Some of the Provisions of the Laws of Massachusetts Affecting Poverty, Vice and Crime.* Cambridge, MA: Printed at the University Press, by Hillard & Metcalf, 1822.

Regis, Pamela. *Describing Early America: Bartram, Jefferson, Crevecoeur, and the Influence of Natural History.* Philadelphia: University of Pennsylvania Press, 1999.

[Report of the] *Penitent Female's Refuge, Boston.* [Boston]: N.p., 1832. Available in the collections of the American Antiquarian Society, Worcester, MA; item 238876.

Rezek, Joseph. *London and the Making of Provincial Literature: Aesthetics and the Transatlantic Book Trade, 1800–1850.* Philadelphia: University of Pennsylvania Press, 2015.

Rifkin, Mark. *When Did Indians Become Straight? Kinship, the History of Sexuality, and Native Sovereignty.* London: Oxford University Press, 2011.

Riley, Stephen T. "Abraham Browne's Captivity by the Barbary Pirates, 1655." *Proceedings of the Colonial Society of Massachusetts* 52 (1980): 31–42.

Roediger, David. *The Wages of Whiteness: Race and the Making of the American Working Class.* New York: Verso, 2007.

Rousseau, Jean-Jacques. *Discours sur l'origine et les fondements de l'inégalité parmi les hommes.* Amsterdam: chez Michel Rey, 1755.

Rubin, Gayle. "Thinking Sex: Notes for a Radical Theory on the Politics of Sexuality." In *Pleasure and Danger: Exploring Female Sexuality,* edited by Carole Vance, 267–293. New York: Routledge, 1984.

Rusert, Britt. *Fugitive Science: Empiricism and Freedom in Early African American Culture.* New York: New York University Press, 2017.

Rush, Benjamin. *Medical Inquiries and Observations upon the Diseases of the Mind.* Philadelphia: Published by Kimber and Richardson, No. 237 Market Street. Merritt, Printed, No. 9, Watkin's Alley, 1812.

———. *Medical Observations and Inquiries into Diseases of the Mind.* Vol. 1 of 4. 2nd ed. Philadelphia: Published by J. Conrad & Co. Chestnut-Street, Philadelphia; M. & J. Conrad and Co., Market-Street, Baltimore; Rapin, Conrad & Co., Washington; Somervell & Conrad, Petersburg; and Bonsall, Conrad & Co., Norfolk; Printed by T&G Palmer, 116, High Street, 1805.

———. "Of the Mode of Education Proper in a Republic," 1798. In *The Selected Writings of Benjamin Rush,* edited by Dagobert Runes, 87–96. New York: Philosophical Library, 2007.

Said, Edward. *Orientalism.* New York: Vintage, 1978.

Schiebinger, Londa. *Nature's Body: Gender in the Making of Modern Science.* Boston: Beacon Press, 1993.

———. *Plants and Empire: Colonial Bioprospecting in the Atlantic World.* Cambridge: Harvard University Press, 2004.

Schiebinger, Londa, and Claudia Swan, eds. *Colonial Botany: Science, Commerce, and Politics in the Early Modern World.* Philadelphia: University of Pennsylvania Press, 2005.

Schneider, Bethany. "Oklahobo: Following Craig Womack's American Indian and Queer Studies." In *After Sex,* edited by Janet Halley and Andrew Parker, 151–168. Durham, NC: Duke University Press, 2011.

Schueller, Malani Johar. *U.S. Orientalisms: Race, Nation, and Gender in Literature, 1790–1890.* Ann Arbor: University of Michigan Press, 2001.

Schueller, Malani Johar, and Edward Watts, eds. *Messy Beginnings: Postcoloniality and Early American Studies.* New Brunswick, NJ: Rutgers University Press, 2003.

Schuller, Kyla. *The Biopolitics of Feeling: Race, Sex, and Science in the Nineteenth Century.* Durham, NC: Duke University Press, 2017.

Scribner, Bob. "Is a History of Popular Culture Possible?" *History of European Ideas* 10 (1989): 175–191.

Sedgwick, Eve Kosofsky. *Epistemology of the Closet.* Berkeley: University of California Press, 1990.

Sewall, Samuel. *Diaries of Samuel Sewall, 1674–1929, vol. 2.* Cambridge: University Press, John Wilson and Son. In *Collections of the Massachusetts Historical Society,* vol. 6, 5th series. Boston: Massachusetts Historical Society, 1879.

Sha'ban, Fuad. *Islam and Arabs in Early American Thought: The Roots of Orientalism in America.* Chapel Hill: University of North Carolina Press, 1991.

Shah, Nayan. *Contagious Divides: Epidemics and Race in San Francisco's Chinatown.* Berkeley: University of California Press, 2001.

Sharp, Daniel. *Report of the Union Committee of the Sunday Schools of the Three Baptist Societies in Boston.* Boston: Printed by Farnham and Badger, Congress Street, 1817.

Shaw, George. *Speculum Linnaeanum.* London: Printed by J. Davis, Chancery Lane, [1790].

Shershow, Scott. "New Life: Cultural Studies and the Problem of the 'Popular.'" *Textual Practice* 12, no. 1 (1998): 23–47.

Shields, David. "Eighteenth-Century Literary Culture." In *A History of the Book in America:* Vol. 1, *The Colonial Book in the Atlantic World,* edited by David Hall and Hugh Amory, 434–476. Chapel Hill: University of North Carolina Press, 2010.

Singy, P. "The Popularization of Medicine in the Eighteenth Century: Writing, Reading, and Rewriting Samuel Auguste Tissot's *Avis au peuple sur sa santé.*" *Journal of Modern History* 82 (2010): 769–800.

Siraisi, Nancy G. *Medieval and Early Renaissance Medicine.* Chicago: University of Chicago Press, 1990.

Slauter, Eric. "Neoclassical Culture in a Society with Slaves: Race and Rights in the Age of Wheatley." *Early American Studies* 2, no. 1 (2004): 81–122.

Slotkin, James. *Readings in Early Anthropology.* New York: Routledge, 1963.

Slotkin, Richard. "Narrative of Negro Crime in New England, 1675–1800." *American Quarterly* 25, no. 1 (March 1973): 3–31.

Smith, Caleb. *The Oracle and the Curse: A Poetics of Justice from the Revolution to the Civil War*. Cambridge, MA: Harvard University Press, 2013.

Smith, Samuel Stanhope. *An Essay on the Causes of the Variety of Complexion and Figure in the Human Species*. Philadelphia: Printed and sold by Robert Aitken, at Pope's Head, Market Street, 1787.

Smith-Rosenberg, Carroll. "Dis-covering the Subject of the 'Great Constitutional Discussion' 1786–1789." *Journal of American History* 79 (1992): 841–873.

Snorton, C. Riley. *Nobody is Supposed to Know: Black Sexuality on the Down Low*. Minneapolis: University of Minnesota Press, 2014.

Somerville, Siobhan. *Queering the Color Line: Race and the Invention of Homosexuality in American Culture*. Durham, NC: Duke University Press, 2000.

Spear, Jennifer. "Indian Women, French Women, and the Regulation of Sex." In *Race, Sex, and Social Order in Early New Orleans*, 17–51. Baltimore: Johns Hopkins University Press, 2009.

Spillers, Hortense. "Mama's Baby, Papa's Maybe: An American Grammar Book." *Diacritics* 17, no. 2 (Summer 1987): 65–81.

Spivak, Gayatri Chakravorty. "Can the Subaltern Speak?" In *Marxism and the Interpretation of Culture*, edited by C. Nelson and L. Grossberg, 271–313. Basingstoke: MacMillan Education, 1988.

Stallybrass, Peter, and Allon White. *The Politics and Poetics of Transgression*. Ithaca, NY: Cornell University Press, 1986.

Stein, Jordan. "Mary Rowlandson's Hunger and the Historiography of Sexuality." *American Literature* 81, no. 3 (2009): 469–495.

Stillingfleet, Benjamin. *Miscellaneous Tracts relating to natural history, husbandry, and physic. To which is added The calendar of flora*. London: R. and J. Dodsley, S. Baker, and T. Payne, 1762.

Stolberg, Michael. "Self-Pollution, Moral Reform, and the Venereal Trade: Notes on the Sources and Historical Context of *Onania* (1716)." *Journal of the History of Sexuality* 9, nos. 1–2 (2000): 37–61.

Stoler, Ann Laura. *Race and the Education of Desire: Foucault's History of Sexuality and the Colonial Order of Things*. Durham, NC: Duke University Press, 1995.

Stone, Lawrence. *The Family, Sex and Marriage in England, 1500–1800*. New York: Harper & Row, 1977.

Storey, John. *Inventing Popular Culture: From Folklore to Globalization*. New York: Blackwell, 2003.

Sweet, John Wood. *Bodies Politic: Negotiating Race in the American North, 1730–1830*. Philadelphia: University of Pennsylvania Press, 2003.

Tawil, Ezra. *The Making of Racial Sentiment: Slavery and the Birth of the Frontier Romance*. Cambridge: Cambridge University Press, 2006.

Terry, Jennifer. *An American Obsession: Science, Medicine and Homosexuality in Modern Society*. Chicago: University of Chicago Press, 1999.

Teute, Fredrike. "The Loves of the Plants; or, The Cross-Fertilization of Science and Desire at the End of the Eighteenth Century." *Huntington Library Quarterly* 63, no. 3 (2000): 319–345.

Thomson, Keith. *A Passion for Nature: Thomas Jefferson and Natural History.* Chapel Hill: Thomas Jefferson Foundation's Monticello Monograph Series, distributed by the University of North Carolina Press, 2008.

Tissot, Samuel Auguste. *Advice to the people in general, with regard to their health.* London [i.e., Boston]: Printed [by Mein and Fleeming], in the year 1767.

———. *Advice to the people in general, with regard to their health.* Philadelphia: Printed [by John Dunlap] for John Sparhawk., 1771.

———. *A treatise on the crime of Onan; illustrated with a variety of cases, together with the method of cure. By M. Tissot, M.D. author of Advice to the people in general with regard to their health.* Translated from the third edition of the original. London: Printed for B. Thomas, in the Strand, 1766.

———. *A Treatise on the Diseases Produced by Onanism.* New York: Published by Collins & Hannay, 1832.

Towner, Lawrence. "True Confessions and Dying Warnings in Colonial New England." In *Past Imperfect: Essays on Histories, Libraries, and the Humanities,* edited by Robert Karrow Jr. and Alfred Young, 56–81. Chicago: University of Chicago Press, 1993.

Traub, Valerie. "Mapping the Global Body." In *Early Modern Visual Culture: Representation, Race, and Empire in Renaissance England,* edited by Peter Erickson and Clark Hulse, 44–97. Philadelphia: University of Pennsylvania Press, 2000.

———. "The Psychomorphology of the Clitoris." *Gay and Lesbian Quarterly* 2, nos. 1–2 (1995): 81–113.

———. *The Renaissance of Lesbianism in Early Modern England.* Cambridge: Cambridge University Press, 2002.

Tyler, Royall. *The Algerine captive, or The life and adventures of Doctor Updike Underhill, six years a prisoner among the Algerines.* Printed at Walpole, NH: By David Carlisle Jr., and sold at his bookstore, 1797.

Vicinus, Martha. "Lesbian History: All Theory and No Facts or All Facts and No Theory." *Radical History Review* 60 (1994): 57–75.

Vitkus, Daniel, ed. *Three Turk Plays: Selimus, Emperor of the Turks; A Christian Turned Turk; and The Renegado.* New York: Columbia University Press, 1999.

———. *Turning Turk: English Theatre and the Multicultural Mediterranean, 1570–1630.* New York: Palgrave, 2003.

Walker, John. *Elements of Geography.* 3rd ed. Dublin: Printed and sold by Thomas Morton Bates, 89, Coombe. nearly opposite Meath-Street, 1797.

Walkowitz, Judith. *City of Dreadful Delight: Narratives of Sexual Danger in Late Victorian London.* Chicago: University of Chicago Press, 1992.

Warner, Michael. "Introduction: Fear of a Queer Planet." *Social Text* 29 (1991): 3–17.

———. "New English Sodom." *American Literature* 64, no. 1 (March 1992): 19–47.

Watt, Tessa. *Cheap Print and Popular Piety, 1550–1640.* Cambridge: Cambridge University Press, 1993.

Weheliye, Alexander. *Habeas Viscus: Racialization Assemblages, Biopolitics, and Black Feminist Theories of the Human.* Durham, NC: Duke University Press, 2014.

Weiner, Mark S. *Black Trials: Citizenship from the Beginnings of Slavery to the End of Caste.* New York: Vintage Reprint, 2006.

Weir, David. *American Orient: Imagining the East from the Colonial Era through the Twentieth Century.* Amherst: University of Massachusetts Press, 2011.

Weiss, Gillian. *Captives and Corsairs: France and Slavery in the Early Modern Mediterranean.* Palo Alto: Stanford University Press, 2013.
Weyler, Karen. "An Actor in the Drama of Revolution: Deborah Sampson, Print, and Performance in the Creation of Celebrity." In *Feminist Interventions in Early American Studies,* edited by Mary Carruth, 183–193. Tuscaloosa: University of Alabama Press, 2006.
Wheelan, Joseph. *Jefferson's War: America's First War on Terror, 1801–1805.* New York: Carrol & Graf, 2004.
Wheeler, Roxann. *The Complexion of Race: Categories of Difference in Eighteenth-Century British Culture.* Philadelphia: University of Pennsylvania Press, 2000.
Whitehill, Walter Muir. *Boston: A Topographical History.* Cambridge, MA: Belknap Press of Harvard University Press, 1959.
Wilf, Steven. *Law's Imagined Republic: Popular Politics and Criminal Justice in Revolutionary America.* Cambridge: Cambridge University Press, 2010.
Williams, Daniel E., ed. *Pillars of Salt: An Anthology of Criminal Execution Narratives.* Madison, WI: Madison House, 1993.
Windus, John. *A journey to Mequinez, the residence of the present emperor of Fez and Morocco: On the occasion of Commodore Stewart's embassy thither for the redemption of the British captives in the year 1721.* London: Printed for Jacob Tonson in *the Strand,* 1721.
Witherspoon, John. *A Series of Letters on Education.* New York: Printed by J. Buel, for C. Davis, 1797.
———. *A sermon on the religious education of children.* Elizabeth-Town, NJ: Printed by Shepard Kollock, 1789.
Wolf, John B. *The Barbary Coast: Algiers under the Turks 1500–1830.* New York: Norton, 1979.
Wollstonecraft, Mary. *Collected Letters of Mary Wollstonecraft.* Edited by Ralph M. Wardle. Ithaca, NY: Cornell University Press, 1979.
———. *A Vindication of the Rights of Woman.* London : Printed for J. Johnson, No. 72, St. Paul's Church Yard, 1792.
Worcester, Noah. *A sermon delivered at Haverhill, New Hampshire, July 28, 1796, at the execution of Thomas Powers, who was executed for a rape, committed at Lebanon, on the 7th of December, 1795.* Haverhill, NH: Printed and sold by N. Coverly, 1796.
Wynter, Sylvia. "Unsettling the Coloniality of Being/Power/Truth/Freedom: Toward the Human, after Man, Its Overrepresentation—an Argument." *CR: The New Centennial Review* 3, no. 3 (Fall 2003): 257–337.
Yeats, Thomas Pattinson. *Institutions of Entomology, being a translation of Linaeus's Ordines et genera insectorum.* London: Printed for R. Horsfield, 1773.
Young, Alfred. *Masquerade: The Life and Times of Deborah Sampson, Continental Soldier.* New York: Knopf, 2004.

INDEX

Ahmed, Sara, 69, 225n17
Allewaert, Monique, 20
anarchia, 139
antivice narrative. *See* cross-dressing narrative
antivice reform, 172, 187, 189, 192, 255n28, 256n38; Boston Female Society for Missionary Purposes, 172, 183, 185–86, 254n27, 257n81; Boston Society for the Moral and Religious Instruction of the Poor, 256n38; Penitent Female's Refuge of Boston, 184–85, 254n27. *See also* Hopkins, Asa; McDowell, John
Ariès, Philippe, 102
Arondekar, Anjali, 15, 106

Baepler, Paul, 226n34, 227n38
Bailey, Ann, 249n51
Bancroft, Aaron: *The importance of a religious education illustrated and enforced* (1793), 130–32
Barbary captivity narrative, 7, 65–68, 71–73, 80, 89, 93, 224n13, 225n21, 226n30, 226n34–35, 231n96. *See also* Brooks, Francis; Browne, Abraham; Dan, Pierre; Davis, William; Gee, Joshua; Sharif, Moulay Ismail ibn
Belcher, Josiah: *The Worst Enemy Conquered* (1698), 115–16
Bennett, Jane, 182, 257n69
Bennett, Judith M., 151
Berman, Jacob, 93
Bersani, Leo, 202–4
Big Dick (1849), 254–55n27

biopolitics, 18, 20, 187–89, 192–93, 197–201, 258n1. *See also* body: biopoliticized
biopower, 20, 187–89, 191–93, 199, 258n1
black feminist theory, 202, 204
blackness: antiblackness, 201; body and, 35; as contagion, 180; criminality and, 23, 102, 105–6, 123, 125, 127, 236n48, 237n52, 239n72, 241n97, 241n103; execution narratives and, 136, 236n48, 239n72; foreignness of, 186–87, 257n81; metaphysics of, 84; minstrelsy and, 181; passion and, 58, 84; restraint and, 125–28; sex economy and, 178; sexual behavior and, 8–9, 23, 28, 58, 102, 105–6, 123; sonic rendering of, 176; spatial dimension of, 9, 177–78, 180, 186; subjectivity and, 201; as "transformable," 125; vice and, 126–28, 180
Blackstone, William, 124
Bleys, Rudi, 97
Blum, Hester, 226n30, 226n35
body: anatomical, 187–88; biopoliticized, 190, 193, 199–200, 204; Christian, 76, 86, 94–96; climate theory and, 37–38, 49, 56–57, 215n17; complexion and, 50–51; disposition of, 50–51; environmental, 6, 21, 41, 49, 60, 107, 135, 176, 187, 190–91, 199, 202, 204–5; humoral, 21, 33–36, 44–45, 49, 187, 209n14; religion and, 79, 115; sexuality and, 31, 53, 107, 190, 192, 200, 205, 233n20; sodomy and, 76, 86, 94–96, 98–99; subjectivity and, 15, 106, 190–92
Boone, Joseph Allen, 89

botany: as cultural idiom, 137–38, 141, 145–46, 153–154, 158; gender and, 138, 244n14, 244–45n17; as sexual science, 137–38, 140, 145, 147–49, 153–54, 161, 243n2, 244n15. *See also* Darwin, Erasmus; Linnaeus, Carolus; Polwhele, Richard

Brooks, Francis: *Barbarian cruelty* (1693), 73, 82–86, 100–1, 230n71

Brown, John: *An Estimate of the Manners and the Principles of the Times* (1757), 54–55

Brown, William: *The Power of Sympathy* (1789), 247n39

Browne, Abraham: *Booke of remembrance of Gods Provydences toward me* (1653–68), 73, 77–79, 100, 227n38, 227n43–44

Browne, Arthur: *Religious education of children recommended* (1739), 109–10

Buffon, Comte de: *Histoire Naturelle* (1750), 35, 39, 48–52, 55–58, 195–96, 216n27, 220–21n82, 221n87

Burke, Peter, 210–11n30

Butt Ender: *Prostitution Exposed* (1839), 255n29, 255–56n36

capital punishment, 123–24. *See also* execution narrative

Case of Young Robinson (1836), 177–78

castration, 88–89

childhood: distinction from youth, 234–35n27; emergence of, 109, 234n27; habit and, 102, 108–11, 125, 131, 134; masturbation and, 111–13, 124, 247n40

Chiles, Katy, 59, 214n11, 219–220n73, 241n99

circumcision. *See* conversion

civilizationalist narratives: stadial theory and, 40, 52–54, 57, 194–95; women and, 56–57, 196, 219n68. *See also* Jefferson, Thomas

climate theory, 7, 21, 30, 35–39, 51, 53, 117–18, 215–16n23, 216n27, 217n35

Cohen, Daniel, 110, 113, 143, 165, 167–68, 172, 235n32, 252n2, 253n22, 254n27

Cohen, Ed, 199–200, 237n58, 258n1

Collection of Voyages and Travels, A (1701–1747), 90–91, 207n3

Colley, Linda, 80, 224n13, 225n19, 226n25, 229n57, 229n66

colonialism: climate theory and, 38–39; contrapuntal reading of, 214n3; natural history and, 24–25, 224n14; philology and, 224n14. *See also* settler colonialism

Comment, Kristin, 249–50n57

complexion, 1, 36, 39–40, 43, 45, 50–51, 56, 92, 179, 183, 209n14, 215n17, 217n35, 218n50, 231n96. *See also* disposition; humoral theory

conversion: Barbary captivity narratives and, 100–1; circumcision and, 76, 97–98, 228n48, 232n108; "turning Turk," 67, 75–76, 79, 100, 223n8, 228n48. *See also* sodomy: Islam and

Coviello, Peter, 61

Crèvecœur, J. Hector St. John de: *Letters from an American Farmer* (1782), 32, 41–44, 217n40

Crewe, Jonathan, 96, 99

criminality: cause of, 134, 235n32; climate theory and, 51; environmental understanding of, 107, 110–11, 134; poverty and, 184, 255n35; race and, 8, 105–106, 108, 116, 123, 131; rehabilitation and, 134–135; sexual behavior and, 109, 116; sex crimes, 20, 114, 119, 239n72; sodomy and, 7–8, 102; vice and,. *See also* blackness; execution narratives; habit; prostitution; rape; vice

Cronon, William: "A Place for Stories" (1992), 20–21

cross-dressing narrative, 23, 143–44, 162, 167, 248n45; *Female Marine* texts, 165–66, 168, 173–76, 178–79, 181, 183, 254n27 (*see also* Negro Hill; prostitution); *The Adventures of Louisa Baker* (1815), 165–68, 172, 253n13, 257n81; *The Adventures of Lucy Brewer* (1815), 175–76, 179–81; *An Affecting Narrative of Louisa Baker* (1816), 177–78, 185. *See also* Mann, Herman; Sampson, Deborah

Curran, Andrew, 38, 213n58, 216n27

Dain, Bruce, 27–28, 213n58

Dan, Pierre: *Histoire de Barbarie et de ses Corsaires* (1649), 63–67, 97, 222–23n1

Dana, James: *The Intent of Capital Punishment* (1790), 123–25

Danforth, Samuel: *The Cry of Sodom Enquired Into* (1674), 10–12, 25, 111, 119, 209–10n22, 210n26–27, 237n56
Darby, Robert, 232n108
Darwin, Erasmus: *The Botanic Garden* (1789), 140, 148, 243n2; *Families of Plants* (1787), 149; *The Loves of the Plants* (1789), 137–38, 148–49, 244n15, 248n47–48
Davidson, Cathy, 150, 167, 247n39
Davis, William, 1–4, 7, 66, 68; *Travels and miserable Captivity of William Davis* (1614, 1704, 1745), 73, 89–98, 101, 207n1, 207n3, 231n96–97
degeneration: cultural, 41; environment and, 39–40, 42–43, 56–57, 118; manners and, 57–58; moral, 42–43; racial, 40–41, 44, 133, 181; stadial theory and, 57, 118. *See also* Buffon, Comte de; Jefferson, Thomas
DeLombard, Jeannine, 13, 136, 168
demography: biopower and, 192–93; natural history and, 194, 197; population theory and, 165, 193–98. *See also* Malthus, Thomas
Description of the nature of slavery among the Moors (1721), 73, 84, 86–89
Dimock, Wai Chee, 170
disposition: 18th C. uses of, 50–51; climate and, 51; criminality and, 128, 131; degeneration and, 42–44; gender and, 48, 148; humoral, 46; of Muslims, 93; sexual, 18, 48, 50–51, 142; race and, 50, 100; temperament and, 36, 43–44; 93, 100, 128. *See also* complexion; habit
Dugaw, Diane, 143, 245–46n25

effeminacy: climate and, 54–55
environment: 18th C. definitions of, 22; antivice reform movements and, 183–85; complexion and, 40; corrupting influence of, 180–81, 184, 187; cross-dressing and, 148; degeneration and, 42–44; early sexuality and, 7, 9, 14–15, 18–19, 30, 43, 94, 205; environmental studies, 20–21; human body and, 6, 21, 33, 36–37, 41, 49, 107, 135, 176, 187, 190, 205 (*see also* body: environmental); human variety and, 6, 22, 29, 33, 35, 58, 60 (*see also* race; racialization; sexual behavior); humanism and, 21;

as milieu, 211–12n39; modern sexuality and, 191–92, 200; natural history and, 14; population theory and, 193, 197 (*see also* biopolitics); religion and, 101; sexual behavior and, 8, 14, 30, 33–34, 43–44, 49–51, 54, 59–61, 106, 113, 200 (*see also* habit); stadial theory and, 52–53; state of society and, 40–42, 57; terminology, 19, 211–12n39; vice and, 43, 172, 183. *See also* body; climate theory; humoral theory; Negro Hill; urbanity
eugenics, 197
execution narratives, 12–13, 102, 105–7, 113–14, 116, 118–19, 121, 126, 128, 135–36, 236n48, 237n57, 238n62. *See also* Frost, Samuel; Mather, Cotton; Mountain, Joseph; Powers, Thomas

Faderman, Lillian, 15
Falconer, William: *Remarks on the Influence of Climate* (1781), 36–37, 51
femininity: artifice and, 178–79; civilization and, 157; modesty and, 137–38; novel reading and, 146, 247n39; perdition of, 161–62; republican womanhood, 146, 245–46n25, 247n38; shame and, 3–4, 8, 46, 48, 208n8; virtue and, 170
Ferguson, Adam: *Essay on the History of Civil Society* (1767), 38, 48, 51–52, 219n58
Forbes, Erin, 132
Foster, Travis, 23
Foucault, Michel, 8, 10, 13, 18, 20, 60–61, 67, 189–90, 192, 194, 198–99, 200–3, 205, 211–12n39, 214n6, 223–24n9, 258n1. *See also* biopolitics; biopower
four-stages theory. *See* stadial theory
Franklin, Benjamin: *Observations concerning the increase of mankind, peopleing of countires, &c.* (1755), 196–98, 259–60n24
Freud, Sigmund, 203–5
Frost, Samuel: *The Confession and Dying Words of Samuel Frost* (1793), 128–34, 242n111

gallows literature. *See* execution narratives
Gee, Joshua: *Narrative of Joshua Gee of Boston, Mass* (1680–87), 73–77, 80, 226n34, 226–27n35–36

gender: botany and, 153, 244n14, 244–45n17; civilization and, 57, 196; climate and, 54–55; habit and, 108; inversion, 160, 251n73; Native Americans and, 156–58, 250n65; racialization and, 3–4, 57; settler colonialism and, 158; theory, 202. *See also* cross-dressing narrative; femininity; Sampson, Deborah; Mann, Herman

Godbeer, Richard, 237n56

Grew, Nehemiah: *Anatomy of Plants* (1682), 243n2

Gustafson, Sandra, 245–246n25

habit, 106–7, 136, 233n12; childhood and, 102, 109–12, 124, 131, 134 (*see also* masturbation); criminality and, 102, 104–9, 123, 125, 134; custom and, 233n12; in execution narratives, 114–15, 118–19, 126, 128, 136; human body and, 107; modes of living, 40–41, 241n99; prison-reform movements and, 134–35; racial status and, 125, 241n99; racialization and, 104–5, 107–9, 114–15, 118, 123, 128, 131, 135, 234n25; of reflection, 132–33; self-governance and, 106, 109, 125; sexual behavior and, 109, 113, 115, 118, 131, 134–35, 239n73; sin and, 11, 104–5, 114, 116–17, 122–26, 130, 134 (*see also* vice); state of society and, 105, 107; of virtue, 124–25, 184

Hakluyt, Richard: *The Principal Navigations, Voiages, Traffiques and Discoueries of the English Nation* (1589–1600), 28, 213n60

Hallock, Thomas, 137–38, 243n3

Hanno, Joseph, 103, 105, 107–8, 233n13

Hervey, James: *Meditations and Contemplations* (1745–1746), 182

Hiltner, Judith, 249n52

Hobson, Barbara Meil, 251–52n2, 253n19, 254n24, 255n35, 256n38

Hopkins, Asa, 169, 170–72, 174, 256n38

Mitchell, John: *Essay Upon the Causes of the different Colours of People in Different Climates* (1744), 48, 218n50
Moore, Lisa, 248n47
Morgensen, Scott, 155, 158
Mountain, Joseph, 123–24, 240n78, 240–41n88; *Sketches of the Life of Joseph Mountain, a Negro. . . For a Rape* (1790), 119–22, 240n84–85
Myles, Anne, 209n20

natural history: Barbary captivity narratives and, 90–91; blackness and, 123; climate theory and, 37; criminal justice and, 135; demography and, 194, 197 (*see also* demography: population theory); environmentalism and, 14, 32–33; externality and, 198; habit and, 113; history of sexuality and, 4; as narrative, 22–23; orientalism and, 224n14; popular narrative and/as, 5–6, 13–14, 22; psychoanalysis and, 204–205; racial difference and, 22, 59, 135, 178, 193; scientific racism and, 187; sexology and, 193, 197; sexual behavior and, 33–34; sodomy and, 67; violence and, 205. *See also* botany; taxonomy
Negro Hill, 8, 164–165, 167, 169–70, 172–73, 175–87, 251–52n2, 254n24, 254–55n27
Nelson, William Max, 197
Nestle, Joan, 206
Nussbaum, Felicity, 48, 54, 219n58, 219n68

Ogilby, John, 84
Onania, or the Heinous Sin of Self-Pollution, 111–12, 235n34
onanism. *See* masturbation
opacity: history of early sexuality and, 141; ignorance and, 154; precipitousness and, 161–62, 247n42; queer modernism and, 251n74; sexual knowledge and, 161
orientalism, 224n14; homoerotics of, 94
Ottoman Empire, 70–72, 81, 229n57

Parker, Patricia, 67, 76, 79, 85, 227n39, 228n48, 228n50
Parrish, Susan Scott, 44–45, 161

pathic subjection, 67–68, 74, 76, 85, 100, 223n8. *See also* sodomy
Pepys, Samuel, 80–81
piracy, 62, 70, 225n19
Polwhele, Richard: *The Unsex'd Females* (1798), 137–38, 149, 242–43n1
Powers, Thomas: *The Narrative and Confession of Thomas Powers, a Negro. . . for Rape* (1796), 125–28, 241n103
Prosser, Jay, 160, 251n73
prostitution, 23, 166, 169–73, 183, 255n29; anti-brothel riots, 172–73; interracial sexual contact and, 179–80; policing of, 184; sexology and, 193, 198, 252n3; vicious poor and, 173, 183–84; workhouses and, 184–85, 254n27. *See also* antivice reform; cross-dressing narrative; sexology; urbanity; vice
psychoanalysis, 173, 203, 205
public health, 186–87, 193, 198. *See also* prostitution; vice

queer theory: antisocial turn, 202, 204; black feminist theory and, 202; psychoanalysis and, 205 (*see also* Bersani, Leo); race and, 202; sex and, 202–3; skepticism of the subject, 191, 201
Quincy, Josiah: *Remarks on Some of the Provisions of the Laws of Massachusetts Affecting Poverty, Vice and Crime* (1822), 255n35

race: ambiguity of, 178; anatomical theories of, 59, 178, 187, 216n27; animality and, 77–78; antivice reform and, 172; biopolitics and, 190; climate theory and, 38, 216n27; criminalization and, 8, 106, 108, 114, 123; Curse of Ham, 27–28, 97, 213n60; domestic heterosexuality and, 234n25; environmental nature of, 6, 19, 33–37, 40–44, 187, 193; execution narratives and, 106, 136, 236n48; habit and, 107–8, 135; humoral theory and, 45; ideology and, 14; mutability of, 125, 214n11; natural history and, 5, 22, 33, 178, 193; racial degeneration, 39–40, 118, 181, 216n27; sex and, 30, 33, 44, 58–60;

race (*continued*)
   sexuality and, 4–7, 14, 26, 29; slavery and, 115; sodomy and, 68, 99; taxonomy and, 2–4, 45–46, 140; terminology, 17–20, 213–14n2. *See also* blackness; natural history: human variety; whiteness
racialization: biopolitics and, 193, 198 (*see also* demography: population theory); of Christianity, 67–68, 79; climate theory and, 118; of crime, 114; disposition and, 44, 50; domestic heterosexuality and, 234n25; environment and, 136; execution narratives and, 128; gender and, 4, 160; governance and, 20; habit and, 106–8, 114–16, 123, 128, 134–35; heterosexuality and, 234n25; incontinence and, 133–34; of Islam, 67–68, 83–84, 101, 231n96–97 (*see also* orientalism); penis and, 97; sciences of, 6; sexual politics of, 5, 20, 25–28, 115, 118, 136, 191; sexual behavior and, 30, 44, 104–6, 109, 134, 136, 160, 167, 196–197, 205; sodomy and, 79, 101, 225n15; of space, 118, 176 (*see also* Negro Hill); stadial theory and, 118, 217n32; terminology, 20; urban, 187
rape: blackness and, 23, 25–26, 123–25, 127, 240n85
Rawlins, John, 228n48
Regis, Pamela, 220n79
Riley, Stephen, 227n43
Rowson, Susanna: *Charlotte Temple* (1791), 253n13
Rush, Benjamin, 139, 146, 247n38

Said, Edward, 69, 214n3, 224n14
Sampson, Deborah, 24, 138–43, 150, 159–62, 243n7, 245n18, 245n21, 245–46n25, 247n42, 249n53, 250–51n70–71, 253n13
sapphism, 8, 23–24, 249–50n57; as exceptional, 141, 150; indigenization of whiteness and, 140, 155, 157; republicanism and, 139–40, 154; settler colonialism and, 158
Schiebinger, Londa, 244n17
Sedgwick, Eve Kosofsky, 154, 192
settler colonialism: botany and, 140; European sexual and gender standards and, 157, 250n65; gendered and sexual violence and, 156–57; indigineity and, 140, 155, 158; natural history and, 24; precipitousness and, 162; sapphism and, 158
Sewall, Samuel, 80–81
sex: botany and, 137–38; climate and, 52–53; criminalization of, 106, 114; cultural knowledge and, 25, 33; difference and, 147; discourse and, 61; ecumenical genealogy of, 11–12; environment and, 6–7, 60; execution narratives and, 236n48; gender and, 247n41; human body and, 8; interiority and, 9; natural history and, 14, 23; popular print and, 13; population theory and, 198; queer theory and, 202; race and, 7, 23, 29–30, 59–60; sexuality and, 9–10, 60–61; sodomy and, 7–8; sonic rendering of, 175–76; subjectivity and, 60–61, 203, 205; terminology, 9, 17–19; "turning Turk," 228n48; visibility of, 7
sexology, 9, 18, 161, 165, 211n38; as biopolitical science, 193, 198; racial difference and, 197; as racial science, 193; sex work and, 173, 252n3; sexuality and, 173, 214n6
sexual behavior: biopolitics and, 200; botany and, 140; contrapuntal reading and, 214n3; Curse of Ham and, 28, 97; disposition and, 50; as embodied phenomenon, 187; environmental nature of, 8, 14, 35, 49, 54, 109, 113, 115, 118, 131, 134–36, 185; habit and, 135; interiority and, 9; natural history and, 32–34; popular print and, 13–14; race and, 4, 49, 59–60, 106, 135, 196; racialization and, 6; self-governance and, 52–54, 135; social hierarchy and, 53–54; spatial dimension of, 9 (*see also* Negro Hill; prostitution; urbanity); weaponization of, 157–58 (*see also* settler colonialism). *See also* masturbation; sodomy; vice
sexuality: behavior modification and, 209n19; biopoliticized body and, 200; as disciplinary, 203; environmental nature of, 6–8, 16, 43–44, 60, 191–92, 200, 205–6; genealogy of, 205; geography and, 48; habit and, 135; history of, 4, 6–7, 15, 60, 160–61; interracial, 234n25; "like race" analogy and, 26–27, 29, 191, 212–13n53; modern, 9, 202–3; natural history and, 4;

of plants, 243n6; politics and, 206; power and, 203–4; racial difference and, 4–8, 15, 20, 62–63, 109; self-governance and, 52–53; sex and, 60–61; sexology and, 18, 214n6; subjectivity and, 9–10, 31, 60–61, 173, 190–92, 200–1, 203; terminology, 8–10, 17–18

Sharif, Moulay Ismail ibn (Muley Ismael, Mully Ishmael), 83–89, 100–1, 229n66, 229n68, 230n71

Slauter, Eric, 233n14

slavery: American, 115, 121, 186, 197; Barbary slavery, 62, 70–72, 81–82, 86–89, 91–92, 95, 100, 222–23n1, 223n2, 228n48; mental, 105, 233n14

Slotkin, Richard, 236n48, 237n52, 238n62, 239n72, 241n97

Smith, Caleb, 121, 238n62

Smith, Samuel Stanhope: *Essay on the Causes of the Variety of Complexion and Figure in the Human Species* (1787), 22, 37, 39–40, 43, 217n35

Smith-Rosenberg, Carroll, 170

Snell, Hannah, 23, 144, 157, 246n25, 246n28

sodomy: barbarism and, 76; Barbary captivity narratives and, 67–69, 75–77, 96, 224n11, 231–32n106; as contrary to Christianity, 101; Christian body and, 86; climate theory and, 7; criminalization of, 102; excess and, 93–94; geography and, 99; history of sexuality and, 207n4; Islam and, 7, 23, 67–69, 74–76, 79, 88, 93; lateral sodomitical frame, 96–97; prosecution of, 232n119; representation of, 89, 232n110; as racialized formation, 7–8, 62, 67–69, 99–102, 225n15; scholarship on, 7, 68–69; sodomitical reading, 96; terminology, 16–17; Turks as sodomites, 1–2, 7, 23, 68, 86, 92–94; unimaginability of, 96, 98

Somerville, Siobhan, 29

Spivak, Gayatri Chakravorty, 200

stadial theory, 40, 49, 52–53, 57, 194–96, 217n32, 219n63, 219n68

state of society, 22; climate and, 37; complexion and, 40–41, 217n35; habits and, 40, 105, 107, 125; manners and, 42, 57–58; natural history and, 41; of northern Africa, 100; slavery and, 91–92; of "Turks," 93.

See also Buffon, Comte de; Crèvecœur, J. Hector St. John de; Jefferson, Thomas; Smith, Samuel Stanhope

Stein, Jordan, 147

Stolberg, Michael, 111

Sweet, John Wood, 217n35

Tawil, Ezra, 48, 218n48, 222n92

taxonomy: binomial, 45, 145, 148, 246n35; botanical, 4, 137–38, 140–41, 144–45, 148, 161, 208n10, 243n2–3, 244–45n17; racial difference and, 2, 59; sexual difference and, 153, 244n13–15. *See also* Buffon, Comte de; Falconer, William; Linnaeus, Carolus

Tenney, Tabitha: *Female Quixotism* (1801), 247n39

Teute, Fredrike, 248n48, 248n50

Tissot, Samuel Auguste: *L'onanisme: Dissertation sur les maladies produites par la masturbation* (1760), 112, 235–36n40

Towner, Lawrence, 236n48

Traub, Valerie, 108, 234n25

Tyler, Royall: *Algerine Captive* (1797), 89, 230n88

urbanity: biopower and, 199; management and, 166–67; morality and, 170; population science and, 168–69, 193–94; poverty and, 167; sexual danger and, 166, 170, 174–75; sexuality and, 166, 252n8; urban planning, 165; youth and, 174–175. *See also* antivice reform; blackness: spatial dimensions of; Negro Hill; prostitution; vice: spatialized understandings of

venereal, 19, 43, 87, 99, 116

venery, 45, 49–50, 166

vice: blackness and, 127, 178, 180; as contagion, 174; criminalization of, 171, 255n35; environment and, 182–83; habit and, 130, 171, 174; lust, 115–16, 127, 135; poverty and, 171–73, 255n35; racial difference and, 105; sexual aberrance and, 105, 126–27; spatialized understandings of, 8–9, 165–66, 170, 172–73, 178. *See also* antivice reform; prostitution

Vicinus, Martha, 151, 154, 243n7

Warner, Michael, 99, 224n11
warrior woman, 245–46n25–26. *See also* cross-dressing narratives
Weyler, Karen, 150
Wheelan, Joseph, 223n2
Wheeler, Roxann, 14, 36, 209n14, 215n17, 215n23, 216n29, 217n32, 218n46, 231n96
Whitehill, Walter, 164–65, 169, 251n1, 254n24
whiteness: 18th C. and, 157–58; Christian, 84; contingency of, 181; feminism and, 160, 250–51n71; indigenization of, 140, 257n81; queer studies and, 191

Williams, Daniel, 114, 120, 127–28, 238n62, 241n103
Windus, John: *Journey to Mequinez* (1721), 86
Witherspoon, John: *Series of Letters on Education* (1797), 110, 113
Wollstonecraft, Mary: *Vindication of the Rights of Woman* (1792), 137–38, 149
Wynter, Sylvia, 202

Young, Alfred: *Masquerade* (2004), 159, 250n70

www.ingramcontent.com/pod-product-compliance
Lightning Source LLC
Chambersburg PA
CBHW030119240426
43673CB00041B/1332